Spring Boot+Vue

框架技术精讲与项目实战

缪勇 施俊◎编著

清華大学出版社
北京

内容简介

本书全面介绍了Spring Boot和Vue全栈开发技术，书中示例所用的Spring版本为5.0，Spring Boot版本为2，Vue版本为3。全书分为三个部分，第一部分（第1～10章）详细介绍了Spring Boot框架，主要内容为Spring Boot核心知识、Spring Boot的Web开发、Spring Boot的数据访问、Spring Boot缓存、消息服务、任务管理、安全管理等；第二部分（第11章）详细介绍了前端框架Vue，主要内容为Vue简介、Vue脚手架、目录结构、初识setup和ref、模板语法、生命周期钩子、reactive函数、初识Vue组件、深入setup、计算属性、条件渲染、列表渲染、watch监视、provide与inject、Vue路由、axios发送请求；第三部分（第12章）使用Spring Boot＋Vue框架，实现了一个前后端分离的电商平台后台管理系统。为方便读者学习和教学，本书提供全程真实课程录像。

本书既适合有一定Java Web基础的开发者阅读，也适合网络开发人员作为技术参考书，同时，也可作为高等院校计算机相关专业学生的课堂教材。

图书在版编目(CIP)数据

Spring Boot＋Vue框架技术精讲与项目实战/缪勇，施俊编著.—北京：清华大学出版社，2022.10 (2024.7重印)
ISBN 978-7-302-61311-4

Ⅰ.①S… Ⅱ.①缪… ②施… Ⅲ.①网页制作工具－程序设计 Ⅳ.①TP393.092.2

中国版本图书馆CIP数据核字(2022)第124607号

责任编辑：颜廷芳
封面设计：刘 键
责任校对：李 梅
责任印制：刘海龙

出版发行：清华大学出版社
网 址：https://www.tup.com.cn，https://www.wqxuetang.com
地 址：北京清华大学学研大厦A座 **邮 编**：100084
社 总 机：010-83470000 **邮 购**：010-62786544
投稿与读者服务：010-62776969，c-service@tup.tsinghua.edu.cn
质量反馈：010-62772015，zhiliang@tup.tsinghua.edu.cn
课件下载：https://www.tup.com.cn，010-83470410
印 装 者：三河市龙大印装有限公司
经 销：全国新华书店
开 本：185mm×260mm **印 张**：22.5 **字 数**：542千字
版 次：2022年12月第1版 **印 次**：2024年7月第3次印刷
定 价：88.00元

产品编号：090116-01

PREFACE 前言

Spring Boot 框架是继 SSM 之后，目前比较主流的 Java EE 企业级框架，适用于搭建各种大型企业级应用系统。Spring Boot 框架通过简化配置，进一步简化 Spring 应用程序的搭建和开发过程。

Vue 是一套构建用户界面的渐进式 JavaScript 框架，在设计上使用 MVVM(Model-View-View Model)模式，其特点是易用、灵活、高效，可应用于构建 UI，也便于与第三方库或既有项目整合。

1. 内容结构

本书全面介绍了 IDEA 开发平台、Spring Boot 框架、Vue 框架等基础知识，最后通过一个前后端分离的电商平台后台管理系统项目，详细介绍了 Spring Boot＋Vue 框架的整合和运用。全书共分 12 章，具体内容如下。

第 1 章介绍 Spring Boot 简介、环境搭建、创建第一个 Spring Boot 项目、使用 Spring Initializer 快速创建项目。

第 2 章介绍 Spring Boot 配置文件、Properties 配置、YAML 配置、Profile 多环境配置、Web 容器配置。

第 3 章介绍日志框架和日志使用。

第 4 章介绍 Spring Boot 对 Web 开发的支持、自定义消息转换器 HttpMessageConverter、Spring Boot 序列化和反序列化 JSON 数据、Spring Boot 对静态资源的访问、Thymeleaf 模板引擎、错误处理、CORS 支持、对 JSP 的支持。

第 5 章介绍配置数据源、使用 JdbcTemplate、使用 Spring Data JPA、整合 MyBatis。

第 6 章介绍整合 Redis、整合 MongoDB。

第 7 章介绍 JCache (JSR-107)规范、缓存抽象与缓存注解、EhCache 2.x 缓存、Redis 缓存。

第 8 章介绍消息服务概述、整合 JMS、整合 AMQP。

第 9 章介绍异步任务、定时任务、邮件任务。

第 10 章介绍 Spring Security 概述、Spring Security 快速入门、用户认证、用户授权。

第 11 章介绍 Vue 简介、Vue 脚手架、目录结构、初识 setup 和 ref、模板语法、生命周期钩子、reactive 函数、初识 Vue 组件、深入 setup、计算属性、条件渲染、列表渲染、watch 监视、provide 与 inject、Vue 路由、axios 发送请求。

第 12 章结合前端 Vue 框架，详细介绍一个前后端分离的电商平台后台管理系统的具体实现过程。

2. 特点和优势

本书作者在Java EE Web领域具有多年的开发和教学经验，熟悉Java开发理论知识体系，本书具有以下特色。

(1) 细致全面：本书的内容编排由开发环境搭建开始，从基本知识入手，由浅入深逐步过渡到高级部分，内容囊括了Spring Boot和Vue框架的重要知识点。注重介绍如何在实际工作中灵活运用基础知识，从而高质量地进行程序开发。

(2) 结合示例：本书在各章知识点的介绍中，都结合了小示例的精讲加以验证。对特别难懂的知识点，通过恰当的示例帮助读者进行分析、理解。

(3) 讲解透彻：本书选取的实战项目是一个前后端分离的电商平台后台管理系统，在讲解的过程中，均按功能分类，便于读者理解每个功能的实现过程。

(4) 实用性强：本书以经验为后盾、以实践为导向、以实用为目标，深入浅出地介绍Spring Boot＋Vue开发中的各种问题。

(5) 课堂实录：采用知识介绍＋课堂实录的方式，提供一套全过程课程录像，便于读者跟进学习，也可以直接用于学校教学。

3. 读者对象

(1) 有一定Java Web框架开发基础，需要对Spring Boot核心技术进一步了解和掌握的程序员。

(2) 高等院校正在学习编程开发的计算机及相关专业的学生。

(3) 公司管理人员或人力资源管理人员。

4. 配套资源

本书附赠完整的学习资源，包括同步教学录像、教学PPT、源代码、素材文件等内容，读者可从清华大学出版社官网(http://www.tup.tsinghua.edu.cn)下载。如有问题或需要答疑，请发送电子邮件至shikham66@163.com，邮件主题为“Spring Boot＋Vue框架技术精讲与项目实战”。

5. 作者及致谢

本书由施俊和缪勇编著。其中，施俊编写第1～10章，缪勇编写第11和12章。施俊对全书进行了审核和统筹，其他参编人员还有刘娇、陆佰林、邱宇、朱锦程、李艳会、王晶晶、游名扬、王梅、王永庆、蒋梅芳、谢伟、纪航、沈勇等。同时扬州国脉通信发展有限责任公司、江苏智途科技股份有限公司也为本书的编写提供了帮助，在此一一致谢。

由于作者水平有限，书中难免存在一些疏漏之处，敬请读者批评指正。

编著者

2022年3月

CONTENTS 目录

第 1 章 Spring Boot 入门 …… 1

1.1 Spring Boot 简介 …… 1

1.2 环境搭建 …… 2

1.2.1 JDK 的下载与安装 …… 2

1.2.2 Maven 的安装与设置 …… 2

1.2.3 IntelliJ IDEA 安装与设置 …… 4

1.3 创建第一个 Spring Boot 项目 …… 6

1.3.1 创建 Maven 工程 …… 6

1.3.2 导入相关依赖包 …… 9

1.3.3 创建启动类和控制器 …… 9

1.3.4 启动项目 …… 10

1.4 使用 Spring Initializer 快速创建项目 …… 11

1.5 小结 …… 14

第 2 章 Spring Boot 基本配置 …… 15

2.1 Spring Boot 配置文件 …… 15

2.2 Properties 配置 …… 16

2.3 YAML 配置 …… 18

2.4 Profile 多环境配置 …… 20

2.5 Web 容器配置 …… 21

2.6 小结 …… 26

第 3 章 Spring Boot 日志 …… 27

3.1 日志框架 …… 27

3.1.1 Spring Boot 的日志 …… 27

3.1.2 Logback 简介 …… 29

3.2 日志使用 …… 29

3.2.1 默认日志格式 …… 29

3.2.2 日志级别 …… 29

3.2.3 日志文件输出 …… 31

3.2.4 自定义日志格式 …… 32

3.2.5 基于 XML 配置日志 …… 33

3.2.6 使用 Log4j2 日志实现 …… 36

3.3 小结 …… 37

第 4 章　Spring Boot 的 Web 开发 …… 38
4.1　Spring Boot 对 Web 开发的支持 …… 38
4.2　自定义消息转换器 HttpMessageConverter …… 39
4.3　Spring Boot 序列化和反序列化 JSON 数据 …… 43
4.4　Spring Boot 对静态资源的访问 …… 44
4.4.1　默认规则 …… 44
4.4.2　自定义规则 …… 46
4.5　Thymeleaf 模板引擎 …… 47
4.5.1　Thymeleaf 简介 …… 48
4.5.2　引入 Thymeleaf …… 48
4.5.3　Thymeleaf 语法规则 …… 49
4.5.4　整合 Thymeleaf …… 53
4.6　错误处理 …… 56
4.6.1　异常处理机制 …… 56
4.6.2　自定义错误页 …… 57
4.7　CORS 支持 …… 62
4.8　对 JSP 的支持 …… 64
4.9　小结 …… 66
第 5 章　Spring Boot 访问 SQL 数据库 …… 67
5.1　配置数据源 …… 67
5.2　使用 JdbcTemplate …… 68
5.2.1　JdbcTemplate 增删改的操作 …… 69
5.2.2　JdbcTemplate 查询的操作 …… 74
5.3　使用 Spring Data JPA …… 77
5.3.1　Spring Data JPA 介绍 …… 77
5.3.2　整合 Spring Data JPA …… 79
5.4　整合 MyBatis …… 84
5.4.1　基于 XML 配置的方式整合 MyBatis …… 85
5.4.2　基于注解的方式整合 MyBatis …… 90
5.5　小结 …… 91
第 6 章　Spring Boot 使用 NoSQL …… 92
6.1　整合 Redis …… 92
6.1.1　Redis 简介 …… 92
6.1.2　Redis 安装 …… 93
6.1.3　Spring Boot 整合 Redis …… 95
6.2　整合 MongoDB …… 98
6.2.1　MongoDB 简介 …… 98
6.2.2　MongoDB 安装 …… 99
6.2.3　Spring Boot 整合 MongoDB …… 101

6.3 小结 …… 109
第 7 章 Spring Boot 与缓存 …… 110
7.1 JCache (JSR-107)规范 …… 110
7.2 缓存抽象与缓存注解 …… 111
7.3 EhCache 2.x 缓存 …… 117
7.4 Redis 缓存 …… 123
7.5 小结 …… 124
第 8 章 Spring Boot 消息服务 …… 125
8.1 消息服务概述 …… 125
8.2 整合 JMS …… 128
8.2.1 JMS 简介 …… 128
8.2.2 Spring Boot 整合 JMS …… 129
8.3 整合 AMQP …… 135
8.3.1 RabbitMQ …… 136
8.3.2 安装 RabbitMQ 以及整合环境搭建 …… 139
8.3.3 Spring Boot 整合 RabbitMQ 实现 …… 142
8.4 小结 …… 149
第 9 章 Spring Boot 任务管理 …… 150
9.1 异步任务 …… 150
9.2 定时任务 …… 155
9.3 邮件任务 …… 157
9.4 小结 …… 164
第 10 章 Spring Boot 安全管理 …… 165
10.1 Spring Security 概述 …… 165
10.2 Spring Security 快速入门 …… 166
10.2.1 入门案例 …… 166
10.2.2 Spring Security 的适配器 …… 168
10.2.3 角色访问控制 …… 169
10.3 用户认证 …… 171
10.3.1 JDBC 身份认证 …… 171
10.3.2 UserDetailsService 身份认证 …… 176
10.4 用户授权 …… 181
10.4.1 用户访问控制 …… 181
10.4.2 用户登录 …… 183
10.4.3 用户退出 …… 186
10.4.4 获取登录用户信息 …… 188
10.4.5 记住我功能 …… 190
10.5 小结 …… 195

第 11 章 Vue 前端框架 …… 196
11.1 Vue 简介 …… 196
11.2 Vue 脚手架 …… 196
11.3 目录结构 …… 198
11.4 初识 setup 和 ref …… 201
11.5 模板语法 …… 203
11.5.1 插值 …… 204
11.5.2 指令 …… 206
11.5.3 用户输入 …… 207
11.5.4 缩写 …… 208
11.6 生命周期钩子 …… 208
11.7 reactive 函数 …… 211
11.8 初识 Vue 组件 …… 212
11.9 深入 setup …… 214
11.10 计算属性 …… 221
11.11 条件渲染 …… 222
11.12 列表渲染 …… 223
11.13 watch 监视 …… 225
11.14 provide 与 inject …… 227
11.15 Vue 路由 …… 229
11.16 axios 发送请求 …… 232
11.17 小结 …… 235
第 12 章 电商平台后台管理系统 …… 236
12.1 需求与系统分析 …… 236
12.2 数据库设计 …… 237
12.3 环境搭建 …… 240
12.3.1 后端程序目录结构 …… 240
12.3.2 编辑 Spring Boot 配置文件 …… 241
12.3.3 创建 MyBatis 配置文件 …… 241
12.3.4 集成 JWT 实现 Token 验证 …… 242
12.3.5 配置跨域 …… 246
12.4 创建实体类 …… 246
12.5 创建 Mapper 接口及映射文件 …… 250
12.6 创建 Service 接口及实现类 …… 261
12.7 创建 Controller 控制器类 …… 272
12.8 前端程序目录结构 …… 290
12.9 登录与管理首页面 …… 292
12.10 商品管理 …… 297
12.10.1 商品列表 …… 297

12.10.2 商品类别 …… 308
12.11 订单管理 …… 314
12.11.1 订单列表 …… 314
12.11.2 创建订单 …… 319
12.12 用户权限管理 …… 328
12.12.1 后台用户管理 …… 328
12.12.2 角色管理 …… 338
12.12.3 前台用户管理 …… 346
12.13 小结 …… 347
参考文献 …… 348

第1章 Spring Boot 入门

互联网的兴起使 Spring 占据了 Java 领域轻量级开发的王者之位，随着 Java 语言的发展，为满足市场开发需求，Spring 推出 Spring Boot 框架以简化 Spring 配置，让用户可以轻松构建独立的运行程序，提高开发效率。本章将介绍 Spring Boot 入门知识，包括：Spring Boot 简介、环境搭建、创建第一个 Spring Boot 项目和使用 Spring Initializer 快速创建项目。

1.1 Spring Boot 简介

Spring Boot 是由 Pivotal 团队提供的全新框架，伴随 Spring 4 而诞生，是一种全新的编程规范，其设计目的是用来简化新 Spring 应用的初始搭建以及开发过程。Spring Boot 在继承 Spring 优点的基础上，简化了框架的使用，使得开发者可以更容易地创建基于 Spring 可"即时运行"的应用和服务。该框架采用"约定优于配置"的方式开发，使得开发人员不再需要定义样板化的配置，通过这种方式可以快速构建 Spring 应用。所以 Spring Boot 本质上是 Spring 框架的另一种表现形式。

Spring Boot 具有如下特征。

- 使用 Spring Boot 可以创建独立的 Spring 应用程序。
- 在 Spring Boot 中直接嵌入了 Tomcat、Jetty、Undertow 等 Web 容器，所以在使用 Spring Boot 做 Web 开发时不需要部署 WAR 文件。
- 通过提供自己的启动器(Starter)依赖，简化项目构建配置。
- 尽量自动配置 Spring 和第三方类库。
- 提供运维特性，如指标信息、健康检查及外部化配置。
- 没有代码生成，也不需要 XML 配置文件，开箱即用。
- 与云计算的天然继承。

Spring Boot 主要有 3 种版本，介绍如下。

- SNAPSHOT：快照版，即开发版。
- CURRENT：最新版，但不一定是稳定版。
- GA：General Availability，正式发布的版本。

在实际开发中，应当选择正式发布的 GA 版本。

1.2 环境搭建

正如学习 Java 一样，首先需要搭建环境，才能真正进行开发和部署。目前常用的 Java 开发工具有 Eclipse、IntelliJ IDEA、Spring Tool Suite 和 MyEclipse 等。本书采用 IntelliJ IDEA 作为 IDE 进行开发。

1.2.1 JDK 的下载与安装

根据 Spring Boot 官方文档，从 Spring Boot 2.2.10 版本开始要求 JDK 8 及以上版本，读者可以自行到 Oracle 官方网站下载相应版本的 JDK，下载网址是 https://www.oracle.com/java/technologies/javase-downloads.html。本书使用的系统为 64 位的 Windows 7，下载的是 jdk-8u281-windows-x64.exe 文件。安装 JDK 后，配置环境变量 JAVA_HOME，值为 D:\Program Files\Java\jdk1.8.0_281。在系统变量里新建 CLASSPATH 变量，变量值为“.;%JAVA_HOME%\lib;%JAVA_HOME%\lib\tools.jar”，然后将“%JAVA_HOME%\bin;%JAVA_HOME%\jre\bin”加入系统的环境变量 path 中。完成配置后，在命令行窗口输入 java -version，会显示 JDK 的版本信息，再在命令行窗口输入 javac，出现用法提示信息，表示 JDK 已经配置成功，如图 1-1 所示。

```
管理员: 命令提示符
Microsoft Windows [版本 6.1.7601]
版权所有 (c) 2009 Microsoft Corporation。保留所有权利。

C:\Users\Administrator>java -version
java version "1.8.0_281"
Java(TM) SE Runtime Environment (build 1.8.0_281-b09)
Java HotSpot(TM) 64-Bit Server VM (build 25.281-b09, mixed mode)

C:\Users\Administrator>javac
用法: javac <options> <source files>
其中, 可能的选项包括:
  -g                         生成所有调试信息
  -g:none                    不生成任何调试信息
  -g:{lines,vars,source}     只生成某些调试信息
  -nowarn                    不生成任何警告
  -verbose                   输出有关编译器正在执行的操作的消息
  -deprecation               输出使用已过时的 API 的源位置
  -classpath <路径>            指定查找用户类文件和注释处理程序的位置
```

图 1-1 查看 JDK 版本测试 JDK 是否配置成功

1.2.2 Maven 的安装与设置

Maven 是 Apache 下的一个开源软件项目管理工具，同时提供出色的项目构建能力。Maven 基于项目对象模型(Project Object Model，POM)的理念，通过一段核心描述信息来管理项目构建、报告和文档信息。使用 Maven 可以对项目的依赖包进行管理，也支持构建脚本的继承。对于一些模块(子项目)较多的项目来说，Maven 是更好的选择，子项目可以继承父项目的构建脚本，从而减少构建脚本的冗余。Maven 本身的插件机制也使其更加强大和灵活，使用者可以配置各种 Maven 插件来完成自己的事情。在 Maven 众多特性中，最

重要的是对依赖包的管理，Maven 将项目所使用的依赖包的信息放到 pom. xml 的 dependencies 节点。

本书使用的 Maven 版本为 3. 8. 1，可到 Maven 官方网站 http://maven. apache. org/download. cgi 进行下载，找到 apache-maven-3. 8. 1-bin. zip 的压缩包链接，下载并解压后得到 apache-maven-3. 8. 1 目录，配置环境变量 MAVEN_HOME，值为 D:\Maven\apache-maven-3. 8. 1。配置好 MAVEN_HOME 后，将%MAVEN_HOME%\bin 加入系统的环境变量 path 中。完成配置后，在命令行窗口输入 mvn -v，稍后可以看到输出的 Maven 版本信息，如图 1-2 所示。

```
管理员: 命令提示符
Microsoft Windows [版本 6.1.7601]
版权所有 (c) 2009 Microsoft Corporation。保留所有权利。

C:\Users\Administrator>mvn -v
Apache Maven 3.8.1 (05c21c65bdfed0f71a2f2ada8b84da59348c4c5d)
Maven home: D:\Maven\apache-maven-3.8.1\bin\..
Java version: 1.8.0_281, vendor: Oracle Corporation, runtime: D:\Program Files\J
ava\jdk1.8.0_281\jre
Default locale: zh_CN, platform encoding: GBK
OS name: "windows 7", version: "6.1", arch: "amd64", family: "windows"

C:\Users\Administrator>_
```

图 1-2　查看 Maven 环境变量

配置 Maven 仓库主要关注两个方面：Maven 下载的类库的来源，即远程 Maven 仓库。下载后的类库需要地方存储，即本地 Maven 仓库。本书放在 D:\Maven\MavenRepository 目录下。修改 Maven 的 settings. xml 文件(该文件存放在 MAVEN_HOME/config 目录下)，打开 settings. xml 文件，添加< localRepository > D:/Maven/MavenRepository </localRepository >，如图 1-3 所示。

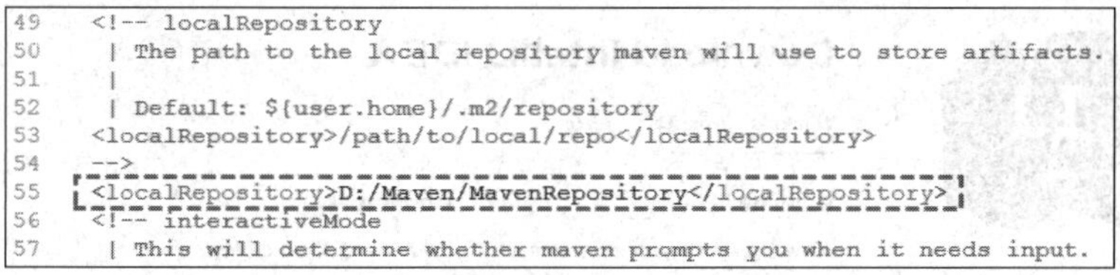

```
<!-- localRepository
 | The path to the local repository maven will use to store artifacts.
 |
 | Default: ${user.home}/.m2/repository
<localRepository>/path/to/local/repo</localRepository>
-->
<localRepository>D:/Maven/MavenRepository</localRepository>
<!-- interactiveMode
 | This will determine whether maven prompts you when it needs input.
```

图 1-3　Maven 本地仓库设置

由于 Apache 官方的仓库位于国外，所以下载速度较慢，这里推荐使用国内阿里云的 Maven 仓库镜像，以提升下载速度，并且修改本地 Maven 仓库的存放地址，修改 Maven 的 settings. xml 文件，在< mirrors ></mirrors >标签中添加 mirror 子节点，配置如下。

```
<mirrors>
  <mirror>
    <id>aliyunmaven</id>
    <mirrorOf>public</mirrorOf>
    <name>阿里云公共仓库</name>
    <url>https://maven.aliyun.com/repository/public</url>
  </mirror>
</mirrors>
```

配置完成,在命令行输入 mvn help：system 进行测试,看到下载链接是 ailiyun 的链接。最后出现 BUILD SUCCESS 的提示,表示配置成功,如图 1-4 所示。

```
管理员: 命令提示符
CLASSPATH=.;D:\Program Files\Java\jdk1.8.0_281\lib;D:\Program Files\Java\jdk1.8.
0_281\lib\tools.jar
MAVEN_PROJECTBASEDIR=C:\Users\Administrator

[INFO] ------------------------------------------------------------------------
[INFO] BUILD SUCCESS
[INFO] ------------------------------------------------------------------------
[INFO] Total time:  02:29 min
[INFO] Finished at: 2021-07-14T17:21:15+08:00
[INFO] ------------------------------------------------------------------------

C:\Users\Administrator>
```

图 1-4　测试阿里云仓库是否配置成功

下载完成后,查看 Maven 的本地仓库 D:\Maven\MavenRepository,就会出现从远程仓库下载的文件。

1.2.3　IntelliJ IDEA 安装与设置

登录 IntelliJ IDEA 的官网(下载地址：https://www.jetbrains.com/idea/download/),如图 1-5 所示。

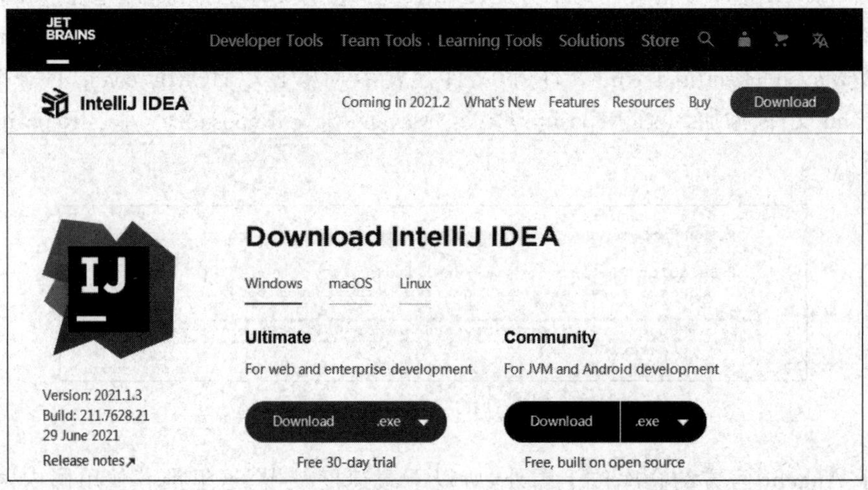

图 1-5　IDEA 官网下载地址

IDEA 主要有旗舰试用版和社区免费版,本书使用的是 ideaIU-2021.1.3 版本,读者可以根据自己的情况下载合适的版本,各版本之间的差异不大,下载完成后安装即可。安装完成后,打开进入 IDEA 欢迎界面,如图 1-6 所示。

Projects 选项说明如下。

- New Project：创建一个新的项目。
- Open：打开或导入一个现有的项目。
- Get from VCS：从版本管理工具内获取项目,可以通过服务器上的项目地址或其他 Git 托管服务器上的项目。

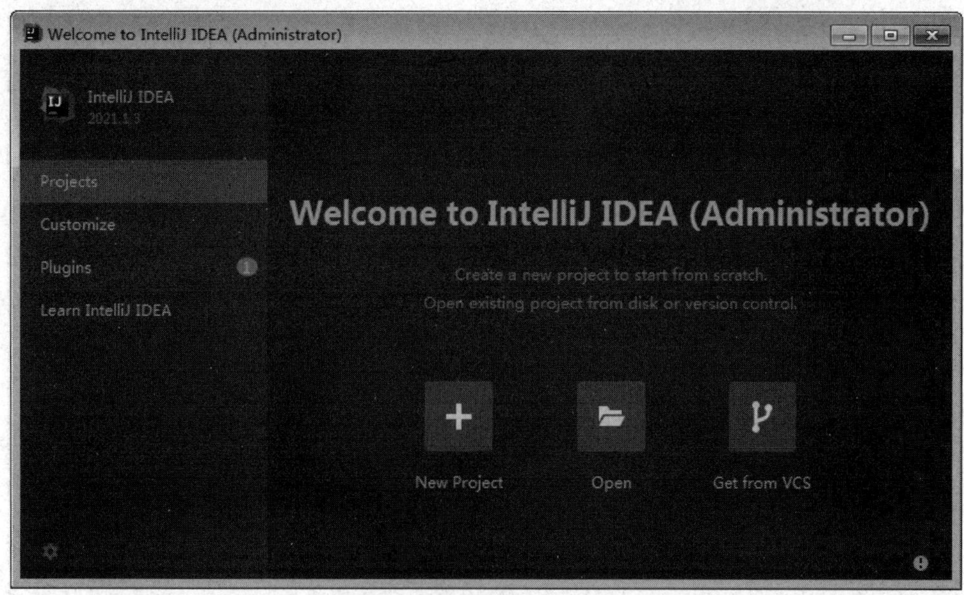

图 1-6 IDEA 的启动界面

在 IDEA 的欢迎界面，单击左侧的 Customize 选项，在右侧的 Color theme 下拉列表框中可以设置 IDEA 的主题，选择相应的颜色模板，这里选择 IntelliJ Light，如图 1-7 所示。

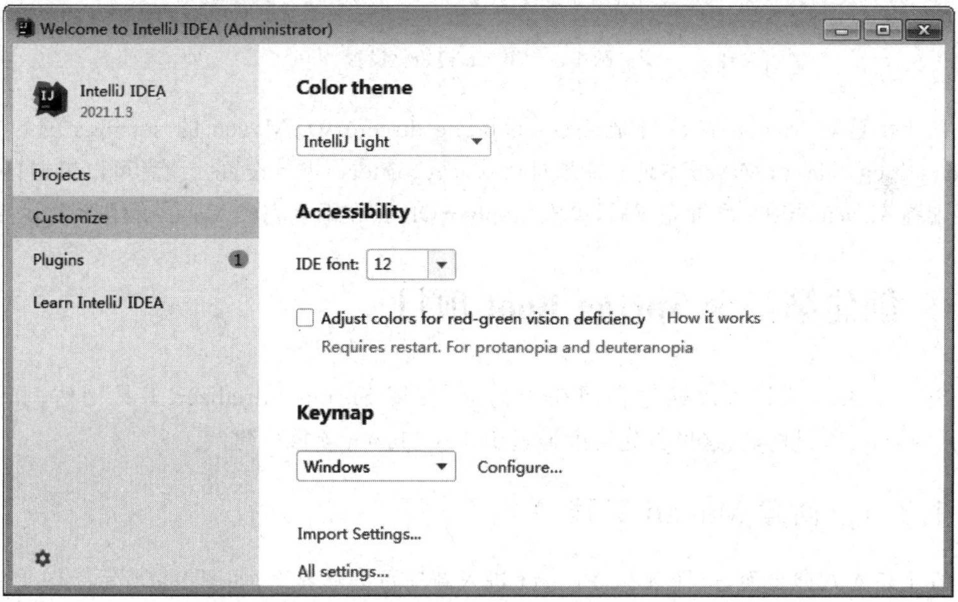

图 1-7 Theme 主题设置

为避免以后每个项目都要配置 Maven，这里在 IDEA 中统一配置 Maven。在 IDEA 欢迎界面的 Customize 选项下，单击右侧下方的 All settings 链接，弹出 Settings 对话框，在界面左侧选择 Build, Execution, Deployment→Build Tools→Maven 选项，在右侧对应的设置界面中进行 Maven 初始化设置，具体如图 1-8 所示。

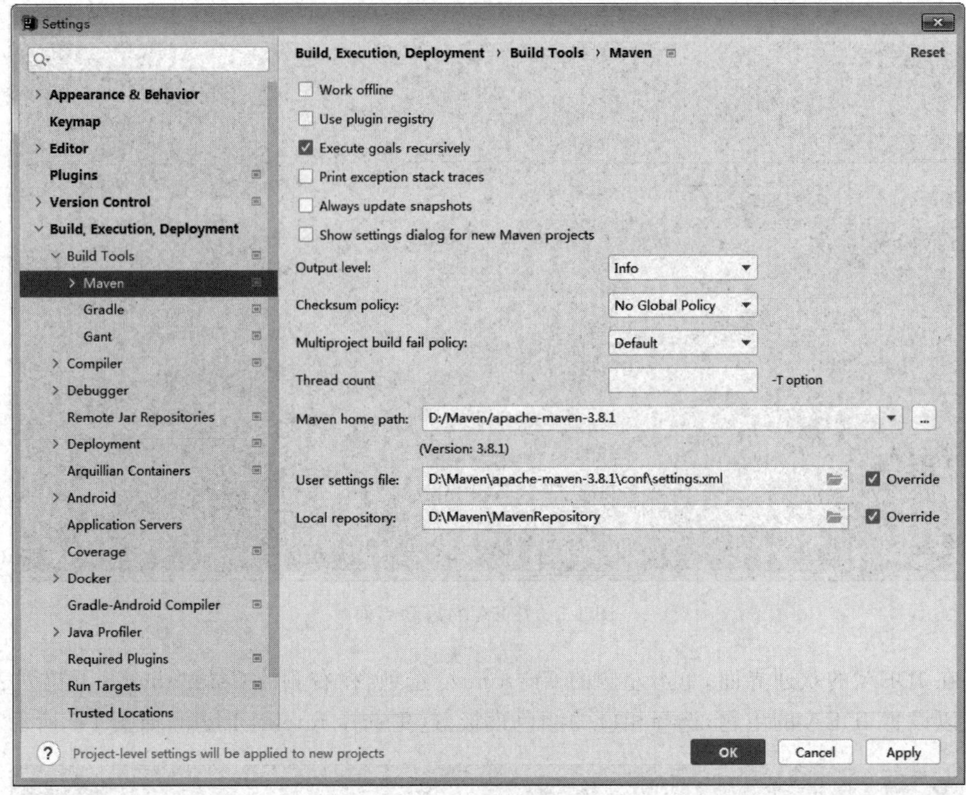

图 1-8　Maven 初始化设置

图 1-8 是对 Maven 安装目录(Maven home directory)、Maven 的 settings 配置文件(User settings file)和 Maven 本地仓库地址(Local repository)进行设置。读者可以根据自己的情况配置 Maven 选项。配置完成后,单击 Apply→OK 按钮即可完成 Maven 的初始化设置。

1.3　创建第一个 Spring Boot 项目

Spring Boot 项目可以通过官网在线构建,使用 Spring Initializer 工具构建,和通过 IDEA Maven 项目构建,这里首先讲解通过 IDEA Maven 项目构建。

1.3.1　创建 Maven 工程

在 IDEA 欢迎界面,选择 New Project 进入新项目创建界面,如图 1-9 所示。图 1-9 中,左侧是可以选择创建的项目类型,包括 Java 项目、Spring 项目、Android 项目、Spring Initializr 项目(即 Spring Boot 项目)、Maven 项目等；右侧是不同类型项目对应的设置界面。这里选择 Maven 选项,右侧选择当前项目的 JDK。

单击图 1-9 中的 Next 按钮进入 Maven 项目创建界面,如图 1-10 所示。在图 1-10 中,Name 用于指定项目名称；Location 用于指定项目的存储路径；单击 Artifact Coordinates,GroupId 表示组织 Id,ArtifactId 表示项目唯一标识符,一般是项目名称；Version 表示项目

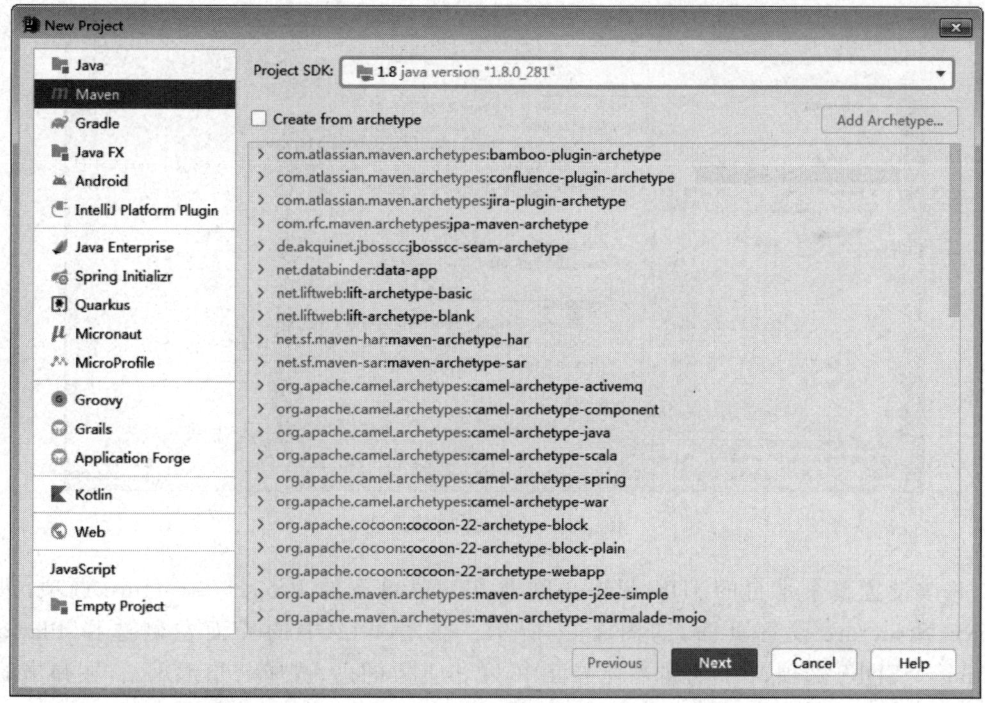

图 1-9　项目类型选择设置

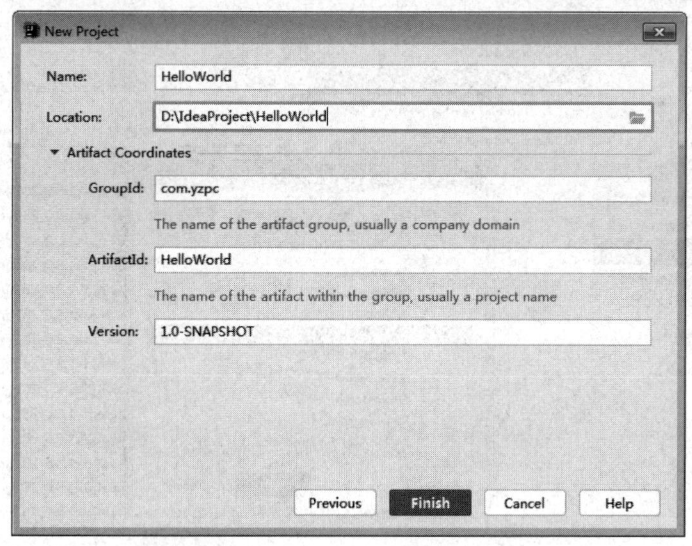

图 1-10　Maven 项目创建界面

版本号。设置好之后，单击 Finish 按钮完成项目的创建。

项目创建完成后，会默认打开创建 Maven 项目生成的 pom. xml 依赖文件，项目结构如图 1-11 所示。

使用 IDEA 开发工具进行 Maven 项目的初始化搭建已经完成，但目前只是一个空的 Maven 项目，要构建 Spring Boot 项目，还需要进行其他设置。

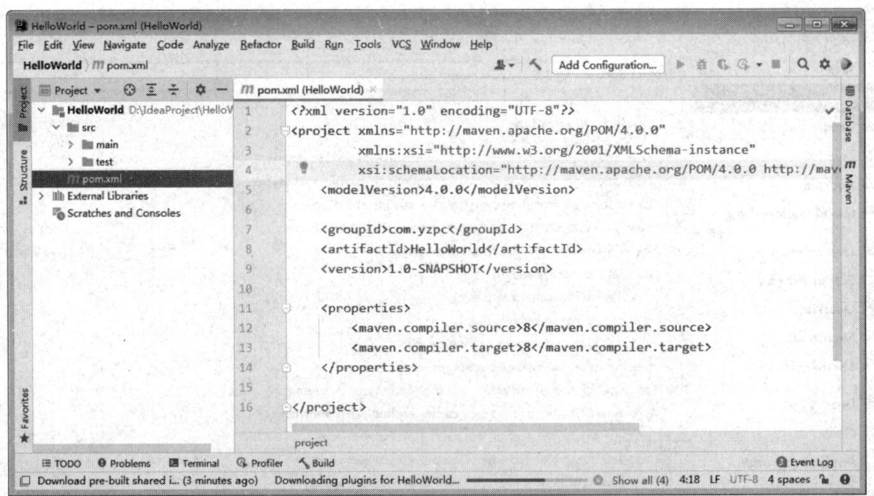

图 1-11　Maven 项目构建

修改设置整个平台的 JDK 版本。单击 File 菜单下的 Project Structure 选项，进入 Project Structure 设置页面，如图 1-12 所示。在图 1-12 中的界面左侧选择 Platform Settings→SDKs 选项，在打开的右侧界面中，单击上方的＋，选择 Add JDK...，在弹出的对话框中，选择 JDK 的路径，这里已经选择前面安装配置的 JDK1.8 版本。

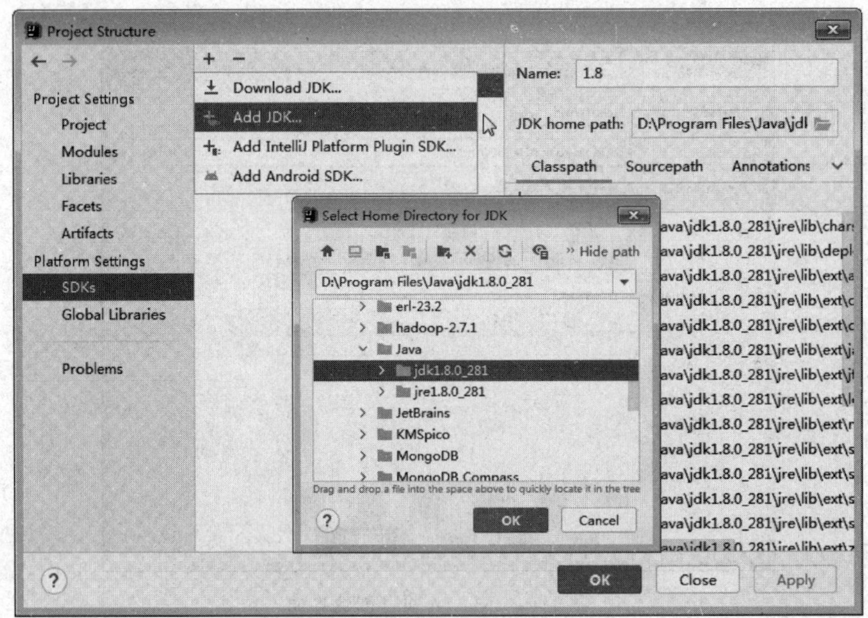

图 1-12　平台的 JDK 设置

设置完成后，在图 1-12 中单击 Apply→OK 按钮，即可完成对平台 JDK 的设置。

IDEA 有两个概念，一个是 Project（工程），另一个是 Module（模块、组件）。可以把 Project 认为是最高的存储目录，在 Project 里又可以创建 Module。IDEA 中的 Project 相当于 Eclipse 系中的 workspace，IDEA 中的 Module 相当于 Eclipse 中的一个项目。而一个

Project 就是由多个 Moudle 组成的整体。在 IDEA 中，并没有对 Project 和 Module 进行强关联和强约束，主要起到项目定义、范围约束、规范类型的效果。

1.3.2 导入相关依赖包

在 HelloWorld 项目下的 pom.xml 文件中添加 Spring Boot 项目和 Web 开发对应的依赖。spring-boot-starter-parent 是一个特殊的 starter，提供一些 Maven 的默认设置，是 Spring Boot 框架集成项目的统一父类管理依赖，添加依赖后可以使用 Spring Boot 的相关特性；<dependencies>标签中添加的 spring-boot-starter-web 依赖是 Spring Boot 框架对 Web 开发集成支持的依赖启动器，代码如下。

```
<!-- 引入 Spring Boot 依赖，继承 Spring Boot 的starter -->
<parent>
    <groupId>org.springframework.boot</groupId>
    <artifactId>spring-boot-starter-parent</artifactId>
    <version>2.5.2</version>
</parent>
<dependencies>
    <!-- 引入 Web 开发依赖启动器，添加 web starter 的依赖 -->
    <dependency>
        <groupId>org.springframework.boot</groupId>
        <artifactId>spring-boot-starter-web</artifactId>
        <version>2.5.2</version>
    </dependency>
</dependencies>
```

pom.xml 文件修改保存后，Maven 将自动在互联网环境下载所需的 jar 文件。

提示：在有些情况下，依赖文件可能无法自动加载，这时需要手动导入依赖文件，具体方法为选择项目，右键选择 Maven→Reload project 进行依赖的重新导入。

1.3.3 创建启动类和控制器

在项目的src/main/java 目录下创建一个名称为 com.yzpc 的包，在该包下新建类 HelloWorldApplication 作为启动类，代码如下。

```
package com.yzpc;
import org.springframework.boot.SpringApplication;
import org.springframework.boot.autoconfigure.SpringBootApplication;
// 声明该类为 Spring Boot 的一个引导类
@SpringBootConfiguration
public class HelloWorldApplication {
    // main 是Java 程序入口
    public static void main(String[] args) {
        // run 方法表示运行Spring Boot 的引导类
        SpringApplication.run(HelloWorldApplication.class,args);
    }
}
```

上述代码中使用@SpringBootApplication注解指定该程序是一个Spring Boot应用，该注解是一个组合注解，相当于@SpringBootConfiguration、@EnableAutoConfiguration和@ComponentScan注解的组合，该注解表明HelloWorldApplication类是Spring Boot项目的主程序启动类，HelloWorldApplication类调用run()方法启动Spring Boot应用。

前面添加了spring-boot-starter-web模块，默认集成了Spring MVC。在com.yzpc包中新建一个名称为MyController的请求处理控制类，并编写一个请求处理方法，代码如下。

```
package com.yzpc;
import org.springframework.web.bind.annotation.GetMapping;
import org.springframework.web.bind.annotation.RestController;
@RestController
public class MyController {
    @GetMapping("/hello")
    public String hello(){
        return "Spring Boot, Hello World! ";
    }
}
```

上述代码中使用@RestController注解是一个组合注解，相当于Spring MVC中的@Controller和@ResponseBody注解的功能之和。在MyController类中，在方法上加入注解@GetMapping("/hello")，此注解是@RequestMapping(method=RequestMethod.GET)的缩写，将HTTP Get映射到方法上。

1.3.4 启动项目

打开HelloWorldApplication入口类，单击类名所在行或main方法所在行前面的绿色按钮，在弹出的选项中选择Run'HelloWorldApplication…'，启动Spring Boot应用程序，如图1-13所示。

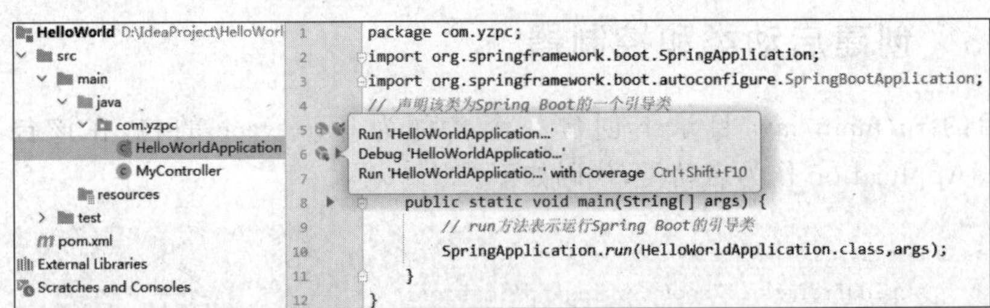

图1-13 启动Spring Boot应用程序

启动后的控制台启动信息，如图1-14所示。

从图1-14中可以看到默认使用的Web容器是Tomcat，项目的端口是8080等信息，在浏览器中输入http://localhost:8080/hello进行测试(hello与MyController类中的@GetMapping("/hello")对应)，可以看到页面输出的内容是Spring Boot，Hello World!，如图1-15所示。

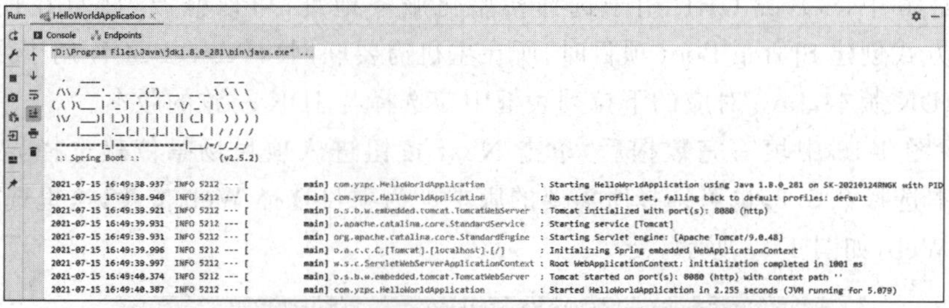

图 1-14 启动后的控制台启动信息

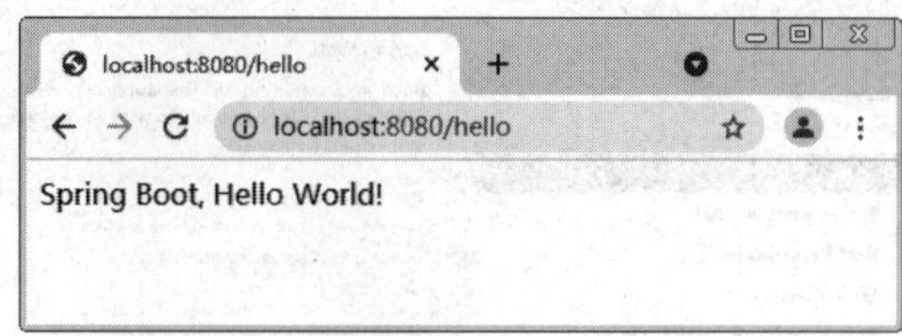

图 1-15 访问 Spring Boot 应用

1.4 使用 Spring Initializer 快速创建项目

除了以上可以使用 Maven 方式创建 Spring Boot 项目外，在实际中更多是通过 IDEA 的 Spring Initializer 方式创建 Spring Boot 项目，创建方式如下。

(1) 在 IDEA 欢迎界面，选择 New Project 进入项目创建界面，在界面左侧选择 Spring Initializer 类型的项目，界面右侧显示项目的配置界面，包括多个选项，如图 1-16 所示。

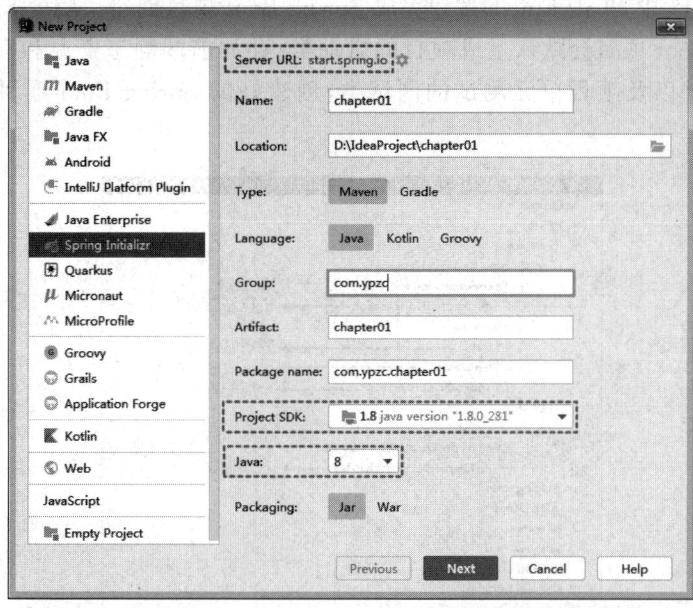

图 1-16 使用 Spring Initializer 类型创建工程

在图 1-16 中，Server URL 用于选择初始化服务地址，可以修改（注意在使用 Spring Initializer 方式创建 Spring Boot 项目时，所在主机需要联网），Project SDK 用于设置创建项目使用的 JDK 版本，Java 对应的下拉列表框中应选择与 JDK 一致的版本。

（2）在图 1-16 中填写完数据后，单击 Next 按钮进入项目场景依赖选择界面，Spring Boot 的版本选择 2.5.2，也可选择所需要的版本。为项目选择 Web 依赖，这里选择 Web 下的 Spring Web，如图 1-17 所示。

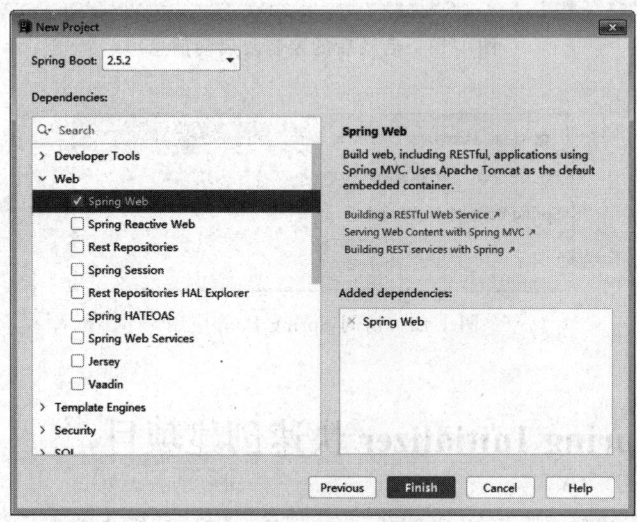

图 1-17　Spring Boot 场景依赖选择界面

选择 Web 依赖后，创建时会自动把选中的依赖添加到项目的 pom.xml 文件中，也可不选择依赖，而后在 pom.xml 中添加所需的依赖。

（3）在图 1-17 中，单击 Finish 按钮完成 Spring Boot 项目创建。使用 Spring Initializer 构建的 Spring Boot 项目会默认生成项目的启动类、存放前端静态资源和模板页面的文件夹、全局配置文件以及项目单元测试的测试类，创建好的 Spring Boot 项目的目录结构如图 1-18 所示。

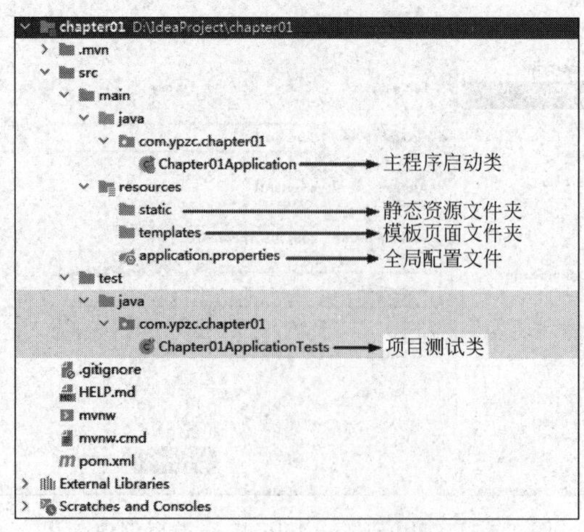

图 1-18　项目目录结构

（4）打开查看项目启动类 Chapter01Application，代码如下。

```
package com.ypzc.chapter01;
import org.springframework.boot.SpringApplication;
import org.springframework.boot.autoconfigure.SpringBootApplication;
@SpringBootApplication
public class Chapter01Application {
    public static void main(String[] args) {
        SpringApplication.run(Chapter01Application.class, args);
    }
}
```

打开查看项目依赖管理文件 pom.xml，代码如下。

```
<?xml version="1.0" encoding="UTF-8"?>
<project xmlns="http://maven.apache.org/POM/4.0.0" xmlns:xsi="http://www.w3.org/2001/
XMLSchema-instance"
    xsi:schemaLocation="http://maven.apache.org/POM/4.0.0 https://maven.apache.org/xsd/
maven-4.0.0.xsd">
    <modelVersion>4.0.0</modelVersion>
    <parent>
        <groupId>org.springframework.boot</groupId>
        <artifactId>spring-boot-starter-parent</artifactId>
        <version>2.5.2</version>
        <relativePath/> <!-- lookup parent from repository -->
    </parent>
    <groupId>com.ypzc</groupId>
    <artifactId>chapter01</artifactId>
    <version>0.0.1-SNAPSHOT</version>
    <name>chapter01</name>
    <description>Demo project for Spring Boot</description>
    <properties>
        <java.version>1.8</java.version>
    </properties>
    <dependencies>
        <dependency>
            <groupId>org.springframework.boot</groupId>
            <artifactId>spring-boot-starter-web</artifactId>
        </dependency>
        <dependency>
            <groupId>org.springframework.boot</groupId>
            <artifactId>spring-boot-starter-test</artifactId>
            <scope>test</scope>
        </dependency>
    </dependencies>
    <build>
        <plugins>
            <plugin>
                <groupId>org.springframework.boot</groupId>
```

```
                <artifactId>spring-boot-maven-plugin</artifactId>
            </plugin>
        </plugins>
    </build>
</project>
```

在项目的 com. yzpc. chapter01 包下创建一个请求控制类 MyController，并编写一个请求处理方法，其代码与 1. 3. 3 小节中的 MyController 类的代码一致。

运行项目的主程序启动类 Chapter01Application，项目成功后，在浏览器地址栏上访问 http://localhost: 8080/hello，运行结果如图 1-15 所示。Spring Boot 项目完全颠覆了对传统 Web 项目的认识，它没有原有的 web. xml 文件，只需几行代码，就可以完成原有 Spring MVC 项目的烦琐配置，甚至不需要配置 Tomcat，直接在内部提供 Tomcat。

1.5 小结

本章首先介绍了 Spring Boot 的概念和特点，然后介绍了 IntelliJ IDEA 的环境搭建，通过创建 Maven 工程开发第一个 Spring Boot 程序，最后通过使用 Spring Initializer 快速构建 Spring Boot 项目，几乎零配置，开发者就可以直接使用 Spring 和 Spring MVC 中的功能。通过本章的学习，读者应对 Spring Boot 有初步的认识，以便为后续学习 Spring Boot 做铺垫。

第2章 Spring Boot基本配置

第 1 章使用 Spring Boot 构建了一个 Web 应用，这个 Web 应用并没有 web. xml，也没有其他配置，如果要对 Web 应用进行配置，或者使用自定义配置，该如何实现？本章将介绍 Spring Boot 的基本配置，包括 Spring Boot 配置文件、Properties 配置、Profile 配置、Web 容器配置和 YAML 配置。

2.1 Spring Boot 配置文件

Spring Boot 中采用了大量的自动化配置，但是对开发者而言，在实际项目中不可避免会有一些需要手动配置。Spring Boot 支持使用 Properties 和 YAML 两种配置方式，Properties 的优先级要高于 YAML（YAML 的文件以. yml 或. yaml 为后缀）。

使用 IntelliJ IDEA 的 Spring Initializr 方式创建 Spring Boot 项目时，IDE 默认会在 src/main/resources 目录下创建一个 application. properties 文件，在这种情况下，配置时需要使用下面的格式（以端口号配置为例）。

```
key = value
```

示例代码如下：

```
server.port = 8088
```

此外，也可以将配置文件 application. properties 的后缀修改为. yml，即文件全名为 application. yml，YAML 文件的好处是采用树状结构，一目了然，其语法结构如下。

```
key: 空格 value
```

端口配置的示例代码如下：

```
server:
   port: 8088
```

Spring Boot 项目中的 application. properties 或者 application. yml 配置文件，可出现在

以下 4 个位置：

(1) 项目根目录的 config 目录；

(2) 项目根目录；

(3) 项目 classpath 下的 config 目录；

(4) 项目 classpath 目录。

如果这 4 个位置中都有 application. properties(. yml)文件，那么加载的优先级从(1)到(4)依次降低，Spring Boot 将按照这个优先级查找配置信息，如图 2-1 所示。

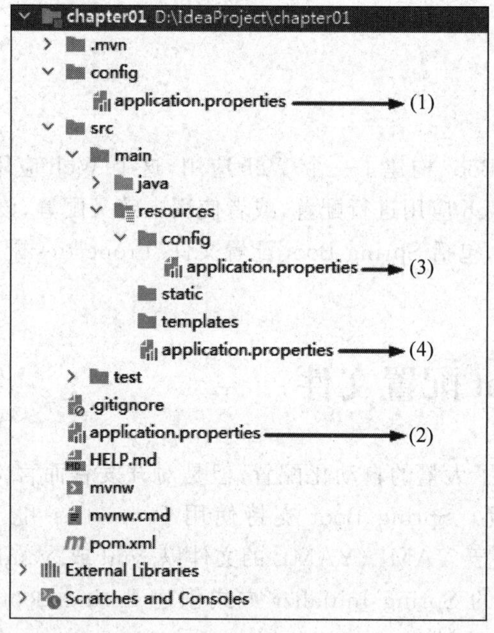

图 2-1　配置文件读取顺序

如果同一配置项出现在多份配置文件中，后读取的值不会覆盖前面读取的值。

2.2　Properties 配置

在 application. properties 文件中定义 Spring Boot 项目的相关属性，这些属性可以是系统属性、环境变量、命令参数等信息，也可以是自定义配置文件名称和位置，示例代码如下。

```
server.address = 80
server.port = 8088
spring.datasource.driver-class-name = com.mysql.cj.jdbc.Driver
spring.config.name = application
```

下面通过实例讲解 application. properties 配置文件的具体使用。

(1) 使用 Spring Initializr 方式创建一个名为 chapter02 的 Spring Boot 项目，包结构为 com. yzpc，在 Dependencies 依赖选择中选择 Web 依赖。

(2) 在 chapter02 项目的 com. yzpc 包下新建一个 entity 包，并在该包下新建 Pet 和

Person 的实体类。

Pet 类的代码如下。

```
package com.yzpc.entity;
public class Pet {
    private String type;
    private String name;
    // 省略属性的getter()方法和setter()方法
    // 省略重写的toString()方法
}
```

Person 类的代码如下。

```
package com.yzpc.entity;
import org.springframework.boot.context.properties.ConfigurationProperties;
import org.springframework.stereotype.Component;
import java.util.*;
@Component
@ConfigurationProperties(prefix = "person")
public class Person {
    private int id;
    private String name;
    private List hobby;
    private String[] family;
    private Map map;
    private Pet pet;
    // 省略属性的getter()方法和setter()方法
    // 省略重写的toString()方法
}
```

@ConfigurationProperties(prefix="person")注解是将配置文件中以 person 开头的属性值通过 setter 方法注入实体类对应属性中，@Component 注解是将当前注入属性值的 Person 类对象作为 Bean 组件放到 Spring 容器中，这样才能被@ConfigurationProperties 注解赋值。

(3) 打开项目 resources 目录下的 application.properties 配置文件，添加配置项及值，用于测试获取配置项的值，代码如下。

```
person.id = 2
person.name = Jenny
person.hobby = read,sport,sing
person.family = mother,father
person.map.key1 = value1
person.map.key2 = value2
person.pet.type = cat
person.pet.name = garfield
```

在配置文件中，通过 person.xx 对 Person 类的相关属性进行设置，这些配置属性会通过

@ConfigurationProperties(prefix = "person")注解注入 Person 实体的相应属性中。

(4) 在项目的测试类 Chapter02ApplicationTests 中引入 Person 实体类 Bean，通过@Autowired 注解将 Person 作为 Bean 注入 Spring 容器，在 contextLoads()方法中输出 Person，测试类的代码如下。

```
package com.yzpc;
import com.yzpc.entity.Person;
import org.junit.jupiter.api.Test;
import org.springframework.beans.factory.annotation.Autowired;
import org.springframework.boot.test.context.SpringBootTest;
@SpringBootTest
class Chapter02ApplicationTests {
    @Autowired
    private Person person;
    @Test
    void contextLoads() {
        System.out.println(person);
    }
}
```

运行 contextLoads()方法，在控制台正确输出了 Person 实体类对象，控制台的输出结果如图 2-2 所示。

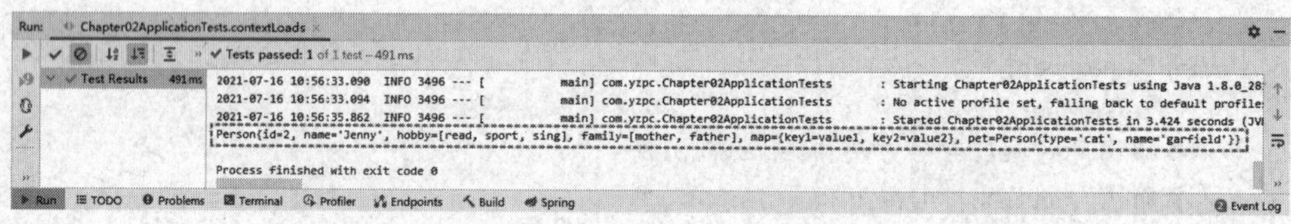

图 2-2 测试方法执行结果

程序运行成功，说明 application.properties 配置文件属性配置正确，并通过相应注解自动完成了属性注入。

2.3 YAML 配置

YAML 格式是 Spring Boot 支持的一种 JSON 超集文件格式，类似 XML、JSON 等。YAML 采用树状结构，一目了然，简洁而强大，通过配置分层、缩进等格式，来增强配置文件的可读性。application.yml(.yaml)配置文件的工作原理和 application.properties 一样，但 Properties 的优先级高于 YAML。

YAML 文件使用缩进控制层级关系。下面介绍 YAML 文件针对不同数据类型配置属性的写法。

(1) value 值为普通数据类型。当 YAML 配置文件中配置的属性值为普通数据类型时，例如数字、字符串、布尔类型时，可直接配置对应的属性值，示例代码如下。

```
server:
  port: 8088
  path: /hello
  tomcat:
    uri-encoding: UTF-8
```

(2) value 值为数组和单列集合。当 YAML 配置文件中配置的属性值为数组或集合类型时,主要有缩进式写法和行内式写法。缩进式写法有两种表现形式,示例代码如下。

```
person:
  hobby:
     - read
     - sport
     - sing
```

或者使用如下示例形式。

```
person:
  hobby:
    read,
    sport,
    sing
```

在 YAML 配置文件中,还可将上述两种缩进式写法简化为行内式写法,行内式写法比缩进式写法更加简洁,示例代码如下。

```
person:
  hobby: [read, sport, sing]
```

使用行内式写法设置属性值时,方括号可以省略,程序会自动匹配校对属性的值。

(3) value 值为 Map 集合和对象。当 YAML 文件配置的属性值为 Map 集合或对象时,YAML 配置文件格式同样可分为缩进式和行内式写法。

缩进式写法的示例代码如下。

```
person:
  map:
    key1: value1
    key2: value2
```

行内式写法的示例代码如下。

```
person:
  map: {key1: value1,key2: value2}      #属性值要用大括号{}包含
```

接下来,在 2.2 节案例的基础上,使用 application.yml 配置文件为 Person 对象赋值,具体步骤如下。

（1）在 chapter02 项目的 resources 目录下，新建一个 application. yml 文件，在该配置文件中设置 Person 对象的属性值，内容如下。

```
person:
  id: 3
  name: 王英瑛
  hobby: [read,sport,sing]
  family: [mother,father]
  map: {key1: value1,key2: value2}
  pet: {type: cat,name: garfield}
```

在 Spring Boot 全局配置文件 application. yml 中配置了 person 的相关属性，这些配置属性将通过@ConfigurationProperties(prefix="person")注解注入 Person 实体类的对应属性中。

（2）打开项目的测试类 Chapter02ApplicationTests，执行测试方法 contextLoads()，查看控制台输出结果，如图 2-3 所示。

```
✔ Tests passed: 1 of 1 test – 315 ms
2021-07-16 11:41:59.501  INFO 7216 --- [           main] com.yzpc.Chapter02ApplicationTests       : Started Chapter02ApplicationTests in 3.364 seconds (
Person{id=3, name='王英瑛', hobby=[read, sport, sing], family=[mother, father], map={key1=value1, key2=value2}, pet=Person{type='cat', name='garfield'}}
```

图 2-3　ContextLoads()测试方法执行结果

注意：需要将之前在 application. properties 配置文件中编写过的配置进行注释，因为 application. properties 配置文件的优先级高于 application. yml 配置文件。

2.4　Profile 多环境配置

在开发 Spring Boot 项目的时候，开发者在项目发布之前，一般需要在开发环境、测试环境和生产环境之间进行切换，不同环境可能需要不同的环境配置，针对这种情况，手动修改配置文件的做法不太现实。通常情况下，可配置多个配置文件，在不同的环境下进行替换。Spring Boot 框架提供了两种多环境配置方式，分别是 Profile 文件多环境配置和@Profile 注解多环境配置。下面介绍使用 Profile 文件进行多环境配置。

在 Spring Boot 项目中，新建几个配置文件，文件名以 application-{profile}. properties 的格式表示，其中{profile}对应具体的环境标识，示例如下。

- application-dev. properties：开发环境。
- application-test. properties：测试环境。
- application-prod. properties：生产环境。

如果想要使用上述对应环境的配置文件，可以在 application. properties 主配置文件中配置 spring. profiles. active 属性激活配置，示例代码如下。

```
spring.profiles.active = dev                #激活相应开发环境配置文件
```

除了在主配置文件配置当前要使用的配置文件外，也可以在将项目打包成 jar 包后启动时，在控制台命令行动态指定激活环境配置，命令如下。

```
java -jar xxx.jar --spring.profiles.active=dev
```

下面通过一个案例来演示 Profile 多环境配置文件的具体使用，步骤如下。

(1) 在 chapter02 项目的 resources 目录下，分别新建 application-dev. properties、application-test. properties 和 application-prod. properties 多环境配置文件，并在各配置文件中对服务端口进行不同的设置，代码如下。

- application-dev. properties

```
server.port=8082
```

- application-test. properties

```
server.port=8084
```

- application-prod. properties

```
server.port=8086
```

上述对不同的运行环境设置了不同的服务端口号。

(2) 打开 resources 目录下的主配置文件 application. properties，配置 spring. profiles. active 属性激活开发环境的配置文件，即 application-dev. properties，代码如下。

```
spring.profiles.active=dev                #指定要激活的多环境配置文件
```

(3) 启动项目的启动类 Chapter02Application，查看控制台输出，如图 2-4 所示。

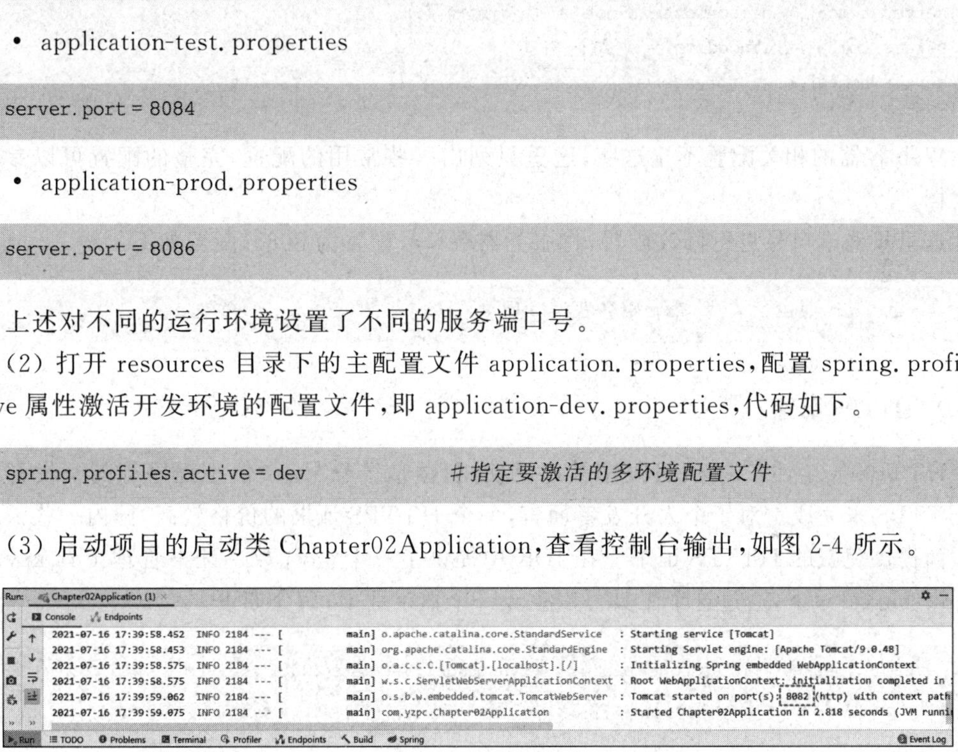

图 2-4 使用 Profile 多环境配置的运行结果

从图 2-4 的运行结果看出，服务启动的端口号为 8082，这与选择激活的 application-dev. properties 文件中的端口号一致，说明配置生效。如果想要其他配置文件激活相应环境，可以在主配置文件中设置相应的配置文件，重启项目查看结果。

2.5 Web 容器配置

使用 Spring Boot 时，首先引人注意的便是其启动方式，人们所熟知的 Web 项目都需要部署到服务容器上，如 Tomcat、Jetty 和 Undertow，然后启动 Web 容器运行系统。在 Spring Boot 项目中，当开发人员添加了 spring-boot-starter-web 依赖后，默认会使用 Tomcat 作为 Web 容器。

1. 常规配置

可以在 application.properties 主配置文件中对 Tomcat 做进一步的配置，示例如下。

```
server.port = # 服务器端口号,默认 8080
server.address = # 服务绑定的网络地址
server.contextPath = # 上下文路径
server.connectionTimeout = # 连接超时时间, -1 无限超时
server.session.timeout = #session 超时时间
server.tomcat.basedir = #Tomcat 的基本目录,未配置时,默认使用系统的临时目录
server.tomcat.redirectContextRoot = # 重定向请求
server.tomcat.uriEncoding = #编码格式
server.tomcat.max-threads = #Tomcat 的最大线程数
```

Web 容器的相关配置不止这些，这里只列出一些常用的配置，完整的配置可以参考官方文档。

这里仅对端口号进行修改配置，将服务器端口号配置为 8088，配置如下。

```
server.port = 8088            #服务器端口号
```

2. HTTPS 配置

HTTPS 安全性较高，在开发中被广泛使用，微信公众号、微信小程序等的开发都要使用 HTTPS 来完成。对于个人开发者而言，一个 HTTPS 证书的价格较高，国内一些云服务器厂商提供免费的 HTTPS 证书。在 JDK 中提供了一个 Java 数字证书管理工具 keytool，在\jdk\bin\目录下，通过这个工具可以生成一个数字证书，命令如下。

```
keytool -genkey -alias tomcathttps -keyalg RSA -keysize 2048 -keystore yang.p12
-validity 365
```

命令的参数意义如下。

-genkey：表示创建一个新的密钥。

-alias：表示 keystore 的别名。

-keyalg：表示使用的加密算法是 RSA。

-keysize：表示密钥的长度。

-keystore：表示生成密钥的存放位置。

-validity：表示密钥的有效时间(单位为天)。

在 cmd 的命令窗口中，执行如上命令，在执行过程中，需要输入密钥口令等信息，根据提示输入即可，如图 2-5 所示。

命令执行完成后，会在当前 bin 目录下生成一个名为 yang.p12 的文件，将这个文件复制到第 1 章的 chapter01 项目的根目录下，而后在 application.properties 中做如下配置。

```
管理员: C:\Windows\system32\cmd.exe
C:\Users\Administrator>d:

D:\Program Files\Java\jdk1.8.0_281\bin>keytool -genkey -alias tomcathttps -keyal
g RSA -keysize 2048 -keystore yang.p12 -validity 365
输入密钥库口令:
再次输入新口令:
您的名字与姓氏是什么?
  [Unknown]:
您的组织单位名称是什么?
  [Unknown]:
您的组织名称是什么?
  [Unknown]:
您所在的城市或区域名称是什么?
  [Unknown]:
您所在的省/市/自治区名称是什么?
  [Unknown]:
该单位的双字母国家/地区代码是什么?
  [Unknown]:
CN=Unknown, OU=Unknown, O=Unknown, L=Unknown, ST=Unknown, C=Unknown是否正确?
  [否]:  y

输入 <tomcathttps> 的密钥口令
        (如果和密钥库口令相同, 按回车):
```

图 2-5 生成数字证书

```
server.ssl.key-store=yang.p12                # key-store 表示密钥文件名
server.ssl.key-alias=tomcathttps             # key-alias 表示密钥别名
server.ssl.key-store-password=123456         # key-store-password 表示在cmd 命令执行过
                                               程中输入的密码
```

配置完成后，启动项目，在 Chrome 浏览器中输入 https://localhost:8088/hello ，可以看到证书是自己生成的，不被浏览器认可，如图 2-6 所示。

图 2-6 证书被拦截

在图 2-6 中，单击左下角的“高级”按钮，出现拦截原因的界面，如图 2-7 所示。

提示：对于不同的浏览器，拦截的形式和提示不一样。

图 2-7　证书拦截原因

单击图 2-7 所示的“继续前往 localhost(不安全)”的链接，成功运行的结果如图 2-8 所示。

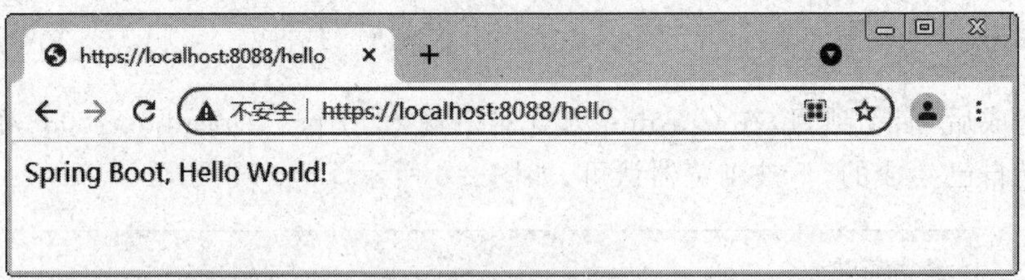

图 2-8　成功运行结果

如果以 HTTP 的形式访问，即在 Chrome 浏览器中输入 http://localhost: 8088/hello，就会访问失败，如图 2-9 所示。

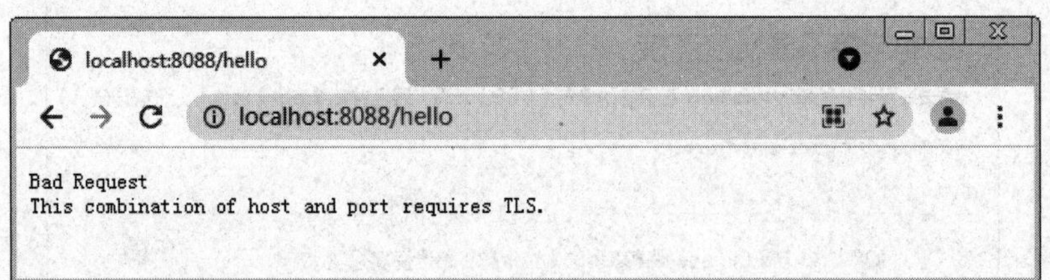

图 2-9　HTTP 访问失败

由于 Spring Boot 不支持同时在配置中启动 HTTP 和 HTTPS，此时可以配置请求重定向，将 HTTP 请求重定向为 HTTPS 请求。在 chapter01 项目的 com. yzpc. chapter01 包下创建一个 TomcatConfig 类来实现 Tomcat 配置，代码如下。

```
package com.ypzc.chapter01;
import org.apache.catalina.Context;
import org.apache.catalina.connector.Connector;
import org.apache.tomcat.util.descriptor.web.SecurityCollection;
import org.apache.tomcat.util.descriptor.web.SecurityConstraint;
import org.springframework.boot.web.embedded.tomcat.
TomcatServletWebServerFactory;
import org.springframework.context.annotation.Bean;
import org.springframework.context.annotation.Configuration;
@Configuration
public class TomcatConfig {
    @Bean
    TomcatServletWebServerFactory tomcatServletWebServerFactory(){
        TomcatServletWebServerFactory factory = new TomcatServletWebServerFactory(){
            @Override
            protected void postProcessContext(Context context) {
                //super.postProcessContext(context);
                SecurityConstraint constraint = new SecurityConstraint();
                constraint.setUserConstraint("CONFIDENTIAL");
                SecurityCollection collection = new SecurityCollection();
                collection.addPattern("/*");
                constraint.addCollection(collection);
                context.addConstraint(constraint);
            }
        };
        factory.addAdditionalTomcatConnectors(createTomcatConnector());
        return factory;
    }
    private Connector createTomcatConnector() {
        Connector connector = new Connector("org.apache.coyote.http11.Http11NioProtocol");
        connector.setScheme("http");
        connector.setPort(8080);
        connector.setSecure(false);
        connector.setRedirectPort(8088);
        return connector;
    }
}
```

这里先配置一个TomcatServletWebServerFactory，然后添加一个Tomcat 的连接方法监听 8080 端口，并将请求转发到 8088 端口上去，重新启动项目，控制台信息如图 2-10 所示。

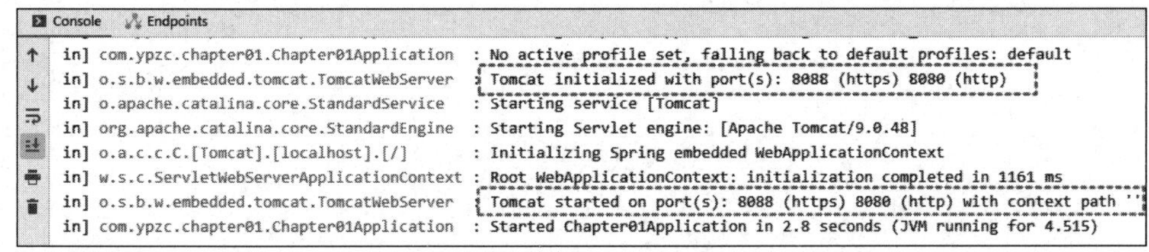

图 2-10 设置重定向端口

在浏览器中输入 http://localhost：8080/hello，会自动重定向到 https://localhost：8088/hello。

2.6 小结

本章主要介绍了 Spring Boot 的常见基础配置，包括 properties 配置文件、YAML 配置文件、Profile 多环境配置、Web 容器配置等，这些配置将是后面章节学习的基础。希望通过本章的学习，读者能够掌握 Spring Boot 的基础配置。

第3章 Spring Boot 日志

日志是追溯系统使用、问题追踪的依据，是一个系统不可缺少的记录事件的组件。在开发阶段，比较容易对系统进行监控，但是对于分布式系统，日志收集对开发者维护系统、Bug定位起着至关重要的作用。通过 Spring Boot 提供的日志抽象层可以方便地管理应用日志的输出。

3.1 日志框架

3.1.1 Spring Boot 的日志

Spring Boot 使用 Commons Logging 进行所有内部日志记录，但保留底层日志实现。项目中只要导入spring-boot-starter.jar 依赖，就会传递导入spring-boot-starter-logging.jar，从 IDEA 的 Maven 面板可以查看依赖关系，如图 3-1 所示。

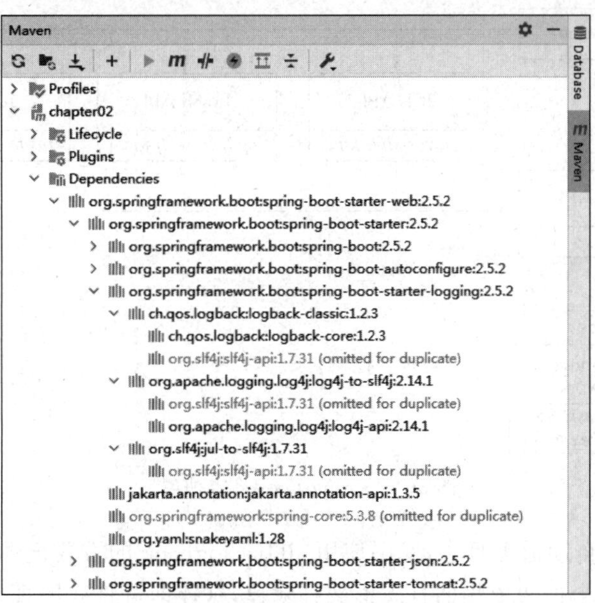

图 3-1 Spring Boot 日志的依赖 JAR 包

spring-boot-starter-logging. jar 依赖如下三个 JAR 包。

- logback-classic. jar：它传递依赖于 logback-core. jar 和 slf4j-api. jar。
- log4j-to-slf4j. jar：它传递依赖于 log4j-api. jar 和 slf4j-api. jar。
- jul-to-slf4j. jar：它传递依赖于 slf4j-api. jar。

市面上常见的日志框架有很多，比如：SLF4J（Simple Logging Facade for Java）、Log4j、Log4j2、Logback、Jakarta Common Logging（JCL）、java. util. logging（JUL）、jboss-logging 等。这些日志框架又可分为日志门面和日志实现，如表 3-1 所示。

表 3-1 日志门面和日志实现

日志门面（日志的抽象层）	日 志 实 现
SLF4J JCL jboss-logging	Log4j JUL Log4j2 Logback

在表 3-1 中，左侧选一个门面（抽象层），右侧选一个实现。Spring Boot 底层是 Spring 框架，Spring 框架默认使用 JCL，spring-boot-starter-logging 采用 SLF4J＋Logback 的组合形式，其中 SLF4J 作为日志门面，Logback 作为日志实现，开发者通常不需要直接操作日志实现的 API。Spring Boot 默认会添加 SLF4J 依赖（slf4j-api. jar）和 Logback 依赖（logback-core. jar 和 slf4j-api. jar）。

Spring Boot 框架需要整合大量第三方框架，而这些框架的底层可能会使用 JCL、Log4j、JUL 等日志。从依赖关系中可以看到，log4j-to-slf4j. jar 用来将 Log4j 日志路由到 SLF4J，jul-to-slf4j. jar 用来将 JUL 日志路由到 SLF4J。Spring Boot 日志抽象层的示意如图 3-2 所示。

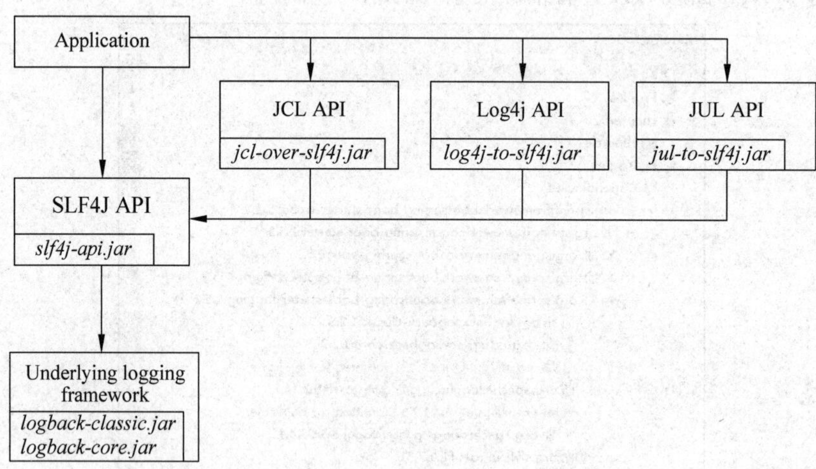

图 3-2 Spring Boot 日志抽象层

Spring Boot 能自动适配日志，底层使用 slf4j＋logback 的方式记录日志，引入其他框架的时候，只需要将这个框架依赖的日志框架移除掉，这样即可将底层日志实现改为使用其他框架。

注意：当把 Spring Boot 应用部署到 Web 服务器或应用服务器上时，JUL 生成的日志不再被路由到 Spring Boot 应用的日志中。

3.1.2 Logback 简介

Logback 日志框架是由 log4j 创始人开发的另一套开源日志组件，是 Spring Boot 默认选择的日志实现，体系非常强大，提供了 logback-core、logback- classic 和 logback-access 三个模块供开发者使用。

- logback-core：是 Logback 的基础模块，是其他两个模块的基础模块。
- logback-classic：可看成 Log4j 的一个改进版本，此外，logback-classic 完整实现 SLF4J API 使开发者可以在其他日志框架(如 Log4j)之间自由切换。
- logback-access：与 Servlet 容器(如 Tomcat)集成，以提供 HTTP 访问日志功能。

3.2 日志使用

3.2.1 默认日志格式

Spring Boot 项目启动后，默认可以看到如图 3-3 所示的界面。

```
  .   ____          _            __ _ _
 /\\ / ___'_ __ _ _(_)_ __  __ _ \ \ \ \
( ( )\___ | '_ | '_| | '_ \/ _` | \ \ \ \
 \\/  ___)| |_)| | | | | || (_| |  ) ) ) )
  '  |____| .__|_| |_|_| |_\__, | / / / /
 =========|_|==============|___/=/_/_/_/
 :: Spring Boot ::                (v2.5.2)

2021-08-04 20:15:30.886  INFO 5760 --- [           main] com.yzpc.Chapter02Application            : Starting Chapter02Application using Java 1.8.0_281 on SK-20210124RNGK with PID
2021-08-04 20:15:30.888  INFO 5760 --- [           main] com.yzpc.Chapter02Application            : The following profiles are active: dev
2021-08-04 20:15:31.905  INFO 5760 --- [           main] o.s.b.w.embedded.tomcat.TomcatWebServer  : Tomcat initialized with port(s): 8082 (http)
2021-08-04 20:15:31.917  INFO 5760 --- [           main] o.apache.catalina.core.StandardService   : Starting service [Tomcat]
2021-08-04 20:15:31.917  INFO 5760 --- [           main] org.apache.catalina.core.StandardEngine  : Starting Servlet engine: [Apache Tomcat/9.0.48]
2021-08-04 20:15:32.034  INFO 5760 --- [           main] o.a.c.c.C.[Tomcat].[localhost].[/]       : Initializing Spring embedded WebApplicationContext
2021-08-04 20:15:32.035  INFO 5760 --- [           main] w.s.c.ServletWebServerApplicationContext : Root WebApplicationContext: initialization completed in 1094 ms
2021-08-04 20:15:32.424  INFO 5760 --- [           main] o.s.b.w.embedded.tomcat.TomcatWebServer  : Tomcat started on port(s): 8082 (http) with context path ''
2021-08-04 20:15:32.435  INFO 5760 --- [           main] com.yzpc.Chapter02Application            : Started Chapter02Application in 2.339 seconds (JVM running for 4.648)
```

图 3-3 Logback 项目启动日志

从图 3-3 可以看到，Spring 的 Logo 部分是 Spring Boot 框架自带的，日志格式大致如下。

- 时间日期：显示日志打印时间，精确到毫秒。
- 日志级别：日志级别分为 FATAL、ERROR、WARN、INFO、DEBUG、TRACE。
- 进程 ID：指的是当前应用对应的 PID。
- 分隔符：用于区分实际日志消息的开始。
- 线程名称：括在方括号中。
- 记录器名称：一般使用类名。
- 日志内容：日志输出的具体内容。

3.2.2 日志级别

在 Spring Boot 默认应用日志配置中，会将日志默认输出到控制台。在默认情况下，只会记录 ERROR、WARN 和 INFO 级别的日志消息，也可以指定日志级别进行日志输出，如果指定了日志级别，那么只会对应输出高于指定级别的日志信息。Spring Boot 中日志级别为：TRACE < DEBUG < INFO < WARN < ERROR。

下面通过示例来讲解 Spring Boot 日志的用法。

(1) 使用 Spring Initializr 方式创建一个名为 chapter03 的 Spring Boot 项目，Group 为 com.yzpc，Package name 为 com.yzpc，在 Dependencies 依赖选择中选择 Spring Web 依赖。

(2) 在项目的测试类Chapter03ApplicationTests 中添加日志记录器Logger，并在 contextLoads()测试方法中通过 Logger 调用相应级别的方法，代码如下。

```
package com.yzpc;
import org.junit.jupiter.api.Test;
import org.slf4j.Logger;
import org.slf4j.LoggerFactory;
import org.springframework.boot.test.context.SpringBootTest;
@SpringBootTest
class Chapter03ApplicationTests {
    //日志记录器
    Logger logger = LoggerFactory.getLogger(getClass());
    @Test
    void contextLoads() {
        //日志的级别,由低到高: trace < debug < info < warn < error
        //可以调整输出的日志级别;日志就只会在这个级别以后的高级别生效
        logger.trace("……这是 trace 日志……");
        logger.debug("……这是 debug 日志……");
        //Spring Boot 默认给我们使用的是info 级别的,没有指定级别的就用Spring Boot 默认规
定的级别
        logger.info("……这是 info 日志……");
        logger.warn("……这是 warn 日志……");
        logger.error("……这是 error 日志……");
    }
}
```

运行测试方法 contextLoads()，控制台输出的日志信息，如图 3-4 所示。

```
2021-08-04 23:49:18.733  INFO 11028 --- [           main] com.yzpc.Chapter03ApplicationTests       : Starting Chapter03ApplicationTests u
2021-08-04 23:49:18.744  INFO 11028 --- [           main] com.yzpc.Chapter03ApplicationTests       : No active profile set, falling back 1
2021-08-04 23:49:20.769  INFO 11028 --- [           main] com.yzpc.Chapter03ApplicationTests       : Started Chapter03ApplicationTests in
2021-08-04 23:49:21.112  INFO 11028 --- [           main] com.yzpc.Chapter03ApplicationTests       : ……这是info日志……
2021-08-04 23:49:21.112  WARN 11028 --- [           main] com.yzpc.Chapter03ApplicationTests       : ……这是warn日志……
2021-08-04 23:49:21.113 ERROR 11028 --- [           main] com.yzpc.Chapter03ApplicationTests       : ……这是error日志……

Process finished with exit code 0
```

图 3-4　默认 INFO 级别日志

从图 3-4 可以看出，Spring Boot 默认只输出 INFO、WARN 和 ERROR 级别的日志，说明 Spring Boot 应用默认设置的日志级别为 INFO。

(3) 改变日志的设置级别。可以通过 debug=true 或 trace=true 等属性(可通过配置文件、命令行参数、系统变量等方式)改变整个 Spring Boot 核心的日志级别；还可通过 logging.level.<logger-name>=<level>属性(可通过配置文件、命令行参数、系统变量等方式)设置指定日志的日志级别，其中<logger-name>代表日志名，通常就是包名或全限定类

名,<level>则可以是 trace、debug、info、warn 和 error 等级别。

在 application. properties 配置文件中,设置日志级别为 trace,设置如下。

```
logging.level.com.yzpc = trace        #设置 com.yzpc 包下的类以trace 级别输出
```

运行测试方法 contextLoads(),控制台输出的日志信息,如图 3-5 所示。

```
2021-08-05 00:50:31.253  INFO 9464 --- [           main] com.yzpc.Chapter03ApplicationTests       : Starting Chapter03ApplicationTests
2021-08-05 00:50:31.255 DEBUG 9464 --- [           main] com.yzpc.Chapter03ApplicationTests       : Running with Spring Boot v2.5.3, Sp
2021-08-05 00:50:31.264  INFO 9464 --- [           main] com.yzpc.Chapter03ApplicationTests       : No active profile set, falling back
2021-08-05 00:50:33.649  INFO 9464 --- [           main] com.yzpc.Chapter03ApplicationTests       : Started Chapter03ApplicationTests i
2021-08-05 00:50:34.001 TRACE 9464 --- [           main] com.yzpc.Chapter03ApplicationTests       : ……这是trace日志……
2021-08-05 00:50:34.001 DEBUG 9464 --- [           main] com.yzpc.Chapter03ApplicationTests       : ……这是debug日志……
2021-08-05 00:50:34.001  INFO 9464 --- [           main] com.yzpc.Chapter03ApplicationTests       : ……这是info日志……
2021-08-05 00:50:34.001  WARN 9464 --- [           main] com.yzpc.Chapter03ApplicationTests       : ……这是warn日志……
2021-08-05 00:50:34.001 ERROR 9464 --- [           main] com.yzpc.Chapter03ApplicationTests       : ……这是error日志……

Process finished with exit code 0
```

图 3-5 设置为 TRACE 级别日志

3.2.3 日志文件输出

默认情况下,Spring Boot 只会将日志消息输出到控制台,并不会将日志写入日志文件。如果要将日志输出到文件,只需在 application. properties 文件或 application. yml 文件内设置如下两个属性之一。

- logging. file. name:设置日志文件,这里可以设置文件的绝对路径,也可设置文件的相对路径,如 logging. file. name=my. log。
- logging. file. path:设置日志目录,设置好目录后,会在设置目录文件夹下创建一个 spring. log 文件,如 logging. file. path=/D:/log。

上述两个属性,如果只设置一个,Spring Boot 应用会默认读取该配置;如果同时设置,则只有 logging. file. name 会生效。

Spring Boot 应用日志文件输出与控制台输出一致。默认只将 INFO、WARN、ERROR 三个级别的日志输出到文件。当日志文件达到 10MB 时,会自动重新使用新的日志文件。另外,日志文件可以通过 logging. file. max-size 属性更改大小设置(在最新的 Spring Boot 中该属性已过时)。

下面通过示例来讲解 Spring Boot 的日志文件输出。

(1) 不指定路径在当前项目根目录下生成日志文件。在 chapter03 项目的 application. properties 配置文件中,设置日志文件的名称,配置如下。

```
logging.file.name = my.log      # 不指定路径在当前项目下生成 my.log 日志
```

运行测试方法 contextLoads(),控制台输出相应日志信息,查看 chapter03 项目的根目录,会出现一个 my. log 的日志文件,打开 my. log 即可看到日志信息,如图 3-6 所示。

(2) 指定日志文件的输出目录。将 application. properties 配置文件的 logging. file. name 的属性行注释掉,添加配置如下。

```
logging.file.path = SpringLogs/   # 指定日志文件的输出目录,使用spring.log 作为默认文件
```

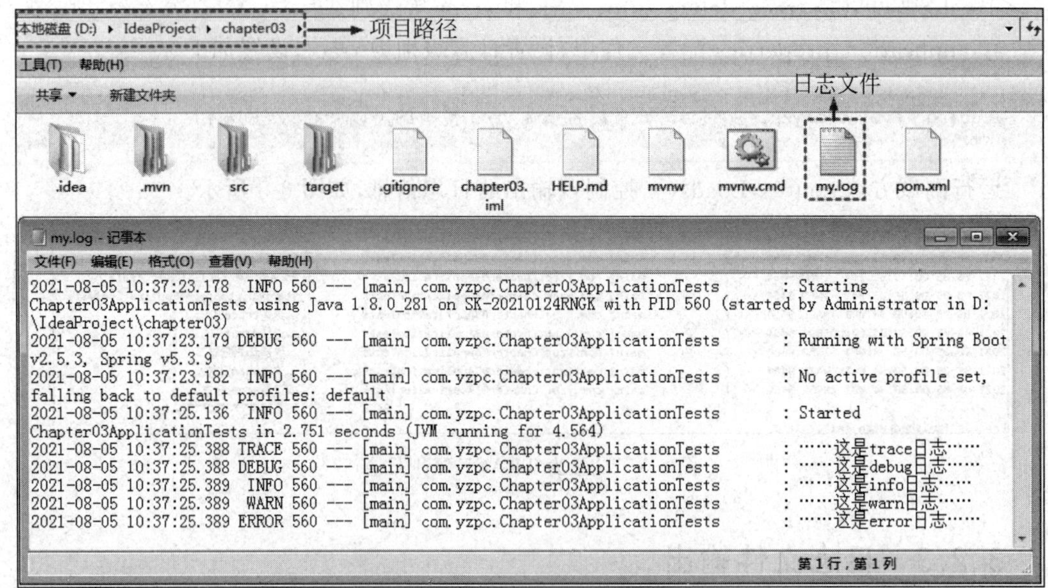

图 3-6 不指定路径生成的日志文件

运行测试方法 contextLoads()，控制台输出相应日志信息，查看项目 chapter03 的根目录，将生成一个 SpringLogs 的文件夹，里面有日志文件 spring.log，如图 3-7 所示。

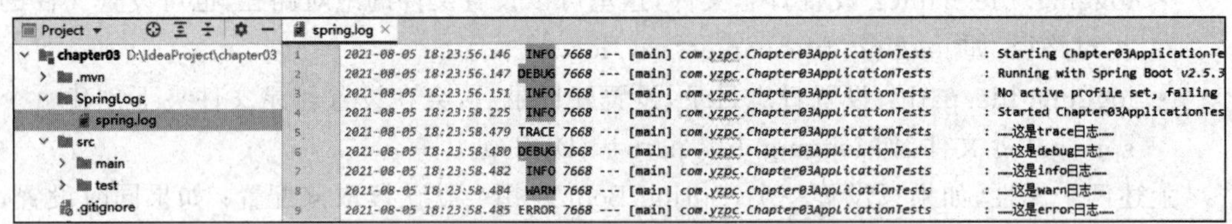

图 3-7 指定日志文件输出目录

（3）自定义路径和文件名生成日志文件。在 application.properties 配置文件中，屏蔽或删除 logging.file.path 属性行，添加 logging.file.name 的属性行，配置如下。

```
logging.file.name = d:/log/my.log   # 可以指定完整的路径和文件名生成日志文件
```

运行测试方法 contextLoads()，控制台输出相应日志信息，查看 d:/log/目录，同样会生成一个 my.log 的日志文件。

3.2.4 自定义日志格式

将日志输出到文件时，可通过 logging.pattern.file 属性和 logging.pattern.console 属性指定日志文件和控制台的输出格式，输出的日志格式如下。

```
%d{yyyy-MM-dd HH:mm:ss.SSS} [%thread] %-5level %logger{50} - %msg%n
```

参数说明如下。

- %d：表示日期时间。
- %thread：表示线程名。
- %-5level：表示级别从左显示 5 个字符宽度。
- %logger{50}：表示 logger 名字最长 50 个字符，否则按照句点分割。
- %msg：表示日志消息。
- %n：表示换行符。

在 chapter03 项目的 application.properties 配置文件中，添加如下配置。

```
logging.pattern.console = %d{yyyy - MM - dd} [%thread] % - 5level %logger{50} - %msg%n
                                          # 在控制台输出的日志的格式
logging.pattern.file = %d{yyyy - MM - dd} --- [%thread] --- % - 5level --- %logger{50}
--- %msg%n                                # 指定文件中日志输出的格式
```

运行测试方法 contextLoads()，控制台输出日志信息以及日志文件中的信息，如图 3-8 所示。

```
2021-08-05 [main] INFO  com.yzpc.Chapter03ApplicationTests - Starting Chapter03ApplicationTests using Java 1.8.0_281 on SK-20210124RNGK
2021-08-05 [main] DEBUG com.yzpc.Chapter03ApplicationTests - Running with Spring Boot v2.5.3, Spring v5.3.9
2021-08-05 [main] INFO  com.yzpc.Chapter03ApplicationTests - No active profile set, falling back to default profiles: default
2021-08-05 [main] INFO  com.yzpc.Chapter03ApplicationTests - Started Chapter03ApplicationTests in 2.972 seconds (JVM running for 5.221)
2021-08-05 [main] TRACE com.yzpc.Chapter03ApplicationTests - ……这是trace日志……
2021-08-05 [main] DEBUG com.yzpc.Chapter03ApplicationTests - ……这是debug日志……
2021-08-05 [main] INFO  com.yzpc.Chapter03ApplicationTests - ……这是info日志……
2021-08-05 [main] WARN  com.yzpc.Chapter03ApplicationTests - ……这是warn日志……
2021-08-05 [main] ERROR com.yzpc.Chapter03ApplicationTests - ……这是error日志……

2021-08-05 --- [main] --- INFO  --- com.yzpc.Chapter03ApplicationTests --- Starting Chapter03ApplicationTests using Java 1.8.0_281
2021-08-05 --- [main] --- DEBUG --- com.yzpc.Chapter03ApplicationTests --- Running with Spring Boot v2.5.3, Spring v5.3.9
2021-08-05 --- [main] --- INFO  --- com.yzpc.Chapter03ApplicationTests --- No active profile set, falling back to default profiles:
2021-08-05 --- [main] --- INFO  --- com.yzpc.Chapter03ApplicationTests --- Started Chapter03ApplicationTests in 3.263 seconds (JVM
2021-08-05 --- [main] --- TRACE --- com.yzpc.Chapter03ApplicationTests --- ……这是trace日志……
2021-08-05 --- [main] --- DEBUG --- com.yzpc.Chapter03ApplicationTests --- ……这是debug日志……
2021-08-05 --- [main] --- INFO  --- com.yzpc.Chapter03ApplicationTests --- ……这是info日志……
2021-08-05 --- [main] --- WARN  --- com.yzpc.Chapter03ApplicationTests --- ……这是warn日志……
2021-08-05 --- [main] --- ERROR --- com.yzpc.Chapter03ApplicationTests --- ……这是error日志……
```

图 3-8　自定义日志输出格式

除了上面介绍过的配置属性外，以下属性用于对日志进行定制。

- logging.exception-conversion-word：记录异常时使用的转换字。
- logging.config：日志配置。
- logging.pattern.dateformat：日志格式内的日期格式。
- logging.charset.console：输出控制台日志时所用的字符集。
- logging.charset.file：输出文件日志时所用的字符集，仅当日志输出到文件时有效。
- logging.pattern.level：输出日志级别时使用的格式(默认为 %5p)。
- PID：当前进程 ID。

3.2.5　基于 XML 配置日志

1. Spring Boot 的默认日志实现

在 Spring Boot 中默认使用 Logback 日志的实现，如图 3-9 所示。

Spring Boot 为 Logback 提供了一些通用的配置文件，这些文件位于 org/springframework/

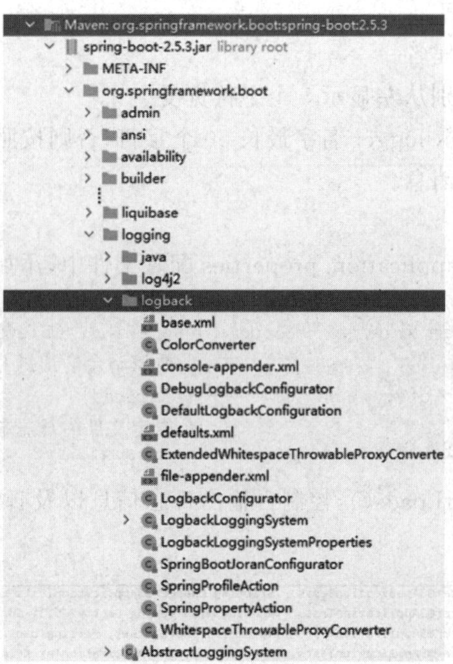

图 3-9 Logback 日志的实现

boot/logging/logback/路径下，其中常用的有如下几个。

- defaults. xml：提供转换规则及各种通用配置。
- console-appender. xml：定义一个 ConsoleAppender，用于将日志输出到控制台。
- file-appender. xml：定义一个 RollingFileAppender，用于将日志输出到文件。

Spring Boot 通过 base. xml 配置文件，加载上述的配置文件来实现日志，代码如下。

```
<?xml version = "1.0" encoding = "UTF - 8"?>
< included >
    < include resource = "org/springframework/boot/logging/logback/defaults. xml" />
    < property name = "LOG_FILE" value = " ${LOG_FILE: - ${LOG_PATH: - ${LOG_TEMP:
 - ${java. io. tmpdir: - /tmp}}}/spring. log}"/>
    < include resource = "org/springframework/boot/logging/logback/
console - appender. xml" />
    < include resource = "org/springframework/boot/logging/logback/
file - appender. xml" />
    < root level = "INFO">
        < appender - ref ref = "CONSOLE" />
        < appender - ref ref = "FILE" />
    </root >
</included >
```

开发者也可通过 XML 文件配置，自定义日志格式及输出，在src/main/resources文件夹下定义 logback. xml 或 logback-spring. xml 作为日志配置，Spring Boot 官方推荐使用带-spring 的配置文件。

2. 控制台输出日志

在 chapter03 项目的 src/main/resources 目录下创建一个 logback-spring. xml 文件，注释掉 application. properties 文件中的属性值，logback-spring. xml 配置文件的设置内容如下。

```
<?xml version = "1.0" encoding = "utf - 8" ?>
< configuration >
    <!-- 控制台输出 -->
    < appender name = "STDOUT" class = "ch.qos.logback.core.ConsoleAppender">
        < encoder class = "ch.qos.logback.classic.encoder.PatternLayoutEncoder">
            <!-- 格式化输出：%date 表示日期，%thread 表示线程名，% - 5level：级别从左显
示 5 个字符宽度，%msg：日志消息，%n 是换行符 -->
            < pattern >%date [%thread] % - 5level %logger{50} - %msg%n
            </pattern >
            < charset > UTF - 8 </charset >
        </encoder >
    </appender >
    <!-- 日志输出级别，等同于在配置文件中设置 logging.pattern.level -->
    < root level = "info">
        < appender - ref ref = "STDOUT" />
    </root >
    < logger name = "com.yzpc" level = "debug" />
</configuration >
```

运行测试方法 contextLoads()，控制台输出日志信息，如图 3-10 所示。

```
2021-08-06 20:13:12,111 [main] INFO  com.yzpc.Chapter03ApplicationTests - Starting Chapter03ApplicationTests using Java 1
2021-08-06 20:13:12,112 [main] DEBUG com.yzpc.Chapter03ApplicationTests - Running with Spring Boot v2.5.3, Spring v5.3.9
2021-08-06 20:13:12,116 [main] INFO  com.yzpc.Chapter03ApplicationTests - No active profile set, falling back to default
2021-08-06 20:13:14,061 [main] INFO  com.yzpc.Chapter03ApplicationTests - Started Chapter03ApplicationTests in 2.611 secc
2021-08-06 20:13:14,319 [main] DEBUG com.yzpc.Chapter03ApplicationTests - ……这是debug日志……
2021-08-06 20:13:14,319 [main] INFO  com.yzpc.Chapter03ApplicationTests - ……这是info日志……
2021-08-06 20:13:14,319 [main] WARN  com.yzpc.Chapter03ApplicationTests - ……这是warn日志……
2021-08-06 20:13:14,319 [main] ERROR com.yzpc.Chapter03ApplicationTests - ……这是error日志……
```

图 3-10 基于 xml 配置的控制台日志

3. 日志文件输出

控制台输出日志文件的形式一般是在开发环境下使用，一般生产环境下需要将日志输出到日志文件进行日志分析，并将日志根据级别输出到不同日志文件中，如果日志文件太大，可以设置日志文件根据大小分隔，配置如下。

```
<?xml version = "1.0" encoding = "utf - 8" ?>
< configuration >
    <!-- 此处省略控制台输出的配置 -->
    <!-- 定义日志文件的存储地址 -->
    < property name = "LOG_HOME" value = "/log" />
    <!-- 按照每天生成日志文件 -->
    < appender name = "FILE" class = "ch.qos.logback.core.rolling.RollingFileAppender">
```

```
        <encoder class="ch.qos.logback.classic.encoder.PatternLayoutEncoder">
            <pattern>%d{yyyy-MM-dd HH:mm:ss.SSS} [%thread] %-5level %logger{50} -
%msg%n</pattern>
            <charset>UTF-8</charset>
        </encoder>
        <rollingPolicy class="ch.qos.logback.core.rolling.TimeBasedRollingPolicy">
            <!-- 日志文件输出的路径和文件名 -->
            <FileNamePattern>${LOG_HOME}/log%d{yyyy-MM-dd}.log
            </FileNamePattern>
            <!-- 日志文件保留天数 -->
            <MaxHistory>30</MaxHistory>
        </rollingPolicy>
        <!-- 日志文件最大的大小 -->
        <triggeringPolicy class="ch.qos.logback.core.rolling.SizeBasedTriggeringPolicy">
            <MaxFileSize>10MB</MaxFileSize>
        </triggeringPolicy>
        <filter class="ch.qos.logback.classic.filter.LevelFilter">
            <level>debug</level>
            <onMatch>ACCEPT</onMatch>
            <onMismatch>DENY</onMismatch>
        </filter>
    </appender>
    <!-- 日志输出级别 -->
    <root level="info">
        <appender-ref ref="STDOUT" />
        <appender-ref ref="FILE" />
    </root>
    <logger name="com.yzpc" level="debug" />
</configuration>
```

在上述配置中包含一些新标签，说明如下。

- file：日志文件位置。
- maxFileSize：设置最大日志文件大小。
- maxHistory：设置日志文件保留的天数。
- fileNamePattern：指定精确到日期的日志切分方式。
- filter：标签中的 level 设置日志级别。

运行测试方法 contextLoads()，在项目所在磁盘的根路径下，会生成一个 log 文件夹，该文件中有按照日期生成的日志文件。

3.2.6 使用 Log4j2 日志实现

Log4j 本身已经非常优秀，Log4j2 完全是 Log4j 的重新设计。若要让 Spring Boot 底层使用 Log4j2，只要在 pom.xml 文件中去除 Logback 的依赖库，并添加 Log4j2 依赖库即可，依赖修改如下。

```
<dependencies>
    <dependency>
        <groupId>org.springframework.boot</groupId>
```

```
            <artifactId>spring-boot-starter-web</artifactId>
            <exclusions>
                <!-- 去除 spring-boot-starter-logging 依赖 -->
                <exclusion>
                    <groupId>org.springframework.boot</groupId>
                    <artifactId>spring-boot-starter-logging</artifactId>
                </exclusion>
            </exclusions>
        </dependency>
        <dependency>
            <groupId>org.springframework.boot</groupId>
            <artifactId>spring-boot-starter-log4j2</artifactId>
        </dependency>
    </dependencies>
```

这里并没有单独去除 Logback 依赖库，Spring Boot 为 Log4j2 也提供了对应的 Starter，所以可以直接去除 Spring Boot 默认的日志 Strater。添加 spring-boot-starter-log4j2.jar 依赖，它会传递性添加它所依赖的 Log4j2 依赖库和 SLF4J 依赖库，从 IntelliJ IDEA 的 Maven 面板上可以看到依赖的 JAR 包，如图 3-11 所示。

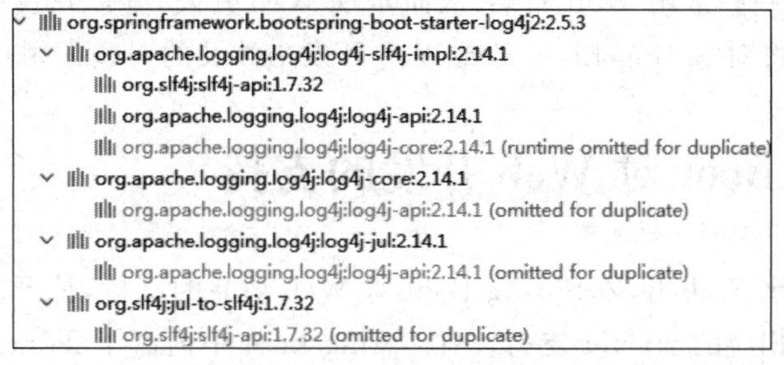

图 3-11 Log4j2 日志的依赖 JAR 包

通过上面的配置，将项目的底层日志框架改成了 Log4j2，这得益于 Spring Boot 日志的抽象机制。如果要对 Log4j2 做自定义的详细配置，则可通过 log4j2.yml 进行配置。

3.3 小结

本章介绍了 Spring Boot 的日志实现机制，首先介绍了 Spring Boot 的日志框架，然后主要就 Spring Boot 默认的日志框架 Logback，介绍了日志的使用、日志级别、日志文件输出、自定义日志格式、日志配置等。

第4章 Spring Boot的Web开发

Web应用开发是现代软件开发中重要的一部分，在一个Web应用中，通常会采用MVC设计模式实现对应的模型、视图和控制器。其中，视图是用户看到并与之交互的界面；后期的Web应用倾向于使用动态模板技术，从而实现前后端分离和页面的动态数据展示。在Web开发中，通常会涉及静态资源的访问支持、视图解析器的配置、转换器、文件上传下载等功能，甚至还需要考虑到与Web服务器关联的Servlet相关组件的定制。Spring Boot框架支持整合一些常用Web框架，从而实现Web开发，并默认支持Web开发中的一些通用功能。本章将对Spring Boot实现Web开发中涉及的一些常用功能进行介绍。

4.1 Spring Boot对Web开发的支持

为了实现并简化Web开发，Spring Boot对Web功能的支持，从开发、测试、部署、运维(安全)等都提供了相应的starter支持。在Spring Boot中内嵌了Tomcat、Jetty、Undertow或Netty等Http服务器的支持。

Spring Boot使用spring-boot-starter-web为Web开发提供支持，spring-boot-starter-web依赖于spring-web和spring-webmvc，其中spring-webmvc代表Spring MVC框架。大多数的Web应用都可以使用spring-boot-starter-web模块进行快速搭建和运行。spring-boot-starter-web主要包括RESTful、参数校验、使用Tomcat作为内嵌容器等功能。也可以使用spring-boot-starter-webflux模块构建响应式的网络应用。

使用Spring进行Web开发时，只需要在项目中引入对应Web开发框架的依赖启动器即可，一旦引入了Web依赖启动器spring-boot-starter-web，那么Spring Boot整合Spring MVC框架默认实现的一些自动配置类就会自动生效，几乎可在无任何额外配置的情况下进行Web开发。

Spring Boot为Spring MVC提供的auto-configuration适用于大多数应用，并在Spring默认功能上添加了以下特性。

(1) 内置两个视图解析器：ContentNegotiatingViewResolver和BeanNameViewResolver。

(2) 对服务器静态资源提供支持，包括对WebJars的支持。

(3) 自动注册Converter、GenericConverter转换器和Formatter格式化器。

(4) 支持使用HttpMessageConverters注册HttpMessageConverter。

（5）自动注册消息代码解析器 MessageCodeResolver。

（6）支持静态的 index. html 首页。

（7）支持定制应用图标 favicon. ico。

（8）自动初始化 Web 数据绑定器 ConfigurableWebBindingInitializer。

Spring Boot 整合 Spring MVC 进行 Web 开发时提供了很多默认配置，而且大多数时候使用默认配置即可满足开发需求。

4.2 自定义消息转换器 HttpMessageConverter

Spring MVC 使用 HttpMessageConverter 接口来转换 HTTP 请求和响应。例如，可以将对象自动转换为 JSON（通过使用 Jackson 库）或 XML（通过使用 Jackson XML 扩展，或者通过使用 JAXB，如果 Jackson XML 扩展不可用）。

1. 常见 JSON 技术

在 JSON 的使用中，有以下几种常见的 JSON 技术。

（1）Jackson。开源的 Jackson 是 Spring MVC 内置的 JSON 转换工具，简单易用且性能相对较高。Jackson 社区相对比较活跃，更新速度比较快。Jackson 对于复杂类型的 JSON 转换 Bean 会出现问题，一些集合 Map、List 的转换出现问题。Jackson 对于复杂类型的 Bean 转换 JSON，转换的 JSON 格式不是标准的 JSON 格式。

（2）Gson。Gson 是目前功能最全的 JSON 解析器，Gson 最初是应 Google 公司内部需求由 Google 自行研发，自从 2008 年 5 月公开发布第 1 版后已被许多公司或用户应用。Gson 的应用主要为 toJson 与 fromJson 两个转换函数，无依赖且不需要额外的 jar 文件，能够直接运行在 JDK 上。Gson 可以完成复杂类型的 JSON 到 Bean 或 Bean 到 JSON 的转换，是 JSON 解析的利器。Gson 在功能上无可挑剔，但是在性能上比 fastjson 稍差。

（3）fastjson。fastjson 是一个 Java 语言编写的高性能 JSON 处理器，由阿里巴巴公司开发。它的特点是无依赖，不需要额外的 jar 文件，能够直接运行在 JDK 上。但是 fastjson 在复杂类型的 Bean 转换 JSON 上会出现一些问题，可能会出现引用的类型不当，导致 JSON 转换出错，因而需要制定引用。fastjson 采用独创的算法，将 parse 的速度提升到极致，超过所有 JSON 库。

2. 默认实现返回 JSON 数据

Spring MVC 中使用消息转换器 HttpMessageConverter 对 JSON 转换提供了很好的支持，在 Spring Boot 中更进一步，对相关配置做了进一步的简化。下面通过默认实现返回 JSON 数据来讲解。

（1）使用 Spring Initializr 方式创建一个名为 chapter04 的 Spring Boot 项目，Group 和 Package name 设为 com. yzpc，在 Dependencies 依赖下选择 Web 节点下的 Spring Web，在 pom. xml 中的 Web 依赖代码如下。

```
<dependencies>
    <dependency>
        <groupId>org.springframework.boot</groupId>
        <artifactId>spring-boot-starter-web</artifactId>
    </dependency>
    ...
<dependencies>
```

这个依赖中默认加入了 jackson-databind 作为 JSON 处理器，此时不需要添加额外的 JSON 处理器就能返回一段 JSON。

（2）在 chapter04 项目的 com.yzpc 包下新建一个 entity 包，并在该包下新建 Person 的实体类，代码如下。

```
package com.yzpc.entity;
import com.fasterxml.jackson.annotation.JsonFormat;
import java.util.Date;
public class Person {
    private String name;
    private String sex;
    private int age;
    @JsonFormat(pattern = "yyyy-MM-dd")
    private Date visitDate;
    //省略 getter、setter 方法
}
```

如果某属性在 JSON 中不输出，可在该属性前添加 @JsonIgnore 的注解。

（3）在 com.yzpc 包下新建一个 controller 包，并在该包下新建 PersonController 类，定义 show()方法返回 Person 对象，代码如下。

```
package com.yzpc.controller;
import com.yzpc.entity.Person;
import org.springframework.web.bind.annotation.GetMapping;
import org.springframework.web.bind.annotation.RestController;
import java.util.Date;
@RestController
public class PersonController {
    @GetMapping("/show")
    public Person show(){
        Person person = new Person();
        person.setName("张三");
        person.setSex("男");
        person.setAge(22);
        person.setVisitDate(new Date());
        return person;
    }
}
```

（4）启动项目，在浏览器中输入 http://localhost:8080/show，即可看到返回了 JSON

数据，如图 4-1 所示。

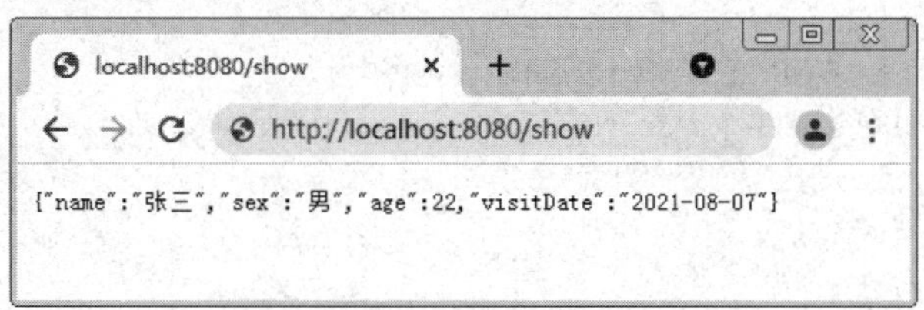

图 4-1　默认实现返回 JSON 数据

这是通过 Spring 中默认提供的 MappingJackson2HttpMessageConverter 来实现的，如果需要添加或自定义转换器，可以使用 Spring Boot 的 HttpMessageConverters，示例代码如下。

```
import org.springframework.boot.autoconfigure.http.HttpMessageConverters;
import org.springframework.context.annotation.*;
import org.springframework.http.converter.*;
@Configuration(proxyBeanMethods = false)
public class MyConfiguration {
    @Bean
    public HttpMessageConverters customConverters() {
        HttpMessageConverter<?> additional = ...
        HttpMessageConverter<?> another = ...
        return new HttpMessageConverters(additional, another);
    }
}
```

上下文中存在的任何 HttpMessageConverter Bean 都将添加到转换器列表中，也可以用同样的方法覆盖默认转换器。

3. 使用 Gson 转换器

常见的 JSON 处理器除了 jackson-databind 之外，还有 Gson 作为 JSON 解析库，这里使用 Gson 来自定义转换器实现 JSON 输出。

(1) 在 chapter04 项目下，使用 Gson 时，需要在 pom.xml 中去除 spring-boot-starter-web 下的默认 jackson-databind 依赖，然后加入 Gson 依赖，这里使用的 Gson 版本为 2.8.7，依赖代码如下。

```
<dependencies>
    <dependency>
        <groupId>org.springframework.boot</groupId>
        <artifactId>spring-boot-starter-web</artifactId>
        <exclusions>
            <exclusion>
                <groupId>com.fasterxml.jackson.core</groupId>
                <artifactId>jackson-databind</artifactId>
            </exclusion>
        </exclusions>
```

```
    </dependency>
    <dependency>
        <groupId>com.google.code.gson</groupId>
        <artifactId>gson</artifactId>
        <version>2.8.7</version>
    </dependency>
</dependencies>
```

由于 Spring Boot 中提供了 Gson 的自动转换类 GsonHttpMessageConvertersConfiguration，因此 Gson 依赖添加成功后，可以像使用 jackson-databind 那样直接使用 Gson。但在 Gson 进行日期转换时，如果对日期数据进行格式化，那么需要开发者自定义 HttpMessageConverter。

(2) 在 com.yzpc 包下新建一个 converter 包，并在该包下新建 GsonConfig 类，代码如下。

```
package com.yzpc.converter;
import com.google.gson.Gson;
import com.google.gson.GsonBuilder;
import org.springframework.context.annotation.Bean;
import org.springframework.context.annotation.Configuration;
import org.springframework.http.converter.json.GsonHttpMessageConverter;
import java.lang.reflect.Modifier;
@Configuration
public class GsonConfig {
    @Bean
    public GsonHttpMessageConverter gsonHttpMessageConverter(){
        // 自己提供一个 GsonGsonHttpMessageConverter 的实例
        GsonHttpMessageConverter converter = new GsonHttpMessageConverter();
        GsonBuilder builder = new GsonBuilder();
        // 设置 Gson 解析时日期的格式
        builder.setDateFormat("yyyy-MM-dd");
        // 设置 Gson 解析时修饰符为protected 的字段被过滤
        builder.excludeFieldsWithModifiers(Modifier.PROTECTED);
        // 创建 Gson 对象放入GsonHttpMessageConverter 的实例并返回converter
        Gson gson = builder.create();
        converter.setGson(gson);
        return converter;
    }
}
```

(3) 修改 Person 类中的 age 字段的修饰符为 protected，代码如下。

```
public class Person {
    private String name;
    private String sex;
    protected int age;
    private Date visitDate;
    // 省略getter、setter 方法
}
```

（4）重新启动项目，在浏览器中访问 http://localhost：8080/show，运行效果如图 4-2 所示。

图 4-2　使用 Gson 转换器返回 JSON

4.3　Spring Boot 序列化和反序列化 JSON 数据

现在 Java 开发大部分都是前后端分离，通过传递 JSON 进行信息传递，不可避免地会遇到对象的序列化和反序列化。JSON 的序列化是指将 Java 中的对象转换成 JSON 字符串，反过来，将 JSON 字符串转换为 Java 中的对象，就是 JSON 的反序列化。下面通过示例来演示 JSON 数据的序列化和反序列化，步骤如下。

（1）在 chapter04 项目中，打开 Chapter04ApplicationTests 项目测试类，添加 toJson()方法实现序列化，代码如下。

```
@SpringBootTest
class Chapter04ApplicationTests {
    …
    @Test
    public void toJson(){
        Map map = new HashMap();
        map.put("id","1001");
        map.put("name","zhangsan");
        Gson gson = new Gson();
        //将 Java 对象转为JSON 字符串,称为 JSON 序列化
        String json = gson.toJson(map);
        System.out.println(json);
    }
}
```

（2）运行测试方法 toJson()，在控制台输出相应的 JSON 格式的字符串，如图 4-3 所示。

```
2021-08-07 14:02:22.127  INFO 5972 --- [           main]
2021-08-07 14:02:22.132  INFO 5972 --- [           main]
2021-08-07 14:02:24.376  INFO 5972 --- [           main]
{"name":"zhangsan","id":"1001"}
```

图 4-3　Java 对象转为 JSON 字符串

（3）在项目测试类中，添加 fromJson()方法实现 JSON 数据的反序列化，代码如下。

```
@Test
public void fromJson(){
```

```
        String json = "{\"name\":\"zhangsan\",\"id\":\"1001\"}";
        Gson gson = new Gson();
        // new TypeToken <要转换的对象类型>(){}.getType()
        Map map = gson.fromJson(json,new TypeToken < Map >(){}.getType());
        System.out.println(map);
    }
```

上述代码中,对象可以是 Map 也可以是具体的 Java Bean,只要 key 与 Java Bean 的属性相同则自动注入。

(4) 运行测试方法 fromJson(),控制台输出如图 4-4 所示。

```
2021-08-07 14:05:32.787  INFO 7784 --- [           main]
2021-08-07 14:05:32.799  INFO 7784 --- [           main]
2021-08-07 14:05:35.001  INFO 7784 --- [           main]
{name=zhangsan, id=1001}
```

图 4-4 JSON 数据的反序列化

4.4 Spring Boot 对静态资源的访问

在 Spring MVC 中,静态资源都需要开发者手动配置静态资源过滤。Spring Boot 中对此也提供了自动化配置,可以简化静态资源过滤配置。

4.4.1 默认规则

默认情况下,Spring Boot 将通过类加载路径下的/static 目录(/public、/resources、/META-INF/resources)或当前应用的根路径来提供静态资源。对大部分开发者而言,只要将 JS 脚本、CSS 样式文件、图片等静态统一放在加载路径下的/static 或/public 目录中即可。

它使用 Spring MVC 中的 ResourceHttpRequestHandler,以便用户可以通过添加自己的 WebMvcConfigurer 并重写 addResourceHandlers()方法来修改该行为。Spring Boot 中对于 Spring MVC 的自动化配置都在 WebMvcAutoConfiguration 类中,因此对于默认的静态资源过滤策略可以从这个类获悉。

在该类中有一个 WebMvcAutoConfigurationAdapter,实现 WebMvcConfigurer 接口,WebMvcConfigurer 接口中有一个添加资源处理的方法 addResourceHandlers(),用来配置静态资源过滤,该方法在 WebMvcAutoConfigurationAdapter 类中得到了实现,部分代码如下。

```
public class WebMvcAutoConfiguration {
  …
  protected void addResourceHandlers(ResourceHandlerRegistry registry) {
      super.addResourceHandlers(registry);
      if (!this.resourceProperties.isAddMappings()) {
            logger.debug("Default resource handling disabled");
      } else {
            ServletContext servletContext = this.getServletContext();
            this.addResourceHandler(registry, "/webjars/**", "classpath:/META - INF/resources/
webjars/");
```

```
            this. addResourceHandler ( registry,  this. mvcProperties. getStaticPathPattern ( ),
(registration -> { registration.addResourceLocations(this.resourceProperties.getStaticLocations
());
                if (servletContext != null) {
                        registration. addResourceLocations ( new  Resource [ ] { new
ServletContextResource(servletContext, "/")});
                }
            });
        }
    }
}
```

上述进行了默认的静态资源过滤配置，其中 this. mvcProperties. getStaticPathPattern()作为参数，得到的路径值在 WebMvcProperties 类的构造方法中进行了定义，代码如下。

```
public class WebMvcProperties {
    private String staticPathPattern;
    public WebMvcProperties() {
        this.staticPathPattern = "/**";
    }
}
```

this. resourceProperties. getStaticLocations()获取到的静态资源定义在 ResourceProperties 的父类 Resources 中进行了定义，代码如下。

```
private static final String[] CLASSPATH_RESOURCE_LOCATIONS =
new String[]{"classpath:/META-INF/resources/", "classpath:/resources/", "classpath:/
static/", "classpath:/public/"};
```

在 addResourceHandlers()方法中，当前面四种资源都未找到，servletContext 不为空时，则静态资源位置设置为/。

Spring Boot 默认会过滤所有的静态资源，所以 /** 访问当前项目下的任何资源，都会去以下 5 个路径下寻找资源，即可将静态资源放到这 5 个位置中的任意一个。

(1) classpath:/META-INF/resources/。

(2) classpath:/resources/。

(3) classpath:/static/。

(4) classpath:/public/。

(5) /：当前项目的根路径。

上面静态资源加载路径的优先级为：/META-INF/resources＞/resources＞/static＞/public。一般情况下，Spring Boot 项目不需要 Webapp(WebContent)目录，故/暂不考虑。

在 chapter04 的项目中，在 resources 目录下分别创建如上 4 个目录，4 个目录中分别放入同名的静态资源 index. html，各目录下的 index. html 静态网页的内容不同，如图 4-5 所示。

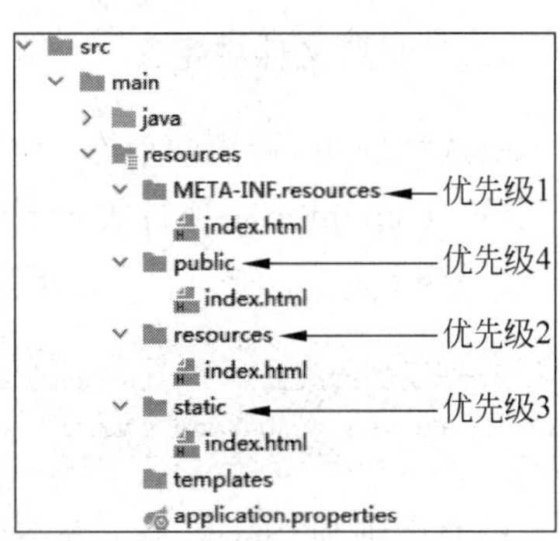

图 4-5　静态资源目录

重新启动项目，在地址栏输入 http://localhost:8080/index.html，即可看到 META-INF/resources/目录下的 index.html，如图 4-6 所示；如果将/META-INF/resources 目录下的 index.html 删除，重启项目，就会访问/resources 目录下的 index.html，如图 4-7 所示，以此类推。

如果使用 IntelliJ IDEA 创建 Spring Boot 项目，会默认创建 static 目录，静态资源一般放在这个目录下。

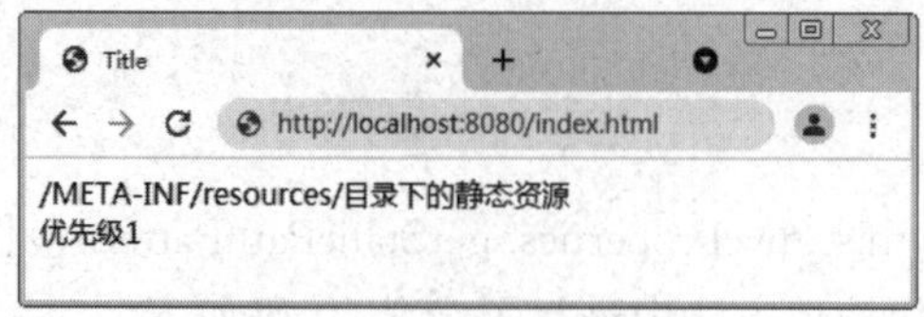

图 4-6 静态资源访问 1

图 4-7 静态资源访问 2

在前面 chapter04 项目的 PersonController 类中，添加方法 hello()，代码如下。

```
@GetMapping("/index.html")
public String hello(){
    return "HELLO!";
}
```

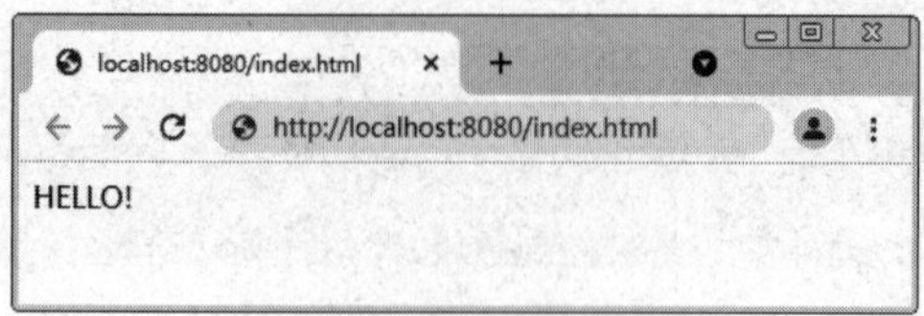

图 4-8 静态映射请求

重新启动该项目，在地址栏输入 http://localhost: 8080/index.html，效果如图 4-8 所示。

从图 4-8 的结果可以看出，静态映射/ ** 请求进来，先去找 Controller 看能不能处理，不能处理的所有请求，交给静态资源处理器。静态资源处理器就去找以上几个目录，如果静态资源能找到则访问，否则就会访问失败。

4.4.2 自定义规则

如果默认的静态资源访问规则不能满足开发需求，也可以自定义规则。例如，可以添加静态资源访问的前缀，自定义静态资源过滤规则有以下两种方法。

1. 在配置文件中定义

默认情况下，资源映射在/ ** 上，可以使用spring.mvc.static-path-pattern 属性对其进行调整，在 chapter04 项目的 application.properties 文件中直接定义过滤规则和静态资源位置，代码如下。

```
spring.mvc.static-path-pattern=/hello/**
spring.web.resources.static-locations=classpath:/static/
```

过滤规则为/hello/ **，静态资源位置为 classpath:/static/。

重新启动项目，在浏览器访问 http://localhost: 8080/hello/index.html，即可看到

static 目录下的 index. html 静态网页文件，如图 4-9 所示。

图 4-9　在配置文件中定义访问规则

2. Java 编码定义

除在配置文件中定义外，也可通过 Java 编码方式来定义，只需要实现 WebMvcConfigurer 接口即可，然后实现该接口的 addResourceHandlers()方法。

在 chapter04 项目的 com. yzpc 包下，新建一个 config 包，在该包中新建 MyWebMvcConfig 类，代码如下。

```
package com.yzpc.config;
import org.springframework.context.annotation.Configuration;
import org.springframework.web.servlet.config.annotation.ResourceHandlerRegistry;
import org.springframework.web.servlet.config.annotation.WebMvcConfigurer;
@Configuration
public class MyWebMvcConfig implements WebMvcConfigurer {
    @Override
    public void addResourceHandlers(ResourceHandlerRegistry registry) {
        registry.addResourceHandler("/hello/**")
                .addResourceLocations("classpath:/public/");
    }
}
```

重新启动项目，在浏览器访问 http://localhost:8080/hello/index. html，可看到 public 目录下的 index. html 静态网页文件，如图 4-10 所示。

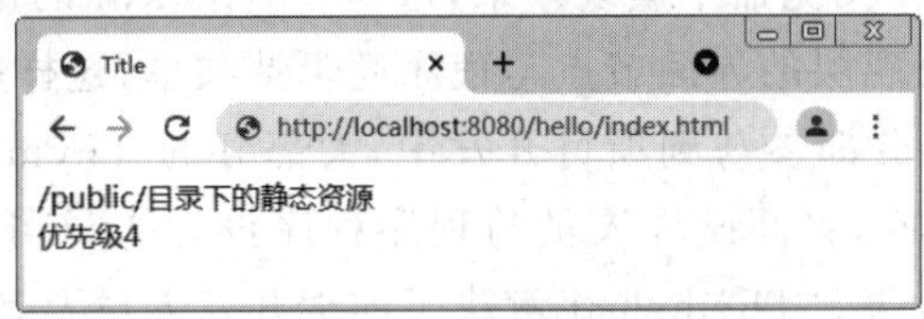

图 4-10　Java 编码定义访问规则

4.5　Thymeleaf 模板引擎

Spring Boot 主要支持 Thymeleaf、Freemarker、Mustache、Groovy 等模板引擎。Spring Boot 实现了与这些前端模板引擎技术的整合支持和自动化配置，Thymeleaf 可以轻易地与 Spring MVC 等 Web 框架进行集成。

4.5.1 Thymeleaf 简介

Thymeleaf 是 Spring Boot 推荐的模板引擎，Spring Boot 为其提供了完美的支持，它是一种现代的基于服务器端的 Java 模板引擎技术，也是一种优秀的面向 Java 的 XML、XHTML、HTML 5 页面模板。Thymeleaf 一般使用在 Web 环境中，可以处理 HTML、XML、JS 等文档，简单来说，Thymeleaf 可以将 JSP 作为 Java Web 应用的表现层，有能力展示与处理数据。

Thymeleaf 中的大多数处理器都是属性处理器，这种页面模板即使在被处理之前，浏览器也可以正确地显示 HTML 模板文件，因为浏览器会简单地忽略其不认识的属性。下面包含 JSP 脚本的 HTML 片段代码，它不能在模板被解析之前通过浏览器直接显示。

```
< input type = "text" name = "userName" value = " ${user.name}"/>
```

上面 value 属性的值是一个 EL 表达式，浏览器不能直接解析并显示它。下面是一段含有 Thymeleaf 语句的 HTML 代码。

```
< input type = "text" name = "userName" value = "用户名" th: value = " ${user.name}"/>
```

以上这段 HTML含有 Thymeleaf 的逻辑，该 THML元素的语法几乎是标准的 HTML 语法，只是增加了th:value 属性，既可以使用浏览器直接打开，也可以放到服务器中用来显示数据，并且样式之间基本不存在差异。在模板被解析之前，浏览器不识别 th:value 属性，那么该属性就会被直接忽略，这样浏览器依旧可以正确显示这些信息，并可打开一个默认值(例如“用户名”)。当该模板被解析之后，th:value 属性指定的表达式解析得到的值又会取代原来的 value 属性值，解析之后 th:value 属性就消失了，该属性对于浏览器就不存在了。

通过对比不难发现，Thymeleaf 就是在标准 HTML 标签中增加一些 th:xxx 属性，在模板被解析之前，这些属性会被浏览器忽略；在被浏览器解析之后，这些属性就完全不存在了，因此它们丝毫不影响在浏览器中呈现效果，这是 Thymeleaf 的优势。

Thymeleaf 可以让表现层的界面节点与程序逻辑被共享，这样的设计，可以让界面设计人员、业务人员与技术人员都参与到项目开发中，大部分的 Thymeleaf 表达式都直接被设置到 HTML 节点中。这样，界面设计人员与程序程序设计人员可以使用同一个模板文件查看静态与动态数据的效果。Thymeleaf 解决了前端开发人员要和后端开发人员配置一样环境的低效。Thymeleaf 通过属性进行模板渲染，不需要引入不能被浏览器识别的新的标签。页面直接作为 HTML 文件，用浏览器打开页面就可看到最终的效果，可以降低前后端人员的沟通成本。

4.5.2 引入 Thymeleaf

在 Spring Boot 项目中使用 Thymeleaf 模板，首先需要在项目中引入 Thymeleaf 依赖，直接在 pom. xml 文件中添加 Thymeleaf 依赖即可，依赖代码如下。

```
<dependency>
    <groupId>org.springframework.boot</groupId>
    <artifactId>spring-boot-starter-thymeleaf</artifactId>
</dependency>
```

此外，需要在 HTML 文件中加入< html lang = "en" xmlns: th = "http://www.thymeleaf.org">命名空间，这样就能完成 Thymeleaf 标签的渲染。

Spring Boot 默认的页面映射路径（即模板文件存放位置）为 classpath:/templates/ *.html，静态文件路径为 classpath:/static/，其中可以存放 CSS、JS 等模板共用的静态文件。

在全局配置文件 application.properties 文件中，可以配置 Thymeleaf 模板解析器属性，一般 Web 项目都会使用下列配置，示例代码如下。

```
spring.thymeleaf.cache = false                          #关闭模板缓存
spring.thymeleaf.encoding = UTF-8                       #模板的编码格式
spring.thymeleaf.mode = HTML5                           #模板的模板模式
spring.thymeleaf.servlet.content-type = text/html       #指定文档类型
spring.thymeleaf.prefix = classpath:/templates/         #指定模板页面存放路径
spring.thymeleaf.suffix = .html                         #指定模板页面的后缀名
```

上述配置中，spring.thymeleaf.cache 表示是否开启 Thymeleaf 模板缓存，默认为 true，在开发过程中，通常会关闭缓存，以保证调试时数据能够及时响应；spring.thymeleaf.prefix 指定了 Thymeleaf 模板页面的存放路径，默认为 classpath: /templates/；spring.thymeleaf.suffix 指定了 Thymeleaf 模板页面的名称后缀，默认为.html。

4.5.3 Thymeleaf 语法规则

使用 Thymeleaf 模板，首先要在< html.../>标签中引入 xmlns:th = "http://www.thymeleaf.org"命名空间，引入 Thymeleaf 模板引擎，然后在其他标签里面使用 th: * 这样的语法。本节将针对 Thymeleaf 常用的标签、表达式进行介绍。

1. 常用标签

在 HTML 页面上使用 Thymeleaf 标签，Thymeleaf 标签能够动态地替换掉静态内容，动态显示页面内容。为了更直观地认识 Thymeleaf，下面展示一个在 HTML 文件中嵌入 Thymeleaf 的页面文件，示例代码如下。

```
<!DOCTYPE html>
<html lang = "en" xmlns: th = "http://www.thymeleaf.org">
<head>
    <meta charset = "UTF-8">
    <link rel = "stylesheet" type = "text/css" media = "all"
          href = "../../css/gtvg.css" th: href = "@{/css/gtvg.css}"/>
    <title>Title</title>
</head>
<body>
```

```
    <p th: text = " ${hello}">您好,欢迎进入 Thymeleaf 学习!</p>
</body>
</html>
```

上述代码中,xmlns: th="http://www.thymeleaf.org"用于引入 Thymeleaf 模板引擎,关键字 th: 标签是 Thymeleaf 模板提供的标签,其中,th:href 用于引入外联样式文件,th:text 用于动态显示标签文本内容。除此之外,Thymeleaf 模板提供了很多 th: 标签,具体如表 4-1 所示。

表 4-1 Thymeleaf 常用标签

th: 标签	使用说明
th:id	用于 id 的声明,类似于 HTML 中的 id 属性
th:insert	页面片段包含(类似 JSP 中 include 标签)
th:replace	页面片段包含(类似 JSP 中 include 标签)
th:each	元素遍历(类似 JSP 中的 c: forEach 标签)
th:if	条件判断,条件成立时显示 th 标签的内容
th:unless	条件判断,条件不成立时显示 th 标签的内容
th:switch	条件判断,进行选择性匹配
th:case	多路选择配合 th: switch 使用
th:object	用于替换对象
th:with	用于定义局部变量
th:attr	通用属性修改
th:attrprepend	通用属性修改,将计算结果追加前缀到现有属性值
th:attrappend	通用属性修改,将计算结果追加后缀到现有属性值
th:value	属性值修改,指定标签属性值
th:href	用于设定链接地址
th:src	用于图片类地址引入
th:text	用于指定标签显示的文本内容
th:utext	用于指定标签显示的文本内容,对特殊标签不转义
th:fragment	声明片段
th:remove	移除片段

表 4-1 中列举的是 Thymeleaf 模板引擎的常用属性,更多属性的介绍,可以查看官方文档或者借助开发工具的快捷提示信息进行了解。

需要说明的是,上述操作是以 HTML 为基础嵌入 Thymeleaf 模板引擎,并使用 th: * 属性进行页面需求开发。这种 Thymeleaf 模板页面虽然与纯 HTML 页面基本相似,但已经不是一个标准的 HTML 5 页面,这是因为在 Thymeleaf 页面中使用的 th: * 属性是 HTML 5 规范所不允许的。如果想要使用 Thymeleaf 模板进行纯 HTML 5 的页面开发,可以使用 data-th-* 属性替换 th: * 属性进行页面开发。例如将上面的示例使用 data-th-* 属性进行修改,示例代码如下。

```
<!DOCTYPE html>
<html lang = "en">
```

```
<head>
    <meta charset="UTF-8">
    <link rel="stylesheet" type="text/css" media="all"
        href="../../css/gtvg.css" data-th-href="@{/css/gtvg.css}"/>
    <title>Title</title>
</head>
<body>
    <p data-th-text="${hello}">您好,欢迎进入 Thymeleaf 学习!</p>
</body>
</html>
```

使用 data-th-* 属性这种方式，不会出现属性的快捷提示，对于开发者来说不太方便，因此在实际开发中，相对推荐使用引入 Thymeleaf 标签的形式进行模板引擎页面的开发。

2. 标准表达式

Thymeleaf 模板引擎提供了多种标准表达式语法，如表 4-2 所示。

表 4-2 Thymeleaf 主要标准表达式

表达式语法	说　明
${...}	变量表达式
*{...}	选择变量表达式
#{...}	消息表达式
@{...}	链接 URL 表达式
~{...}	片段表达式

表 4-2 列举了 Thymeleaf 模板引擎最常用的简单表达式语法，并对这些语法进行功能说明。Thymeleaf 还提供了其他更多的语法支持，例如文本表达式、算式表达式、布尔表达式、比较表达式等，读者可查看具体的官方文档说明，下面对常用的表达式进行介绍。

(1) 变量表达式。变量表达式 ${...}主要用于获取上下文中的变量值，示例代码如下。

```
<p th:text="${title}">标题</p>
```

上述示例使用了 Thymeleaf 模板的变量表达式 ${...}用来动态获取 p 标签中的内容，如果当前程序没有启动或者当前上下文中不存在 title 变量，该片段会显示 p 标签的默认值“标题”；如果当前上下文中存在 title 变量并且程序已经启动，当前 p 标签中的默认文本将会被 title 变量的值所替换，从而达到模板引擎页面数据动态替换的效果。

Thymeleaf 为变量所在域提供了一些内置对象，具体介绍如下。

- #ctx：上下文对象。
- #vars：上下文变量。
- #locale：上下文区域设置。
- #request：(仅在 Web 上下文中)HttpServletRequest 对象。
- #response：(仅在 Web 上下文中)HttpServletResponse 对象。
- #session：(仅在 Web 上下文中)HttpSession 对象。

• #servletContext：(仅在 Web 上下文中)ServletContext 对象。

结合上述内置对象的说明，如果要在 Thymeleaf 模板引擎页面中动态获取当前国家信息，可以使用 #locale 内置对象，示例代码如下。

```
The locale country is: <span th: text = " ${#locale.country}"> US </span>.
```

使用 th:text=" ${#locale.country}"动态获取当前用户所在国家信息，其中<span>标签内默认内容为 US，程序启动后通过浏览器查看当前页面时，Thymeleaf 会通过浏览器语言设置来识别当前用户所在国家信息，从而实现动态替换。

(2) 选择变量表达式。选择变量表达式 *{...}和变量表达式 ${...}用法类似，*{...}一般用于从被选定对象而不是上下文中获取属性值，如果没有选定对象，则和变量表达式一样，示例代码如下。

```
<div th: object = " ${session.user}">
    <p> Name: <span th: text = " ${#object.firstName}"> Zhang </span>.</p>
    <p> Surname: <span th: text = " ${session.user.lastName}"> Sanfeng </span>.</p>
    <p> Nationality: <span th: text = " *{nationality}"> China </span>.</p>
</div>
```

${#object.firstName}变量表达式使用 Thymeleaf 模板提供的内置对象 object 获取当前上下文对象中的 firstName 属性值；${session.user.lastName}变量表达式获取当前 user 对象的 lastName 属性值；*{nationality}选择变量表达式获取当前指定对象 user 的 nationality 属性值。

(3) 消息表达式。消息表达式#{...}主要用于 Thymeleaf 模板页面国际化内容的动态替换和展示，使用消息表达式#{...}进行国际化设置时，还需要提供一些国际化配置文件。

(4) 链接 URL 表达式。链接 URL 表达式@{...}一般用于页面跳转或资源的引入，在 Web 开发中占据着非常重要的地位，示例代码如下。

```
<a href = "details.html"
th: href = "@{http://localhost: 8080/order/details(orderId = ${o.id})}"> view </a>
<a href = "details.html"
th: href = "@{/order/details(orderId = ${o.id})}"> view </a>
```

上述代码中，链接 URL 表达式 @{...}分别编写了绝对地址链接地址和相对链接地址。在有参数表达式中，需要按照@{路径(参数名称=参数值，参数名称=参数值...)}的形式编写，同时该参数的值可以使用变量表达式来传递动态参数值。

(5) 片段表达式。片段表达式~{...}是一种用来将标记片段移动到模板中的方法。其中，最常见的用法是使用 th：insert 或 th：replace 属性插入片段，示例代码如下。

```
<div th: insert = "~{thymeleafDemo: : title}"></div>
```

上述代码中，使用 th:insert 属性将 title 片段模板应用到该<div>标签中。thymeleafDemo 为模板名称，Thymeleaf 会自动查找 classpath:/resources/templates/目录下的 thymeleafDemo

模板,title 为声明的片段名称。

4.5.4 整合 Thymeleaf

下面通过示例来介绍 Spring Boot 整合 Thymeleaf 模板引擎。

(1) 创建 Spring Boot 项目,引入 Thymeleaf 依赖。使用 Spring Initializr 方式创建名称为 chapter04thymeleaf 的 Spring Boot 项目,Group 和 Package name 为 com. yzpc,在 Dependencies 依赖中选择 Web 节点下的 Spring Web 依赖和 Template Engines 节点下的 Thymeleaf 依赖,单击 Finish 按钮,如图 4-11 所示。

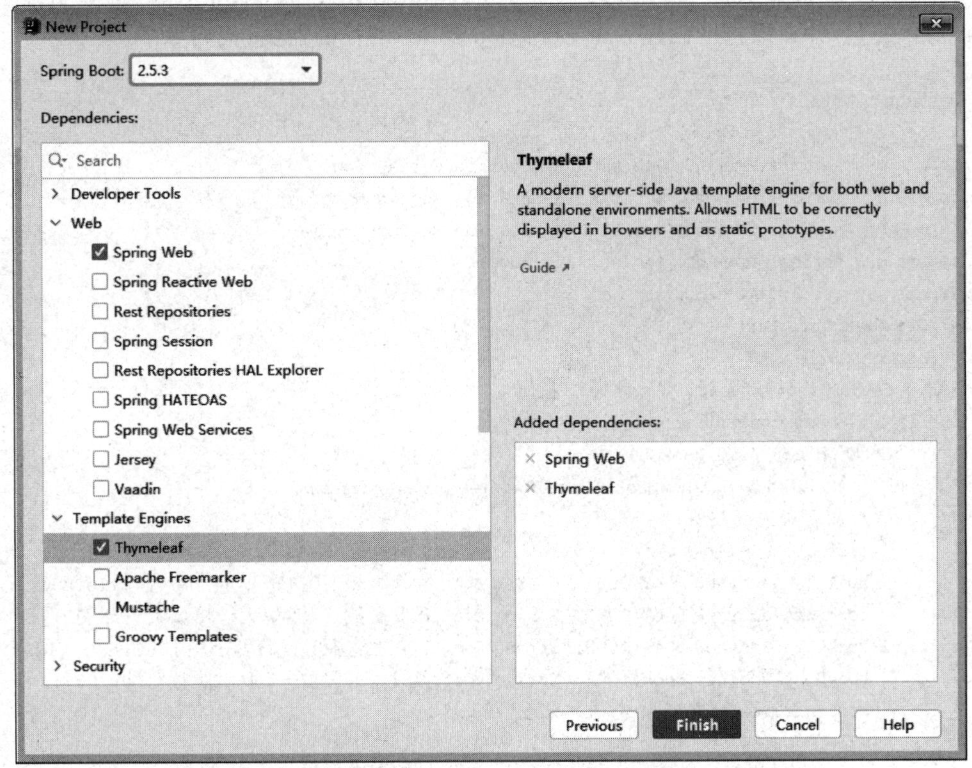

图 4-11 项目引入 Thymeleaf 依赖

(2) 编写配置文件。在 application. properties 全局配置文件中,对 Thymeleaf 模板页面的数据缓存进行设置,代码如下。

```
# Thymeleaf 页面缓存设置默认为true ,开发中,为方便调试设置 false
spring.thymeleaf.cache = false
```

在开发过程中,一般将 Thymeleaf 的模板缓存设置为关闭,否则,在程序修改之后可能不会及时显示修改后的内容。

(3) 创建实体类。在项目的 src/main/java/目录下的 com. yzpc 包中创建名为 pojo 的包,并在该包下创建一个 Book 的实体类,代码如下。

```
package com.yzpc.pojo;
public class Book {
    private int id;
    private String name;
    private String publisher;
    private double price;
    //省略带参数的构造方法
    //省略属性的getter、setter 方法
}
```

(4) 创建控制器类。在项目的 src/main/java/目录下的 com. yzpc 包中创建名为 controller 的包,并在该包下创建一个 BookController 的控制器类,代码如下。

```
package com.yzpc.controller;
import com.yzpc.pojo.Book;
import org.springframework.web.bind.annotation.GetMapping;
import org.springframework.web.bind.annotation.RequestMapping;
import org.springframework.web.bind.annotation.RestController;
import org.springframework.web.servlet.ModelAndView;
import java.util.ArrayList;
import java.util.List;
@RestController
@RequestMapping("/book")
public class BookController {
    @GetMapping("/showBooks")
    public ModelAndView showBooks(ModelAndView modelAndView){
        //加载对象或数值
        List<Book> bookList = new ArrayList<Book>();
        bookList.add(new Book(1001,"Spring Boot 编程思想","电子工业出版社",118.00));
        bookList.add(new Book(1002,"Spring Boot 开发实战","清华大学出版社",59.90));
        bookList.add(new Book(1003,"Spring Boot 2 实战之旅","清华大学出版社",79.00));
        bookList.add(new Book(1004,"Spring Boot 实战派","电子工业出版社",109.00));
        modelAndView.addObject("bookList",bookList);
        modelAndView.addObject("title","图书信息列表");
        modelAndView.setViewName("index");    //指定视图
        return modelAndView;
    }
}
```

控制器中的 showBooks()方法返回一个 ModelAndView 对象,该对象既包含视图信息,也包含模型数据信息,这样 Spring MVC 就可以使用视图对模型数据进行解析。

(5) 创建模板页面。在项目的 src/main/resources 路径下的 templates 目录中创建一个用于显示图书信息的 Thymeleaf 模板引擎页面 index.html,代码如下。

```
<!DOCTYPE html>
<html lang="en" xmlns:th="http://www.thymeleaf.org">
<head>
    <meta charset="UTF-8">
```

```
    <title th:text="${title}">Title</title>
</head>
<body>
    <h2 th:text="${title}">图书列表</h2>
    <table border="1">
        <tr>
            <td>ID号</td>
            <td>书名</td>
            <td>出版社</td>
            <td>价格</td>
        </tr>
        <tr th:each="book: ${bookList}">
            <td th:text="${book.id}">1000</td>
            <td th:text="${book.name}">Spring Boot</td>
            <td th:text="${book.publisher}">高等教育出版社</td>
            <td th:text="${book.price}">99.00</td>
        </tr>
    </table>
</body>
</html>
```

先通过 xmlns:th="http://www.thymeleaf.org" 命名空间引入 Thymeleaf 模板引擎，使用 th:text、th:each 引入后台动态传递过来的 title 标题信息和图书列表信息，并遍历显示图书信息。

(6) 效果测试。直接用浏览器打开 index.html 页面，就是前端所能看到的静态效果，如图 4-12 所示。

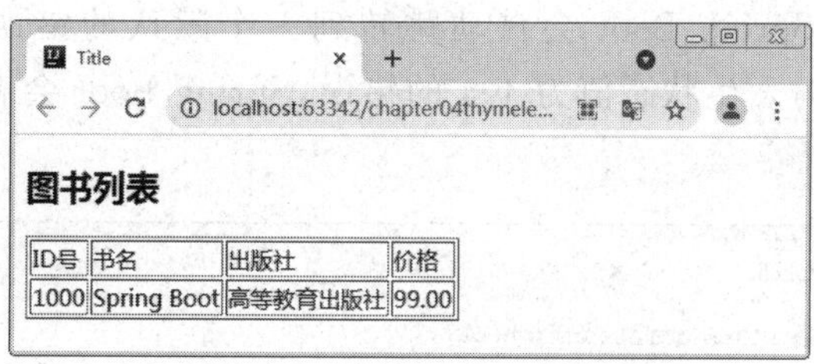

图 4-12 页面前端静态效果

启动项目，在浏览器访问 http://localhost:8080/book/showBooks，进入图书信息列表的页面，运行效果如图 4-13 所示。

图 4-13 图书信息列表页面

从图 4-13 的运行结果可以看出，图书信息列表页面 index. html 显示正常，说明 th: * 相关属性引入的动态数据生效，而不是页面的静态效果。这说明 Spring Boot 整合 Thymeleaf 成功，完成了动态数据的显示。

4.6 错误处理

在项目的开发过程中，不管是对底层数据库的操作，还是业务层、控制层的处理都不可避免地会遇到各种可预知的、不可预知的异常需要处理。

4.6.1 异常处理机制

Spring Boot 的错误处理是对 Spring MVC 异常处理的自动配置。因此 Spring Boot 同样可支持两种错误处理机制。

- 以 Spring Boot 提供的自动配置为基础，通过提供一些配置信息来改变 Spring Boot 默认的错误处理行为。
- 使用@ResponseStatus、@ExceptionHandler、@ControllerAdvice 等基于 AOP 的异常处理机制，这是直接基于 Spring MVC 异常处理机制进行错误处理。

Spring Boot 默认提供了一个/error 映射来处理所有的错误，错误默认由 BasicErrorController 类来处理，并将其注册为 Servlet 容器的全局错误页面。对于程序客户端，会生成一个具有错误详情、HTTP 状态和异常信息的 JSON 响应信息。对于普通浏览器，会生成一个 whitelabel 错误页面，以 HTML 格式呈现同样的错误信息。例如，在遇到不存在的访问请求时，Spring Boot 会自动跳转到一个默认的错误页面，如请求 http://localhost: 8080/hello 时发生状态值为 404 的错误，Spring Boot 会有一个默认的页面展示给用户，如图 4-14 所示。

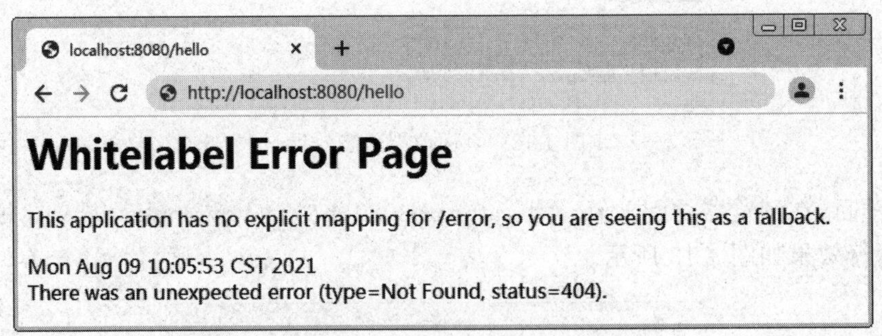

图 4-14 发生状态值为 404 的默认错误页面

虽然 Spring Boot 提供了默认的错误页面映射，但是在实际应用中，错误页面对用户来说并不友好，需要对其进行自定义，添加 view 解析为 error。要完全替代默认行为，可以实现 ErrorController 并注册该类型的 Bean 定义，或提供自定义的 ErrorAttributes 实现类，通过该类中的 getErrorAttributes( ) 方法向错误属性中添加异常信息。Spring Boot 为 ErrorAttributes 提供的实现类为 DefaultErrorAttributes，它生成的错误属性包含如下异常信息。

- timestamp：错误发生时间。
- status：HTTP 状态码。
- error：错误原因。
- message：异常消息。
- path：错误发生时请求的 URL 路径。
- exception：异常的类名。
- errors：BindingResult 异常里的各种错误(如果这个错误是由异常引起的)。
- trace：异常跟踪信息(如果这个错误是由异常引起的)。

4.6.2　自定义错误页

Spring Boot 可指定自定义的错误页面来代替原有的默认错误页面，只要将错误页面放在/error 目录下即可。自定义的错误页面既可以是静态的 HTML 页面，也可以是动态的页面模板，应该将静态的 HTML 错误页面放在静态资源目录(如 classpath:/static/等)下的/error 目录中，而将动态的页面模板放在/resources/templates/error 目录下，且动态页面模板的优先级更高。接下来通过实例介绍在 Spring Boot 中，如何自定义错误页面。

1. 简单配置

按照前面新建 chapter04thymeleaf 项目的方法，新建 chapter04error 的项目，在 src/main/resources/static 目录下创建 error 目录，在该 error 目录中创建错误展示页面 404. html 和 500. html，不同的错误直接展示不同的页面，在两个页面的 body 标签部分，分别显示“404 错误!”和“500 错误!”。

在项目的 src/main/java/目录下的 com. yzpc 包中创建名为 controller 的包，并在该 controller 包中新建 HelloController 的控制器类，代码如下。

```
@RestController
public class HelloController {
    @GetMapping("/hello")
    public String hello(){
        int i = 1/0;     //抛出异常
        return "Hello";
    }
}
```

启动项目，在浏览器中输入一个不存在的路径时，就会展示 404. html 页面的内容，如图 4-15 所示。若在浏览器中输入正确请求，但请求对应的方法会抛出异常，就会展示 500. html 中的内容，如图 4-16 所示。

这种定义均使用静态 HTML 页面，无法向用户展示完整的错误信息，若采用视图模板技术，则可向用户展示更多的错误信息。这里以 Thymeleaf 为例，Thymeleaf 页面模板默认位于 classpath:/templates/目录下，并在该目录下创建 error 目录，在该 error 目录下创建错误展示页面 4xx. html、5xx. html，以 4xx. html 页面为例，代码如下。

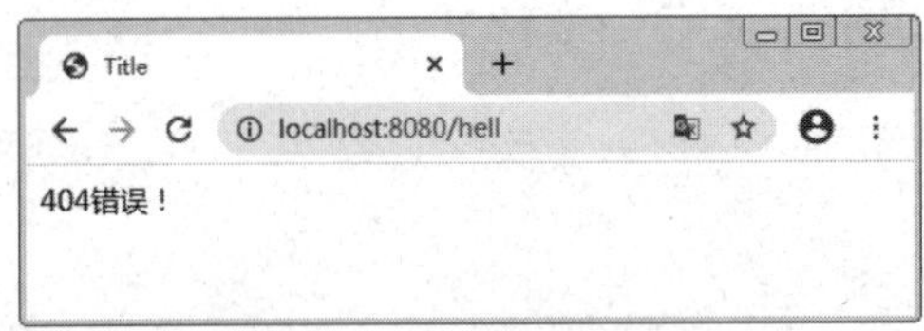

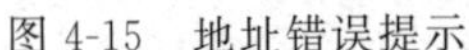
图 4-15　地址错误提示

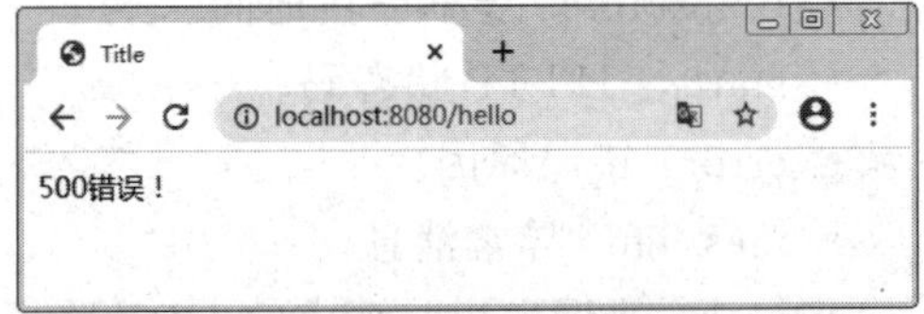

图 4-16　程序异常提示

```
<!DOCTYPE html>
<html lang="en" xmlns:th="http://www.thymeleaf.org">
<head>
    <meta charset="UTF-8">
    <title>Title</title>
</head>
<body>
    <h3>错误信息</h3>
    <table border="1">
        <tr><td>timestamp</td><td th:text="${timestamp}">错误</td></tr>
        <tr><td>status</td><td th:text="${status}">错误</td></tr>
        <tr><td>error</td><td th:text="${error}">错误</td></tr>
        <tr><td>message</td><td th:text="${message}">错误</td></tr>
        <tr><td>path</td><td th:text="${path}">错误</td></tr>
        <tr><td>trace</td><td th:text="${trace}">错误</td></tr>
    </table>
</body>
</html>
```

Spring Boot 在这里一共返回了 6 条错误相关信息，分别是 timestamp、status、error、message、path 和 trace。5xx.html 的页面内容与 4xx.html 页面的内容一致。

在 application.properties 配置文件中，添加配置，代码如下。

```
# 指定启动错误页面(这是默认值)
server.error.whitelabel.enabled=true
# 指定一直显示 message
server.error.include-message=always
# 指定包含异常类
server.error.include-exception=true
# 指定一直包含 BindingErrors
server.error.include-binding-errors=always
# 指定一直包含异常跟踪栈
server.error.include-stacktrace=always
```

先删除/resources/static/error 目录下的 404.html 和 500.html 页面文件。启动 chapter04error 项目，在浏览器中输入一个不存在的路径，例如 http://localhost:8080/hel，结果如图 4-17 所示。如果访问一个会抛出异常的请求，例如 http://localhost:8080/hello，结果如图 4-18 所示。

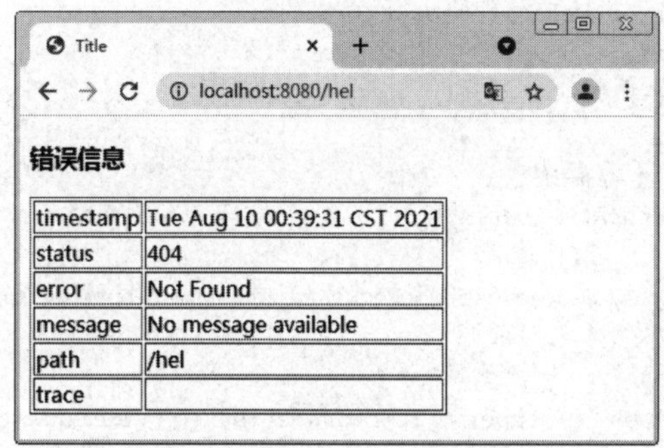

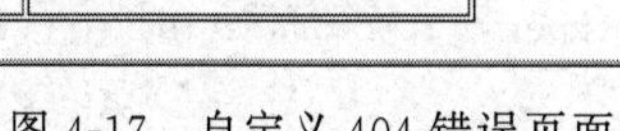
图 4-17　自定义 404 错误页面

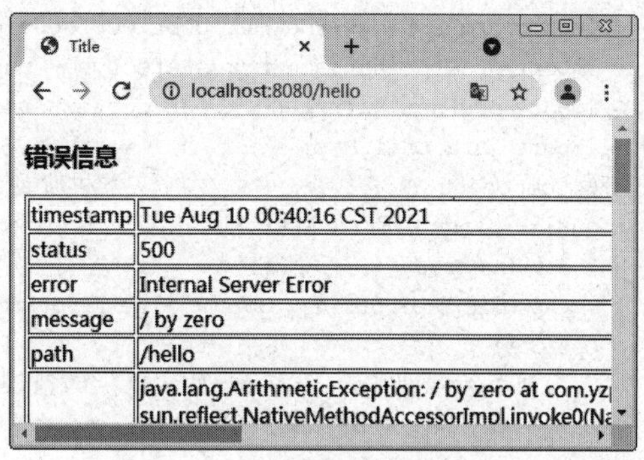

图 4-18　自定义 500 错误页面

提示：若定义了多个错误页面，则 404. html、500. thml 页面的优先级高于 4xx. html、5xx. html 页面，即当前是 404 错误，则优先展示 404. html 而不是 4xx. html；动态页面的优先级高于静态页面，即若 resources/templates 和 resources/static 下同时定义了 4xx. html，则优先展示 resources/templates 下的 4xx. html。

2. 复杂配置

简单配置的方式不够灵活，只能定义 HTML 页面，展示 Spring Boot 默认的返回信息，无法处理 JSON 的定制，无法返回自定义业务数据，Spring Boot 中支持对 Error 信息的定制。

(1) 自定义 Error 数据。自定义 Error 数据就是对返回的数据进行自定义。通过查看 BasicErrorController 类的 errorHtml()和 error()方法，发现都是通过 getErrorAttributes()方法获取 error 信息。该方法最终会调用到 DefaultErrorAttributes 类的 getErrorAttributes()方法，DefaultErrorAttributes 类是在 ErrorMvcAutoConfiguration 类中默认提供的。ErrorMvcAutoConfiguration 类的 errorAttributes()方法源码如下。

```
@Bean
@ConditionalOnMissingBean(value = {ErrorAttributes.class}, search = SearchStrategy.CURRENT)
public DefaultErrorAttributes errorAttributes() {
    return new DefaultErrorAttributes();
}
```

当系统没有提供 ErrorAttributes 时，才会采用 DefaultErrorAttributes。自定义错误提示时，只需要提供一个 ErrorAttributes 即可，而 DefaultErrorAttributes 是 ErrorAttributes 的子类。

在项目的 src/main/java/目录下的 com. yzpc 包中创建名为 customerror 的包，并在该包中新建一个 MyErrorAttribute 类并继承 DefaultErrorAttributes，重写 getErrorAttributes()方法，代码如下。

```
package com.yzpc.customerror;
import org.springframework.boot.web.error.ErrorAttributeOptions;
```

```
import org.springframework.boot.web.servlet.error.DefaultErrorAttributes;
import org.springframework.stereotype.Component;
import org.springframework.web.context.request.WebRequest;
import java.util.Map;
@Component                    //将该类注册到Spring 容器中
public class MyErrorAttribute extends DefaultErrorAttributes {
    @Override
    public Map<String, Object> getErrorAttributes(WebRequest webRequest, ErrorAttributeOptions
options) {
        //获取Spring Boot 默认提供的错误信息
          Map < String, Object > errorAttributes = super.getErrorAttributes(webRequest,
options);
        errorAttributes.put("myerror","程序出错,错误如下");    //自定义Error
        errorAttributes.remove("error");                         //移除默认的error 属性
        return errorAttributes;
    }
}
```

修改 4xx.html 和 5xx.html 页面，将“错误信息”的标题行，使用自定义的 myerror 的错误信息提示，代码如下。

```
<h3 th:text="${myerror}">错误信息</h3>
<!-- 其余代码一致,此处省略 -->
```

重新启动项目，在浏览器地址栏中访问 http://localhost:8080/hel，如图 4-19 所示。

图 4-19 自定义 Error 数据显示

访问一个不存在的路径或者抛出异常的访问请求，就可以看到自定义的标题信息，并且可以看到默认的 error 数据被移除。

(2) 自定义 Error 视图。自定义 Error 视图是展示给用户的页面。通过查看 BasicErrorController 类的 errorHtml()方法，调用了 ErrorViewResolver 提供的 resolveErrorView()方法，用来获取一个 ModelAndView 实例。通过 ErrorMvcAutoConfiguration 类的源码可以看到默认的 ErrorViewResolver 是 DefaultErrorViewResolver 类型，ErrorMvcAutoConfiguration 类的部分源码如下。

```
@Bean
@ConditionalOnBean({DispatcherServlet.class})
```

```
@ConditionalOnMissingBean({ErrorViewResolver.class})
DefaultErrorViewResolver conventionErrorViewResolver() {
    return new DefaultErrorViewResolver(this.applicationContext, this.resources);
}
```

默认使用 DefaultErrorViewResolver，如果想要自定义 Error 视图，只需要提供自己的 ErrorViewResolver 即可。

在项目的 src/main/java/目录下的 com. yzpc. customerror 包中，新建 MyErrorViewResolver 类，代码如下。

```
package com.yzpc.customerror;
import org.springframework.boot.autoconfigure.web.servlet.error.ErrorViewResolver;
import org.springframework.http.HttpStatus;
import org.springframework.stereotype.Component;
import org.springframework.web.servlet.ModelAndView;
import javax.servlet.http.HttpServletRequest;
import java.util.Map;
@Component                    //将该类注册到 Spring 容器中
public class MyErrorViewResolver implements ErrorViewResolver {
    @Override                 //Map 参数是默认的Error 信息
    public ModelAndView resolveErrorView(HttpServletRequest request, HttpStatus status,
Map<String, Object> model) {
        ModelAndView mav = new ModelAndView("error");
        mav.addObject("myerror","Error 视图,程序出错,错误如下");
        mav.addAllObjects(model);
        return mav;
    }
}
```

在 resolveErrorView()方法中，返回一个 ModelAndView，在 ModelAndView 中设置 Error 视图和 Error 数据，说明也可通过实现 ErrorViewResolver 接口来实现 Error 数据的定义，但如果只想自定义 Error 数据，还是建议继承 DefaultErrorAttributes。

在/resources/templates 目录下新建 error. html 页面，页面代码与前面 4xx. html 的代码一致。将 MyErrorAttribute 类前的@Component 注解注释掉或删除 MyErrorAttribute 类。

重新启动项目，无论发生 4xx 错误，还是发生 5xx 错误，都将显示 error. html 页面。如图 4-20 和图 4-21 所示。

图 4-20 自定义 Error 视图 1

图 4-21 自定义 Error 视图

4.7 CORS 支持

CORS(Cross-origin resource sharing)即跨域资源共享，是由 W3C 制定的一种跨域资源共享技术标准，其目的是解决前端的跨域请求，它允许浏览器向跨域服务器发送 Ajax 请求，打破了 Ajax 只能访问本站内的资源限制，CORS 在很多地方都有使用，微信支付的 JS 支付就是通过 JS 向微信服务器发送跨域请求。开放 Ajax 访问可被跨域访问的服务器可大大减少后台开发工作，明确前后台分工，下面通过示例来介绍 CORS。

(1) 新建一个 Spring Boot 工程 chapter04cors，Group 和 Package name 为 com.yzpc，在 Dependencies 依赖中选择 Web 节点下的 Spring Web 依赖，单击 Finish 按钮。

在 application.properties 配置文件中设置端口为 8081，如下所示。

```
server.port = 8081
```

(2) 在 chapter04cors 工程中，在 src/main/java/目录下的 com.yzpc 包中创建名为 controller 的包，并在该包下创建一个 HelloController 的控制器类，代码如下。

```
package com.yzpc.controller;
import org.springframework.web.bind.annotation.CrossOrigin;
import org.springframework.web.bind.annotation.GetMapping;
import org.springframework.web.bind.annotation.RestController;
@RestController
public class HelloController {
    @GetMapping("/hello")
    //使用 @CrossOrigin 注解允许跨域
    @CrossOrigin(origins = "http://localhost: 8082",maxAge = 1800, allowedHeaders = "*")
    public String hello(){
        return "hello cors!";
    }
}
```

@CrossOrigin 注解中的 origins 表示支持的域，这里表示来自 http://localhost: 8082 域的请求是支持跨域的；maxAge 表示探测请求的有效期，默认属性值为 1800 秒；allowedHeaders 表示允许的请求头，* 表示所有的请求头都被允许。

这种配置方式可以控制到每个方法上，也可不在每个方法上添加@CrossOrigin 注解，而是采用全局配置。在项目的 src/main/java/目录下的 com.yzpc 包中，创建名为 config 的包，并在该包下创建一个 MyWebMvcConfig 的类，并实现 WebMvcConfigurer 接口，重写其中的 addCorsMappings()方法，代码如下。

```
package com.yzpc.controller.config;
import org.springframework.context.annotation.Configuration
import org.springframework.web.servlet.config.annotation.CorsRegistry;
import org.springframework.web.servlet.config.annotation.WebMvcConfigurer;
```

```
@Configuration
public class MyWebMvcConfig implements WebMvcConfigurer {
    @Override
    public void addCorsMappings(CorsRegistry registry) {
        registry.addMapping("/**")
                .allowedOrigins("http://localhost:8082")
                .allowedHeaders("*")
                .allowedMethods("*")
                .maxAge(30 * 1000);
    }
}
```

在 addCorsMappings()方法中，addMapping 表示对哪种格式的请求路径进行跨域处理；allowedOrigins 表示支持的域；allowedHeaders 表示允许的请求头部信息，默认允许所有；allowedMethods 表示允许的请求方法；* 表示支持所有的请求方法；maxAge 表示探测请求的有效期。

上面两种配置方式@CrossOrigin 注解配置和全局配置，选择其中一种即可，启动 chapter04cors 工程。

(3) 选中 chapter04cors 工程，右键选择 New 菜单下的 Module，新建一个名为 cors 的 Module，过程与新建工程类似。在 cors 中，在 src/main/resources/static 目录下，新建 index. html 网页文件和一个 js 目录，并将 jquery-3. 6. 0. js 文件放入 js 目录下，如图 4-22 所示。

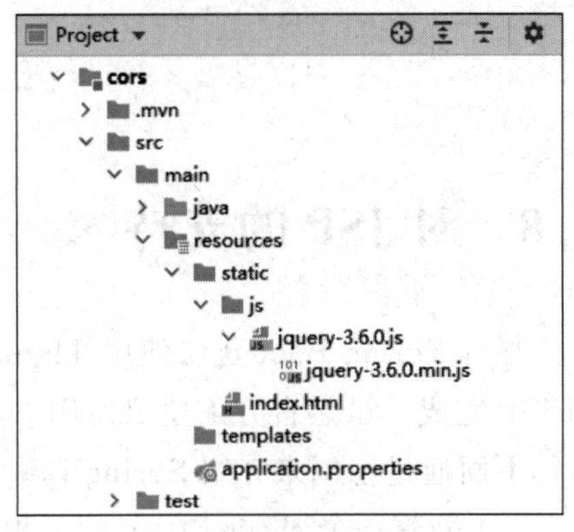

图 4-22 cors 文件目录

静态网页文件 index. html 的代码如下。

```
<!DOCTYPE html>
<html lang="en">
<head>
<meta charset="UTF-8">
<title>跨域访问示例</title>
<script src="js/jquery-3.6.0.js"></script>
<script>
    function getData() {
        $.get('http://localhost:8081/hello',function (msg) {
            $("#app").html(msg);
        })
    }
</script>
</head>
<body>
    <div id="app"></div>
    <input type="button" value="跨域请求" onclick="getData()">
```

```
</body>
</html>
```

修改 application. properties 配置文件，将端口改为 8082，如下所示。

```
server.port = 8082
```

启动 CorsApplication，在浏览器中输入 http://localhost：8082/index.html，页面只显示一个“跨域请求”按钮，单击该按钮，就能看到 hello cors！的请求结果，如图 4-23 所示。

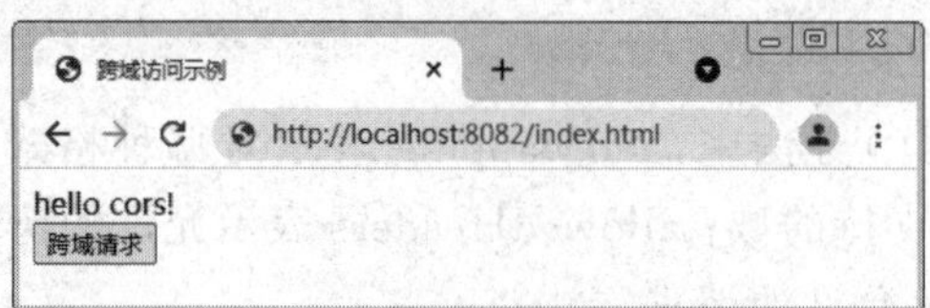

图 4-23　cors 跨域请求

4.8　对 JSP 的支持

尽管 Spring Boot 建议使用 Thymeleaf 来完成动态页面，但也有部分 Java Web 应用使用 JSP 完成。虽然官方不建议使用 JSP 作为渲染页面，但也可以使用。

下面通过示例来介绍 Spring Boot 如何集成 JSP 技术。

（1）新建一个 chapter04jsp 的工程，Group 和 Package name 为 com. yzpc，在 Dependencies 依赖中选择 Web 节点下的 Spring Web 依赖，单击 Finish 按钮。

（2）在 pom. xml 文件中，添加 Tomcat、Servlet 和 JSTL 依赖，代码如下。

```
<!-- 添加 Spring Boot 对JSP 支持 -->
<dependency>
    <groupId>org.apache.tomcat.embed</groupId>
    <artifactId>tomcat-embed-jasper</artifactId>
</dependency>
<!-- Servlet 支持开启 -->
<dependency>
    <groupId>javax.servlet</groupId>
    <artifactId>javax.servlet-api</artifactId>
</dependency>
<!-- 使用JSTL 标签 -->
<dependency>
    <groupId>javax.servlet</groupId>
    <artifactId>jstl</artifactId>
</dependency>
```

（3）在 application. properties 配置文件，设置项目的上下文路径和页面文件的前缀与后缀，内容如下。

```
server.servlet.context-path=/jsp
spring.mvc.view.prefix=/WEB-INF/jsp/
spring.mvc.view.suffix=.jsp
```

(4) 在项目中的 src/main/java 路径下的 com.yzpc 的包中新建 pojo 包,并在该包下创建一个 Book 的实体类,代码如下。

```
package com.yzpc.pojo;
public class Book {
    private int id;
    private String name;
    private String publisher;
    private double price;
    //省略带参数的构造方法
    //省略属性的getter、setter 方法
}
```

(5) 在项目的 src/main/java/目录下的 com.yzpc 包中创建名为 controller 的包,并在该包下创建一个 BookController 的控制器类,代码如下。

```
@Controller
public class BookController {
    @GetMapping("/")
    public String showBooks(Model model){
        List<Book> bookList = new ArrayList<Book>();
        bookList.add(new Book(1001,"Spring Boot 编程思想","电子工业出版社",118.00));
        bookList.add(new Book(1002,"Spring Boot 开发实战","清华大学出版社",59.90));
        bookList.add(new Book(1003,"Spring Boot 2 实战之旅","清华大学出版社",79.00));
        bookList.add(new Book(1004,"Spring Boot 实战派","电子工业出版社",109.00));
        model.addAttribute("bookList",bookList);
        model.addAttribute("title","图书信息列表");
        return "index";
    }
}
```

(6) 在项目的 src/main 目录下创建目录 webapp/WEB-INF/jsp/,并在该目录下新建 index.jsp 文件,代码如下。

```
<%@ page language="java" contentType="text/html; UTF-8"
pageEncoding="UTF-8" %>
<%@ taglib prefix="c" uri="http://java.sun.com/jsp/jstl/core" %>
<!DOCTYPE html>
<html lang="en">
<head>
    <meta charset="UTF-8">
    <title>Title</title>
</head>
<body>
```

```
<h2>${title}</h2>
<table border="1">
    <tr>
        <td>ID号</td>
        <td>书名</td>
        <td>出版社</td>
        <td>价格</td>
    </tr>
    <c:forEach var="book" items="${bookList}">
        <tr>
            <td>${book.id}</td>
            <td>${book.name}</td>
            <td>${book.publisher}</td>
            <td>${book.price}</td>
        </tr>
    </c:forEach>
</table>
</body>
</html>
```

(7) 启动项目,在浏览器访问 http://localhost:8080/jsp/,效果如图 4-24 所示。

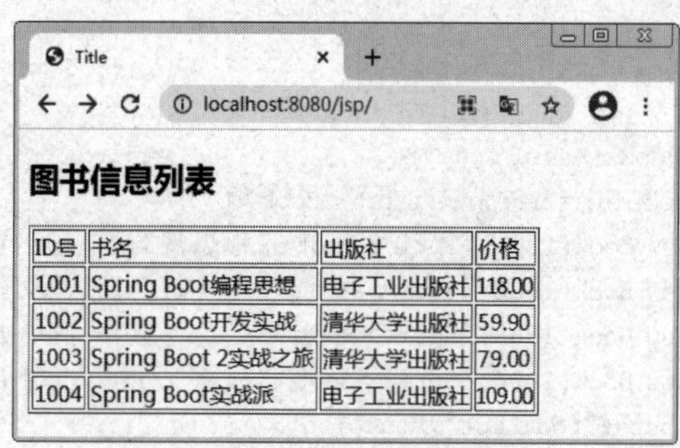

图 4-24　图书列表显示

Spring Boot 还有很多模板可以使用,如果不是必需,建议不要使用。

4.9 小结

本章首先介绍了 Spring Boot 对 Web 开发的支持,然后介绍了自定义消息转换器、序列化和反序列化、对静态资源的访问规则、Thymeleaf 模板引擎。同时还介绍了错误处理、CORS 支持和对 JSP 的支持等 Web 应用开发的常用功能。

第5章 Spring Boot访问SQL数据库

在项目开发中，通常会对数据库中的数据进行操作，Spring Boot 在简化项目开发以及实现自动化配置的基础上，对关系型数据库和非关系型数据库的访问操作都提供了非常好的整合支持，本章将对 Spring Boot 访问 SQL 数据库进行介绍。

5.1 配置数据源

数据库分为两种，即关系型数据库和非关系型数据库。关系型数据库是指通过关系模型组织数据的数据库，并且可以利用外键等保持一致性，如 MySQL、Oracle、SQL Server 等；而非关系型数据库其实不像是数据库，更像是一种以 key-value 模式存储对象的结构，如 Redis、MangoDB 等。

Spring Framework 为访问 SQL 数据库提供了广泛的支持。从直接使用 JdbcTemplate 进行 JDBC 访问到完全的对象关系映射(object relational mapping)技术，比如 Hibernate、MyBatis。Spring Data 提供了更多级别的功能，直接从接口创建的 Repository 实现，并使用了约定从方法名生成查询。本节介绍 Spring Boot 如何使用 MySQL、SQL Server、Oracle 等不同的关系型数据库。

1. 使用 MySQL 数据库

在 Spring Boot 中使用 MySQL 数据库，这里以 MySQL 8.0 版本为例，在 pom.xml 文件中添加依赖，并在 application.properties 文件中配置数据源信息。

(1) 在 pom.xml 文件中添加 MySQL 的依赖，代码如下。

```
<dependency>
    <groupId>mysql</groupId>
    <artifactId>mysql-connector-java</artifactId>
    <scope>runtime</scope>
</dependency>
```

(2) 在 application.properties 配置文件中配置数据源信息，代码如下。

```
# 数据库驱动
spring.datasource.driver-class-name=com.mysql.cj.jdbc.Driver
```

```
# 数据库地址
spring.datasource.url = jdbc: mysql://localhost: 3306/test?serverTimezone = UTC
# 数据库用户名
spring.datasource.username = root
# 数据库密码
spring.datasource.password = 123456
```

2. 使用 SQL Server 数据库

在 Spring Boot 中使用 SQL Server 数据库。

(1) 在 pom.xml 文件中添加 SQL Server 依赖,代码如下。

```
<dependency>
    <groupId>com.microsoft.sqlserver</groupId>
    <artifactId>mssql-jdbc</artifactId>
    <scope>runtime</scope>
</dependency>
```

(2) 在 application.properties 配置文件中配置数据源信息,代码如下。

```
spring.datasource.driver-class-name = com.microsoft.sqlserver.jdbc.SQLServerDriver
spring.datasource.url = jdbc: sqlserver://localhost: 1433; DatabaseName = test
spring.datasource.username = sa
spring.datasource.password = 123456
```

3. 使用 Oracle 数据库

在 Spring Boot 中使用 Oracle 数据库。

(1) 在 pom.xml 文件中添加 Oracle 依赖,代码如下。

```
<dependency>
    <groupId>com.oracle.database.jdbc</groupId>
    <artifactId>ojdbc8</artifactId>
    <scope>runtime</scope>
</dependency>
```

(2) 在 application.properties 配置文件中配置数据源信息,代码如下。

```
spring.datasource.driver-class-name = oracle.jdbc.driver.OracleDriver
spring.datasource.url = jdbc: oracle: thin: @localhost: 1521: test
spring.datasource.username = admin
spring.datasource.password = 123456
```

5.2 使用 JdbcTemplate

JDBC(Java DataBase Connectivity)是用于连接数据库的规范,也就是用于执行数据库

SQL 语句的 Java API。实际上，JDBC 由一组用 Java 语言编写的类和接口组成，为大部分关系型数据库提供访问接口。JDBC 需要每次进行数据库连接，然后处理 SQL 语句、传值、关闭数据库，如果都由开发人员编写代码，则很容易出错，可能会出现使用完成之后，数据库连接忘记关闭的情况。这容易导致连接被占用而降低性能，为了减少这种可能的错误，减少开发人员的工作量，Spring 框架设计了 JdbcTemplate 模板框架。

JdbcTemplate 是 Spring 对 JDBC 的封装，目的是让 JDBC 更加易于使用，完成所有的 JDBC 底层工作。因此，对于数据库的操作，不再需要每次都进行连接、打开、关闭数据库。通过 JdbcTemplate 不需要进行全局修改，就可以轻松地应对开发人员常常要面对的增删改查的操作。更关键的是，JdbcTemplate 对象也是通过自动配置机制注册到 IoC 容器中，自动化配置类是 JdbcTemplateAutoConfiguration。在 JdbcTemplate 中，提供了大量的查询和更新数据库的方法，Spring JDBC 使用这些方法来操作数据库，使用 JdbcTemplate 类可实现对数据表的数据进行增删改查操作。

5.2.1 JdbcTemplate 增删改的操作

update()方法可以完成对数据库中数据表的增加、修改和删除操作。在 JdbcTemplate 类中，update()方法中存在多个重载的方法，其常用方法具体介绍如下。

- int update(String sql)：该方法是最简单的 update 方法重载形式，可以直接传入 SQL 语句并返回受影响的行数。
- int update(PreparedStatementCreator psc)：该方法执行从 PreparedStatementCreator 返回的语句，然后返回受影响的行数。
- int update(String sql, PreparedStatementSetter pss)：该方法通过 PreparedStatementSetter 设置 SQL 语句中的参数，并返回受影响的行数。
- int update(String sql, Object...args)：该方法使用 Object...args 设置 SQL 语句中的参数，要求参数不能为空，并返回受影响的行数。

下面通过示例来介绍使用 JdbcTemplate 的 update()方法实现对 MySQL 数据库中的数据表进行增删改的操作，步骤如下。

(1) 使用 SQLyog 客户端工具(也可使用其他客户端工具)，在 MySQL 数据库中，新建一个 chapter05 的数据库，并在其中新建 user 数据表，输入几条记录，其表结构如图 5-1 所示。

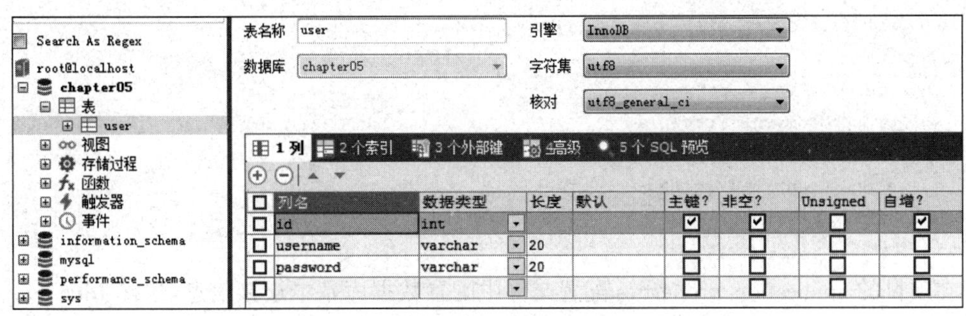

图 5-1 user 数据表的表结构

（2）新建一个 Spring Boot 工程 chapter05jdbctemplate，Group 和 Package name 为 com.yzpc，在 Dependencies 依赖中选择 Web 节点下的 Spring Web 依赖和 SQL 节点下的 JDBC API、MySQL Driver 依赖，单击 Finish 按钮，如图 5-2 所示。

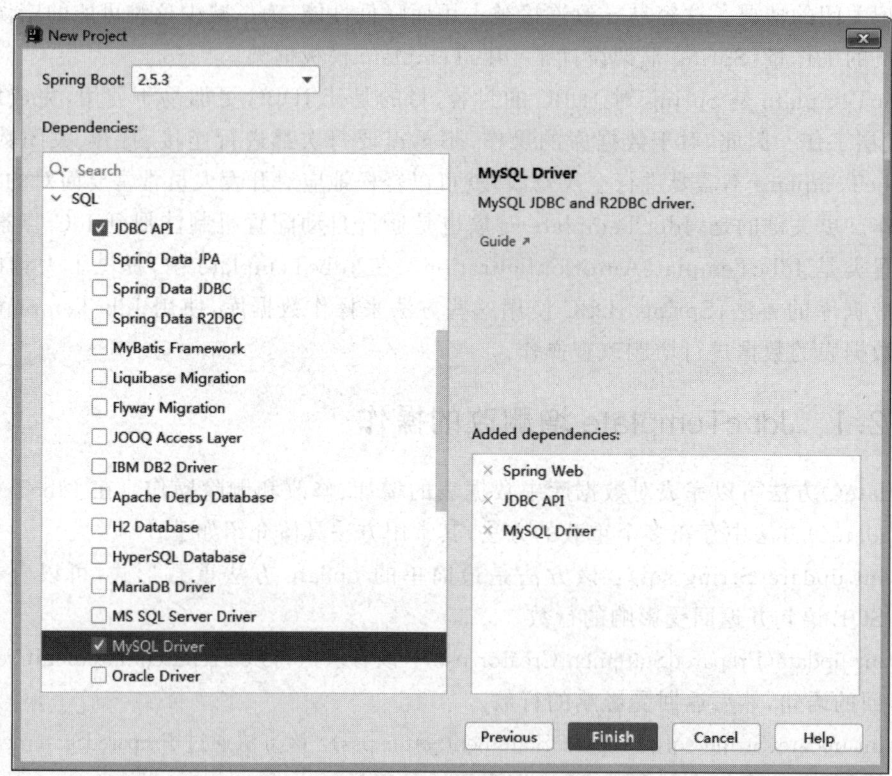

图 5-2 添加 SQL 相关依赖

pom. xml 文件中添加的依赖代码如下。

```
<dependency>
    <groupId>org.springframework.boot</groupId>
    <artifactId>spring-boot-starter-jdbc</artifactId>
</dependency>
<dependency>
    <groupId>org.springframework.boot</groupId>
    <artifactId>spring-boot-starter-web</artifactId>
</dependency>
<dependency>
    <groupId>mysql</groupId>
    <artifactId>mysql-connector-java</artifactId>
    <scope>runtime</scope>
</dependency>
```

在项目的 application. properties 配置文件中配置数据库基本连接信息，这样 JdbcTemplate 才能正常连接到数据库，代码如下。

```
spring.datasource.driver-class-name=com.mysql.cj.jdbc.Driver
spring.datasource.url=jdbc:mysql://localhost:3306/chapter05?&serverTimezone=UTC
spring.datasource.username=root
spring.datasource.password=123456
```

(3) 在项目 src/main/java/路径下的 com.yzpc 包中，新建一个 pojo 包，并在该包中新建 User 实体类，代码如下。

```
package com.yzpc.pojo;
public class User {
    private int id;
    private String username;
    private String password;
    //此处省略构造方法
    //此处省略相应属性的 setter/getter 方法
    //重写 toString 方法
    @Override
    public String toString() {
        return "User{" + "id=" + id + ", username='" + username + '\'' + ", password='" +
password + '\'' + '}';
    }
}
```

(4) 在项目 src/main/java/ 路径下的 com.yzpc 包中，新建一个 dao 包，并在该包中新建 UserDao 接口，并在接口中定义添加、修改、删除用户的方法，代码如下。

```
package com.yzpc.dao;
import com.yzpc.pojo.User;
public interface UserDao {
    public int addUser(User user);
    public int updateUser(User user);
    public int deleteUser(int id);
}
```

在 com.yzpc.dao 包中，新建一个 impl 包，在该包中创建 UserDao 接口的实现类 UserDaoImpl，并在类中实现添加、修改和删除的方法，代码如下。

```
package com.yzpc.dao.impl;
import com.yzpc.dao.UserDao;
import com.yzpc.pojo.User;
import org.springframework.beans.factory.annotation.Autowired;
import org.springframework.jdbc.core.JdbcTemplate;
import org.springframework.stereotype.Repository;
@Repository
public class UserDaoImpl implements UserDao {
    @Autowired
    private JdbcTemplate jdbcTemplate;
    @Override
```

```
    public int addUser(User user) {
        return jdbcTemplate.update("insert into user(username,password) values(?,?)",user.
getUsername(),user.getPassword());
    }
    @Override
    public int updateUser(User user) {
        return jdbcTemplate.update("update user set username = ?,password = ? where id = ?",
user.getUsername(),user.getPassword(), user.getId());
    }
    @Override
    public int deleteUser(int id) {
        return jdbcTemplate.update("delete from user where id = ?",id);
    }
}
```

在 UserDaoImpl 类中对 JdbcTemplate 的引用，使用@Autowired 注解实现 UserDaoImpl 对 JdbcTemplate 类的依赖注入。

(5) 在 src/main/java/ 路径下的 com.yzpc 包中，新建一个 controller 包，在该包中新建 UserController 类，并在该类中添加增加、删除、修改的方法，代码如下。

```
package com.yzpc.controller;
import com.yzpc.dao.UserDao;
import com.yzpc.pojo.User;
import org.springframework.beans.factory.annotation.Autowired;
import org.springframework.web.bind.annotation.GetMapping;
import org.springframework.web.bind.annotation.PathVariable;
import org.springframework.web.bind.annotation.RestController;
@RestController
public class UserController {
    //使用@Autowired注解实现UserController 对UserDao 的依赖注入
    @Autowired
    private UserDao userDao;
    @GetMapping("/addUser")
    public String addUser(){
        int result = userDao.addUser(new User("yzpc1","yzpc1"));
        String message;
        if (result > 0) {
            message = "成功往数据表中插入了 " + result + " 条数据!";
        }else {
            message = "插入数据失败!";
        }
        System.out.println(message);
        return message;
    }
    @GetMapping("/updateUser")
    public String updateUser(){      //修改的id记录要存在
        int result = userDao.updateUser(new User(10004,"sj","sj"));
        String message;
```

```
        if (result > 0) {
            message = "成功在数据表中修改了 " + result + " 条数据!";
        }else {
            message = "修改数据失败!";
        }
        System.out.println(message);
        return message;
    }
    @GetMapping("/deleteUser/{id}")
    public String deleteUser(@PathVariable("id") int id){
        int result = userDao.deleteUser(id);
        String message;
        if (result > 0) {
            message = "成功在数据表中删除了 " + result + " 条数据!";
        }else {
            message = "删除数据失败!";
        }
        System.out.println(message);
        return message;
    }
}
```

(6) 启动项目,在浏览器中访问 http://localhost: 8080/addUser,如果插入数据成功,控制台输出相应的提示信息,如图 5-3 所示。在浏览器的页面上也输出同样的提示信息,如图 5-4 所示。

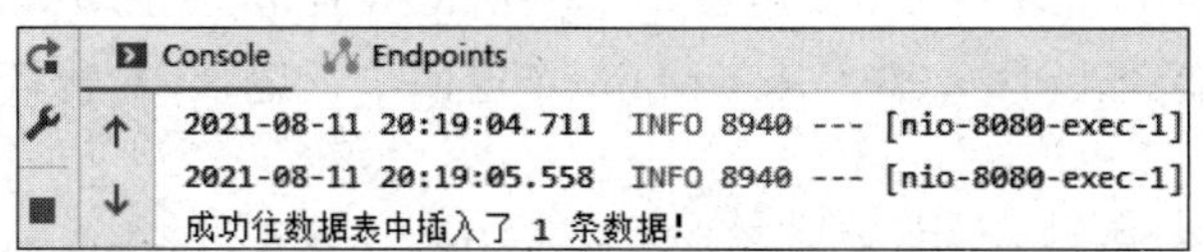

图 5-3 添加记录控制台运行效果

图 5-4 添加记录页面效果

访问 http://localhost: 8080/updateUser,进行修改记录,控制台运行效果,如图 5-5 所示。修改后的数据表中的数据,如图 5-6 所示。

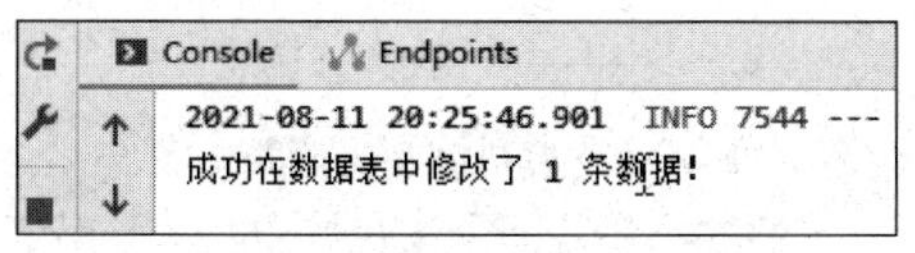

图 5-5 修改记录控制台运行效果

id	username	password
10001	admin	admin
10002	yzpc	yzpc
10003	my	my
10004	sj	sj

图 5-6 修改数据后的数据表

访问 http://localhost: 8080/deleteUser/10004,进行删除记录,如果数据表中存在要

删除的数据，则删除该数据，页面输出"成功在数据表中删除了 1 条数据!"的信息，如图 5-7 所示。如果传递一个不存在的 id(如 10006)参数，则页面输出"删除数据失败!"的信息，如图 5-8 所示。

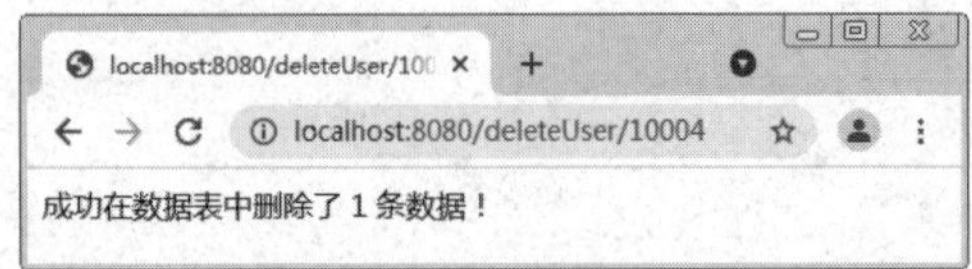

图 5-7 删除成功的运行效果

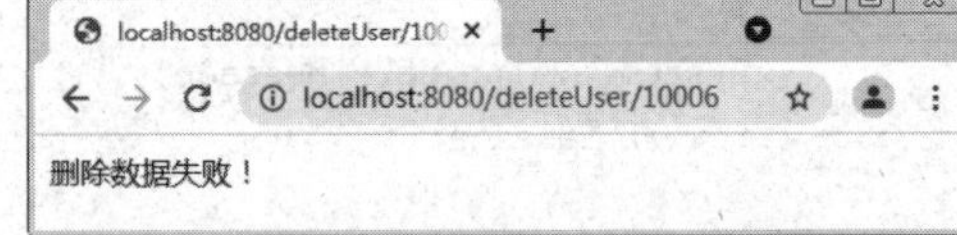

图 5-8 删除失败的运行效果

可以在运行相应的响应请求后，查看数据中的数据表记录，查看运行效果。

5.2.2 JdbcTemplate 查询的操作

JdbcTemplate 对 JDBC 的流程做了封装，提供了大量的 query()方法来处理各种对数据库表的查询操作，常用的 query()方法如下。

- List query(String sql, PreparedStatementSetter pss, RowMapper rowMapper)：该方法根据 String 类型参数提供的 SQL 语句创建 PreparedStatement 对象，通过 RowMapper 将结果返回到 List 中。
- List query(String sql,Object[] args,RowMapper rowMapper)：该方法使用 Object[]的值来设置 SQL 中的参数值，采用 RowMapper 回调方法可以直接返回 List 类型的数据。
- queryForObject(String sql,Object[] args,RowMapper rowMapper)：该方法将参数绑定到 SQL 语句中，通过 RowMapper 返回单行记录，并转换为一个 Object 类型返回。
- queryForList(String sql, Object[] args, class<T> elementType)：该方法可以返回多行数据的结果，但必须是返回列表，elementType 参数返回的是 List 元素类型。

下面尝试从 user 表中查询数据，在 UserDao 接口中增加按照 id 查询用户的方法和查询所有用户的方法，在 UserDaoImpl 中具体实现有两个方法，其实现步骤如下。

(1) 在 UserDao 接口中，分别创建一个通过 id 查询单个用户信息和查询所有用户信息的方法，代码如下。

```
public User getUserById(int id);
public List<User> getAllUser();
```

(2) 在 UserDao 接口的实现类 UserDaoImpl 中，实现接口中的方法，并使用 query()方法分别进行查询，代码如下。

```
//实例化 BeanPropertyRowMapper 对象,将结果集通过 Java 的反射机制映射到Java 对象中
RowMapper<User> rowMapper = new BeanPropertyRowMapper<User>(User.class);
@Override
public User getUserById(int id) {
    String sql = "select * from user where id = ?";
```

```
        return jdbcTemplate.queryForObject(sql,rowMapper, id);
    }
    @Override
    public List < User > getAllUser() {
        String sql = "select * from User";
        return jdbcTemplate.query(sql, rowMapper);
    }
```

在 UserDaoImpl 实现类的方法中，BeanPropertyRowMapper 是 BeanMapper 接口的实现类，它可以自动地将数据表中的数据映射到用户定义的类中(需要用户自定义类中的字段要与数据表中的字段相对应)。BeanPropertyRowMapper 对象创建后，在 getUserById()方法中通过 queryForObject()方法返回一个 Object 类型的单行记录，而在 getAllUser()方法中通过 query()方法返回一个结果集合。

(3) 在 UserController 类中，添加查询方法 getUserById ()、getAllUser()来进行查询，代码如下。

```
    @GetMapping("/getUserById")
    public User getUserById(int id){
        User user;
        user = userDao.getUserById(id);
        if (user!= null) {
            System.out.println("查询 id 为" + id + "的 User 对象为: " + user);
        }
        return user;
    }
    @GetMapping("/getAllUser")
    public List < User > getAllUser(){
        List < User > users = userDao.getAllUser();
        if (users!= null) {
            System.out.println("所有 User 对象列表为: ");
            for (User user : users) {      // 循环输出集合中对象
                System.out.println(user);
            }
        }
        return users;
    }
```

(4) 重新运行项目，访问 http://localhost:8080/getUserById?id=10001，根据 id 查询 id 值为 10001 的用户信息，控制台输出结果如图 5-9 所示。

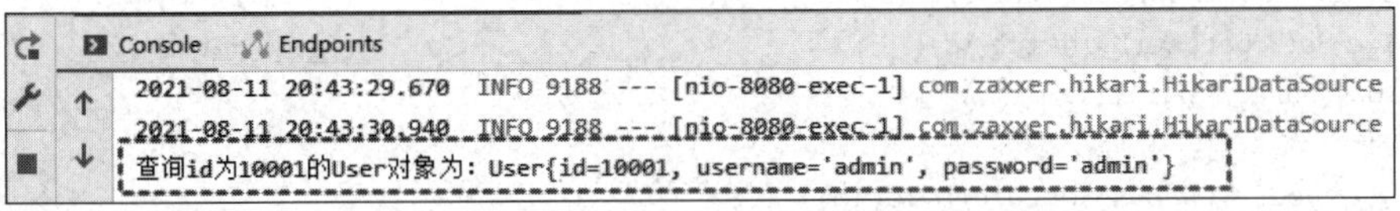

```
Console   Endpoints
2021-08-11 20:43:29.670  INFO 9188 --- [nio-8080-exec-1] com.zaxxer.hikari.HikariDataSource
2021-08-11 20:43:30.940  INFO 9188 --- [nio-8080-exec-1] com.zaxxer.hikari.HikariDataSource
查询id为10001的User对象为: User{id=10001, username='admin', password='admin'}
```

图 5-9　查询单个用户运行结果

访问 http://localhost: 8080/getAllUser，查询所有用户信息，控制台输出结果如图 5-10

所示。

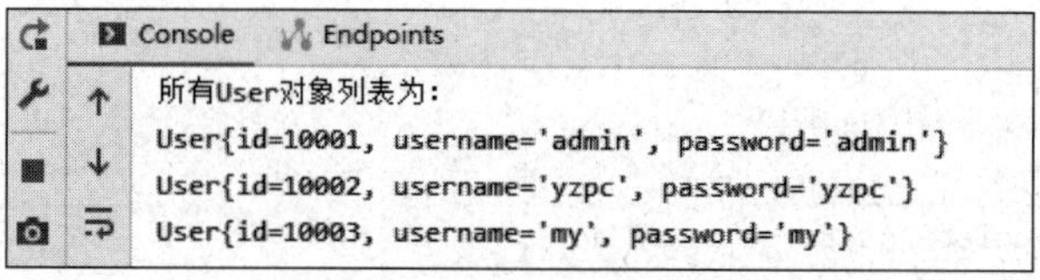

```
所有User对象列表为:
User{id=10001, username='admin', password='admin'}
User{id=10002, username='yzpc', password='yzpc'}
User{id=10003, username='my', password='my'}
```

图 5-10 查询所有用户运行结果

如果查询一个数据库中不存在的数据，例如查询 id 值为 10006 的信息，访问 http://localhost:8080/getUserById?id=10006，控制台将抛出错误异常的提示信息，如图 5-11 所示。

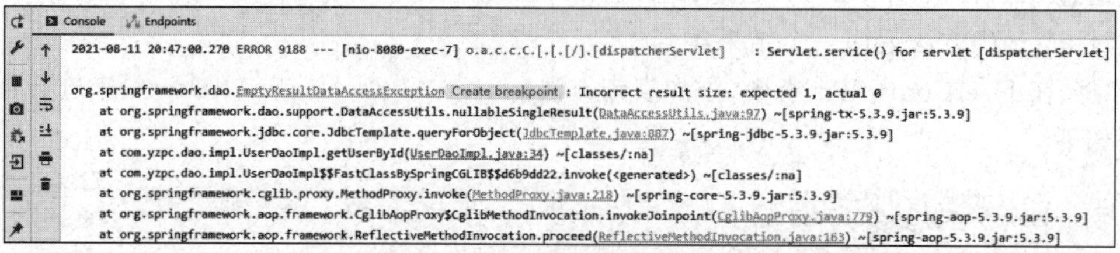

```
2021-08-11 20:47:00.270 ERROR 9188 --- [nio-8080-exec-7] o.a.c.c.C.[.[.[/].[dispatcherServlet]    : Servlet.service() for servlet [dispatcherServlet]

org.springframework.dao.EmptyResultDataAccessException Create breakpoint : Incorrect result size: expected 1, actual 0
    at org.springframework.dao.support.DataAccessUtils.nullableSingleResult(DataAccessUtils.java:97) ~[spring-tx-5.3.9.jar:5.3.9]
    at org.springframework.jdbc.core.JdbcTemplate.queryForObject(JdbcTemplate.java:887) ~[spring-jdbc-5.3.9.jar:5.3.9]
    at com.yzpc.dao.impl.UserDaoImpl.getUserById(UserDaoImpl.java:34) ~[classes/:na]
    at com.yzpc.dao.impl.UserDaoImpl$$FastClassBySpringCGLIB$$d6b9dd22.invoke(<generated>) ~[classes/:na]
    at org.springframework.cglib.proxy.MethodProxy.invoke(MethodProxy.java:218) ~[spring-core-5.3.9.jar:5.3.9]
    at org.springframework.aop.framework.CglibAopProxy$CglibMethodInvocation.invokeJoinpoint(CglibAopProxy.java:779) ~[spring-aop-5.3.9.jar:5.3.9]
    at org.springframework.aop.framework.ReflectiveMethodInvocation.proceed(ReflectiveMethodInvocation.java:163) ~[spring-aop-5.3.9.jar:5.3.9]
```

图 5-11 异常信息显示

查看 DataAccessUtils 源代码，nullableSingleResult 在查到空集合的时候，会默认抛出 EmptyResultDataAccessException 异常，代码如下。

```
@Nullable
public static <T> T nullableSingleResult(@Nullable Collection<T> results) throws
IncorrectResultSizeDataAccessException {
    if (CollectionUtils.isEmpty(results)) {
        throw new EmptyResultDataAccessException(1);
    } else if (results.size() > 1) {
        throw new IncorrectResultSizeDataAccessException(1, results.size());
    } else {
        return results.iterator().next();
    }
}
```

修改 getUserById()方法代码，在方法体内捕获 EmptyResultDataAccessException 异常，代码修改如下。

```
@GetMapping("/getUserById")
public User getUserById(int id){
    User user = null;
    try{
        user = userDao.getUserById(id);
        if (user!= null) {
            System.out.println("查询 id 为" + id + "的 User 对象为: " + user);
        }
```

```
    }catch (EmptyResultDataAccessException e){
        System.err.println("未查询到 id 为 " + id + " 的记录!");
        return null;
    }
    return user;
}
```

重新启动项目,测试方法和前面一致。同样,对于 getAllUser()方法,去执行查询操作时,假设数据表中没有数据,也会抛出 EmptyResultDataAccessException 异常,在 getAllUser()方法体内捕获异常的方法和在 getUserById()方法体中基本一致。

5.3 使用 Spring Data JPA

5.3.1 Spring Data JPA 介绍

Spring Data 是 Spring 的一个子项目,旨在统一和简化各类型数据的持久化存储方式,而不局限于是关系型数据库还是非关系型数据库。其主要目标是封装众多底层数据存储的不同操作方式,对外统一相同的接口,让数据操作变得方便快捷。

JPA(Java Persistence API)即 Java 的持久层的 API,用于对象的持久化,它是一个非常强大的 ORM 持久化的解决方案。JPA 通过注解或 XML 描述"对象—关系(表)"的映射关系,并将内存中的实体对象持久化到数据库。JPA 是一个规范化接口,封装了 Hibernate 的操作作为默认实现,让用户不通过任何配置即可完成数据库的操作。

Spring Data 在 JPA 规范的基础上,充分利用其优点,提出了 Spring Data JPA 模块对具有 ORM 关系的数据进行持久化操作。JPA、Spring Data 和 Hibernate 的关系如图 5-12 所示。

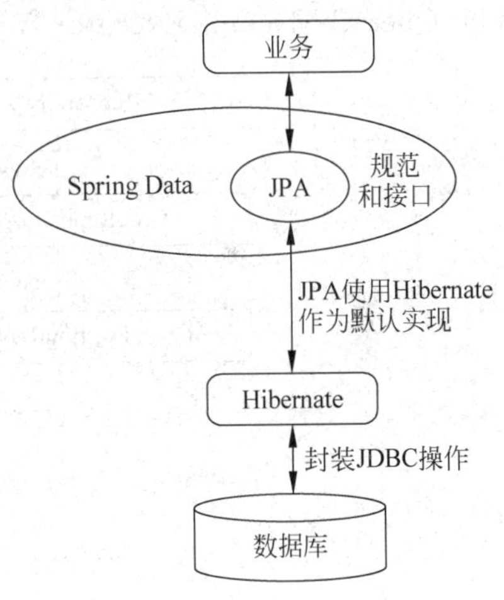

图 5-12 JPA、Spring Data 和 Hibernate 的关系

Spring Data JPA 是 Spring 在 ORM 框架、JPA 规范的基础上封装的一套新的 JPA 应用规范,也是要靠 Hibernate 等 ORM 框架实现的一种解决方案。它提供了增删改查等常用功能,可以使用较少的代码实现数据操作,同时还易于扩展。Spring Data JPA 是 Spring 提供的一套对 JPA 操作更加高级的封装,是在 JPA 规范下的专门用来进行数据持久化的解决方案。

Spring Data JPA 通过基于 JPA 的 Repository 极大地简化了 JPA 的写法,在几乎不写实现的情况下,实现数据库的访问和操作。使用 Spring Data JPA 建立数据访问层十分方便,只需要定义一个继承 JpaRepository 接口的接口即可。继承了JpaRepository 接口的自定义访问接口,具有 JpaRepository 接口的所有数据访问操作方法,JpaRepository 接口的源代码如下。

```
@NoRepositoryBean
public interface JpaRepository < T, ID > extends PagingAndSortingRepository < T, ID >,
QueryByExampleExecutor < T > {
    List < T > findAll();                                     //查找所有实体
    List < T > findAll(Sort var1);                            //排序、查找所有实体
    List < T > findAllById(Iterable < ID > var1);             //返回指定一组 ID 的实体
    < S extends T > List < S > saveAll(Iterable < S > var1);  //保存集合
    void flush();                                             //执行缓存与数据库同步
    < S extends T > S saveAndFlush(S var1);                   //强制执行持久化
    void deleteInBatch(Iterable < T > var1);                  //删除一个实体集合
    void deleteAllInBatch();                                  //删除所有实体
    T getOne(ID var1);                        //返回 ID 对应的实体.若不存在,则返回空值
    //查询满足 Example 的所有对象
    < S extends T > List < S > findAll(Example < S > var1);
    //查询满足 Example 的所有对象,并且进行排序返回
    < S extends T > List < S > findAll(Example < S > var1, Sort var2);
}
```

上述代码中,T 表示实体对象；ID 表示主键,且必须实现序列化；Example 对象是 JPA 提供用来构造查询条件的对象。JpaRepository 接口继承了 PagingAndSortingRepository 接口、QueryByExampleExecutor 接口,JpaRepository 接口的继承关系如图 5-13 所示。

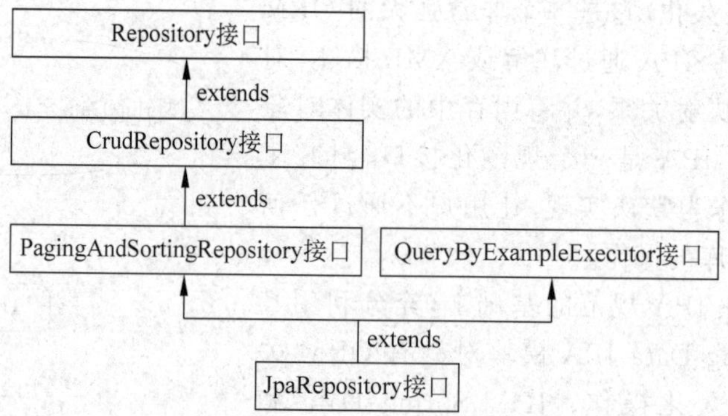

图 5-13 JpaRepository 接口的继承关系

JpaRepository 接口继承关系中涉及的接口说明如下。

- Repository 接口是 Spring Data JPA 提供的用于自定义 Repository 接口的顶级父接口,该接口中没有声明任何方法。
- CrudRepository 接口是 Repository 的继承接口之一,包含了一些基本的 CRUD 方法。
- PagingAndSortingRepository 接口继承 CrudRepository 接口的同时,提供了分页和排序两个方法。
- QueryByExampleExecutor 接口是进行条件封装查询的顶级父接口,允许通过 Example 实例执行复杂条件查询。

5.3.2 整合 Spring Data JPA

下面通过示例来介绍 Spring Boot 整合 Spring Data JPA，步骤如下。

（1）新建一个 Spring Boot 工程 chapter05jpa，Group 和 Package name 为 com.yzpc，在 Dependencies 依赖中选择 Web 节点下的 Spring Web 依赖和 SQL 节点下的 Spring Data JPA、MySQL Driver 依赖，单击 Finish 按钮，如图 5-14 所示。

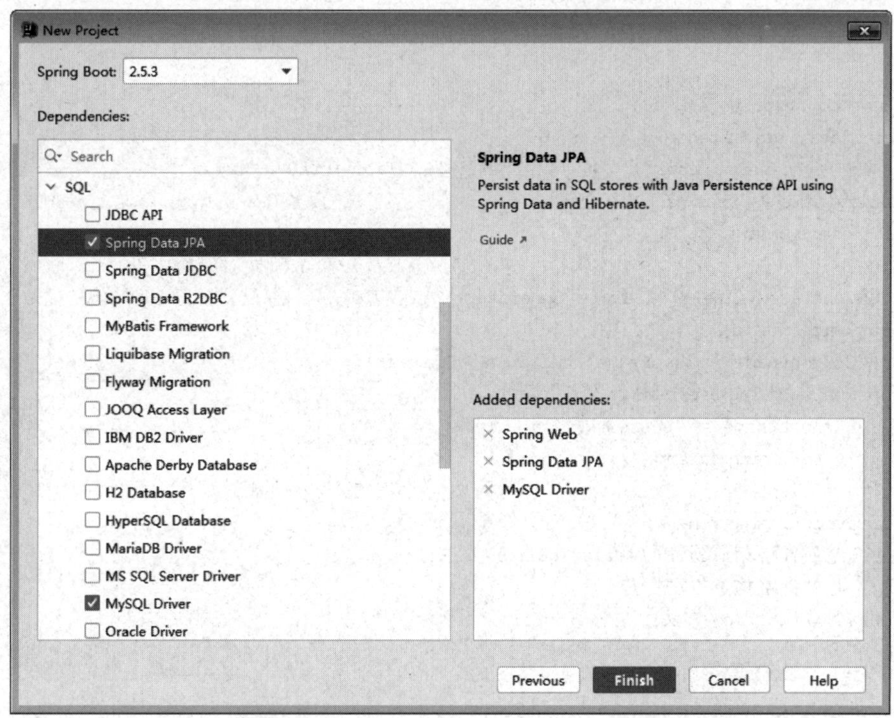

图 5-14　添加 SQL 的 Spring Data JPA 等相关依赖

pom.xml 中自动添加的依赖代码如下。

```
<dependency>
    <groupId>org.springframework.boot</groupId>
    <artifactId>spring-boot-starter-data-jpa</artifactId>
</dependency>
<!-- 省略Web 中Spring Web 依赖和SQL 中的MySQl Driver 依赖代码 -->
```

（2）在 application.properties 配置文件中加入数据库配置，与 5.2.1 小节第(2)步中介绍的一致。新建一个 chapter05jpa 的数据库，设置 JPA 的基本配置，代码如下。

```
# 省略数据库配置的设置,参看 5.2.1 小节第(2)步,数据库使用chapter05jpa
spring.jpa.database = mysql    # JPA 对应的数据库类型
spring.jpa.show-sql = true     # 在控制台打印SQL
```

```
# update: 项目启动,加载 Hibernate 时,根据实体类更新数据库中的表
# validate: 在加载Hibernate 时,验证创建数据表的结构
# create: 每次加载Hibernate 时,重新创建数据表结构,造成数据丢失
# create - drop: 加载Hibernate 时创建,退出时删除表结构
# none: 启动时不做任何操作
spring.jpa.hibernate.ddl - auto = create
```

(3) 在项目的 src/main/java/路径下的 com.yzpc 包中,新建一个 pojo 包,并在该包中新建一个 User 实体类,添加 JPA 对应的注解进行映射配置,代码如下。

```
package com.yzpc.pojo;
import javax.persistence.*;
// @Entity(name = "user") 等同于 @Entity 和@Table(name = "user" )
@Entity(name= "user" )                                    // 设置 ORM 实体类,并指定映射的表名
public class User {
    @Id                                                   //表示映射对应的主键
    @GeneratedValue(strategy = GenerationType.AUTO)      //设置主键自增策略
    private int id;
    @Column(name = "username",nullable = false)
    private String username;
     @Column(name = "password")
    private String password;
    @Transient                                            //表示生成数据库中的表时,该属性被忽略
    private String remark;
    // 此处省略相应属性的 setter/getter 方法
    // 此处省略构造方法
    // 重写 toString 方法
}
```

(4) 在项目的 src/main/java/路径下的 com.yzpc 包中,新建一个 repository 包,并在该包中新建一个用于对数据库表 user 进行操作的接口 UserRepository,代码如下。

```
package com.yzpc.repository;
import com.yzpc.pojo.User;
import org.springframework.data.jpa.repository.JpaRepository;
import org.springframework.data.jpa.repository.Query;
public interface UserRepository extends JpaRepository< User, Integer >{
    //根据 id 查询用户
    @Query(value = "SELECT * FROM user u where u.id = ?1",nativeQuery = true)
    public User findUserById(int id);
    //根据 username 用户名查询用户
    @Query(value = "SELECT * FROM user where user.username = ?1", nativeQuery = true)
    public User findUserByName(String username);
    //查找 id 最大的用户
    @Query(value = "select * from user where id = (select max(id) from user )",nativeQuery =
true)
    public User findMaxIdUser();
    // JpaRepository 接口提供了很多方法,可直接调用
}
```

（5）在项目的 src/main/java/路径下的 com. yzpc 包中，新建一个 controller 包，并在该包中新建一个用于控制的类 UserController，代码如下。

```
package com.yzpc.controller;
…
@RestController
public class UserController {
    @Autowired
    UserRepository repository;
    @GetMapping("/saveUser")
    public void saveUser() {
        repository.save(new User("admin","admin"));
        repository.save(new User("yzpc","yzpc"));
        repository.save(new User("my","my"));
    }
    @GetMapping("/addUser")
    public void addUser(String username,String password) {
        repository.save(new User(username,password));
    }
    @GetMapping("/updateUser")
    public void updateUser(int id,String username,String password) {
        User user =  new User(id,username,password);
        repository.save(user);
    }
    @GetMapping("/deleteUser")
    public void deleteUser(int id) {
        repository.deleteById(id);
    }
    @GetMapping("/getUserById")
    public Optional<User> getUserById(int id) {
        Optional<User> user = repository.findById(id);
        System.out.println(user);
        return user;
    }
    @GetMapping("/getAllUser")
    public List<User> getAllUser() {
        List<User> users = repository.findAll();
        if (users!= null) {
            System.out.println("所有 User 对象列表为：");
            for (User user : users) {       // 循环输出集合中对象
                System.out.println(user);
            }
        }
        return users;
    }
    // 调用在 UserRepository 接口中自定义的方法
    @GetMapping("/findUserById")
    public User findUserById(int id) {
        User user = repository.findUserById(id);
```

```
        System.out.println(user);
        return user;
    }
    @GetMapping("/findUserByName")
    public User findUserByName(String username) {
        User user = repository.findUserByName(username);
        System.out.println(user);
        return user;
    }
    @GetMapping("/findMaxIdUser")
    public User findMaxIdUser(){
        User user = repository.findMaxIdUser();
        System.out.println(user);
        return user;
    }
}
```

(6) 启动项目,在控制台输出 Hibernate 的相关语句,如图 5-15 所示。

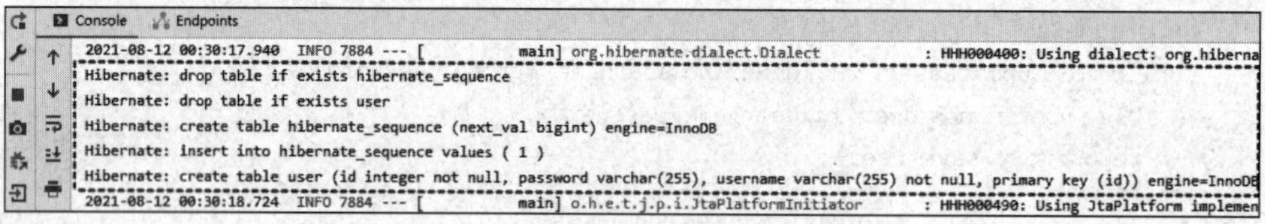

```
2021-08-12 00:30:17.940  INFO 7884 --- [           main] org.hibernate.dialect.Dialect            : HHH000400: Using dialect: org.hiberna
Hibernate: drop table if exists hibernate_sequence
Hibernate: drop table if exists user
Hibernate: create table hibernate_sequence (next_val bigint) engine=InnoDB
Hibernate: insert into hibernate_sequence values ( 1 )
Hibernate: create table user (id integer not null, password varchar(255), username varchar(255) not null, primary key (id)) engine=InnoDB
2021-08-12 00:30:18.724  INFO 7884 --- [           main] o.h.e.t.j.p.i.JtaPlatformInitiator       : HHH000490: Using JtaPlatform implemen
```

图 5-15 创建数据表的 Hibernate 语句

在 application. properties 配置文件中 spring. jpa. hibernate. ddl-auto 的值设置为 create,即通过 Hibernate 实现数据表单的创建。通过 SQLyog 工具,查看数据库,发现 user 数据表已经创建,但并没有记录,下面分别来测试相应的方法实现。

访问 http://localhost: 8080/saveUser,在该请求中 repository 对象三次调用 save()方法,而这个 save()方法是系统提供的,不需要写插入的 SQL 代码,就可实现数据的插入功能。查看控制台,发现有三条 insert 的插入语句,如图 5-16 所示;查看数据表,如图 5-17 所示。

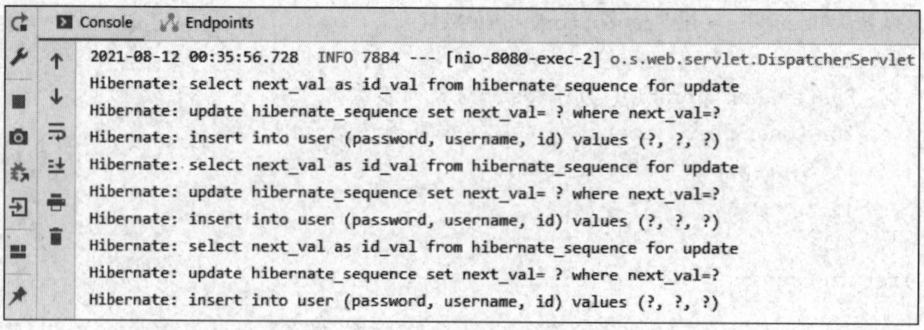

```
2021-08-12 00:35:56.728  INFO 7884 --- [nio-8080-exec-2] o.s.web.servlet.DispatcherServlet
Hibernate: select next_val as id_val from hibernate_sequence for update
Hibernate: update hibernate_sequence set next_val= ? where next_val=?
Hibernate: insert into user (password, username, id) values (?, ?, ?)
Hibernate: select next_val as id_val from hibernate_sequence for update
Hibernate: update hibernate_sequence set next_val= ? where next_val=?
Hibernate: insert into user (password, username, id) values (?, ?, ?)
Hibernate: select next_val as id_val from hibernate_sequence for update
Hibernate: update hibernate_sequence set next_val= ? where next_val=?
Hibernate: insert into user (password, username, id) values (?, ?, ?)
```

图 5-16 插入语句执行

访问 http://localhost:8080/addUser?username=sj&password=sj,查看控制台输出,可以看到一条 insert 的插入语句,查看数据表,发现增加了一条记录,如图 5-18 所示。

id	password	username
1	admin	admin
2	yzpc	yzpc
3	my	my
(NULL)	(NULL)	(NULL)

图 5-17　插入数据后的 user 表

id	password	username
1	admin	admin
2	yzpc	yzpc
3	my	my
4	sj	sj
(NULL)	(NULL)	(NULL)

图 5-18　增加一条记录后的 user 表

访问 http://localhost:8080/updateUser?id=4&username=sss&password=sss，查看数据表发现 id 为 4 的记录的 username 和 password 字段都已经修改为 sss，如图 5-19 所示。

访问 http://localhost:8080/deleteUser?id=4，查看数据表，id 为 4 记录已经被删除，如图 5-20 所示。

id	password	username
1	admin	admin
2	yzpc	yzpc
3	my	my
4	sss	sss
(NULL)	(NULL)	(NULL)

图 5-19　修改一条记录后的 user 表

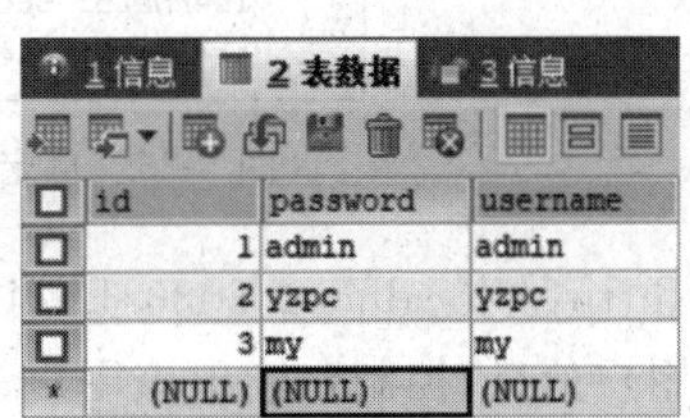

id	password	username
1	admin	admin
2	yzpc	yzpc
3	my	my
(NULL)	(NULL)	(NULL)

图 5-20　删除一条记录后的 user 表

访问 http://localhost:8080/getUserById?id=3，网页页面输出 id 为 3 的记录，如图 5-21 所示；查看控制台，输出相关信息，如图 5-22 所示。

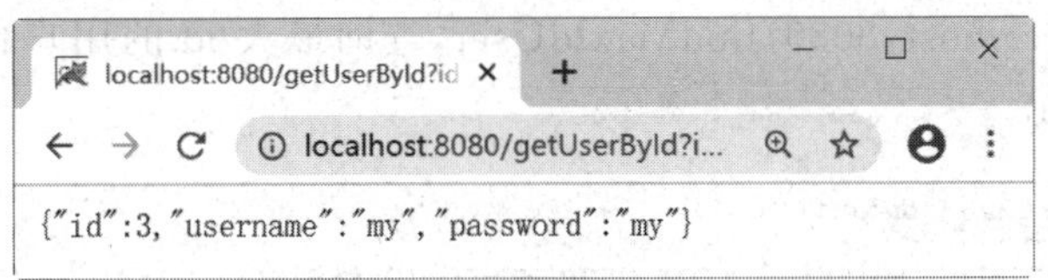

图 5-21　网页显示查询的用户

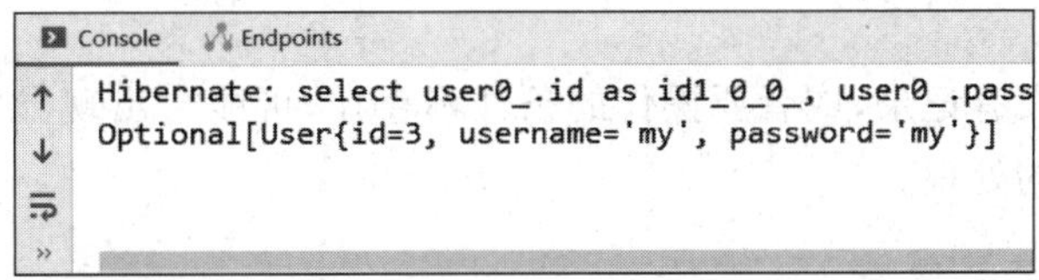

图 5-22　控制台输出查询的用户

访问 http://localhost:8080/getAllUser，查询所有用户信息，网页页面显示查询的所有用户，如图 5-23 所示；控制台输出查询的所有用户，如图 5-24 所示。

访问 http://localhost:8080/findUserById?id=2，这里调用的是 UserRepository 接口中的自定义的方法，查询 id 为 2 的用户信息，控制台输出该用户，如图 5-25 所示。

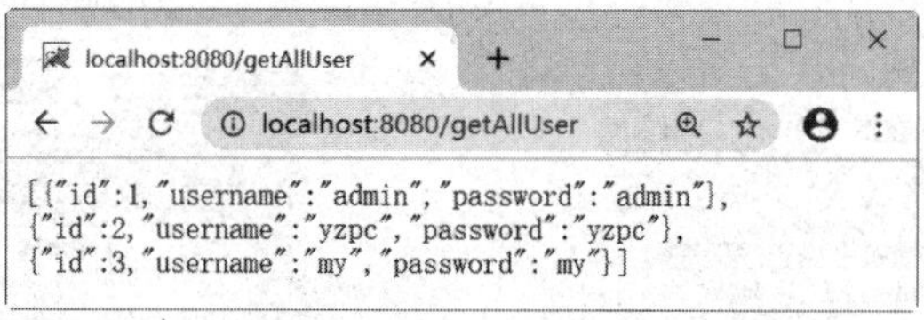
localhost:8080/getAllUser
localhost:8080/getAllUser
[{"id":1,"username":"admin","password":"admin"},
{"id":2,"username":"yzpc","password":"yzpc"},
{"id":3,"username":"my","password":"my"}]

图 5-23　网页显示查询的所有用户

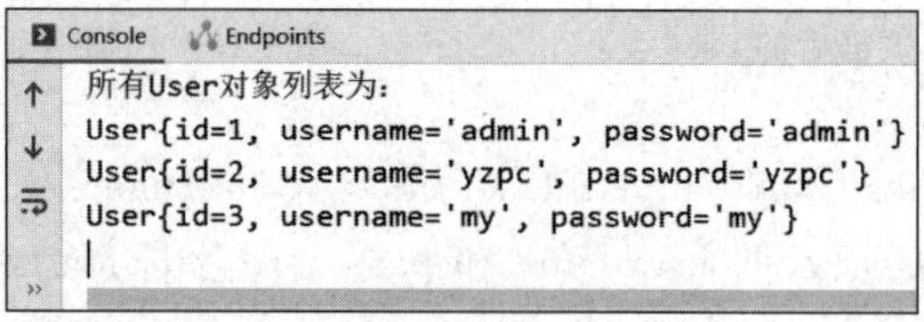
Console
Endpoints
所有User对象列表为:
User{id=1, username='admin', password='admin'}
User{id=2, username='yzpc', password='yzpc'}
User{id=3, username='my', password='my'}

图 5-24　控制台输出查询的所有用户

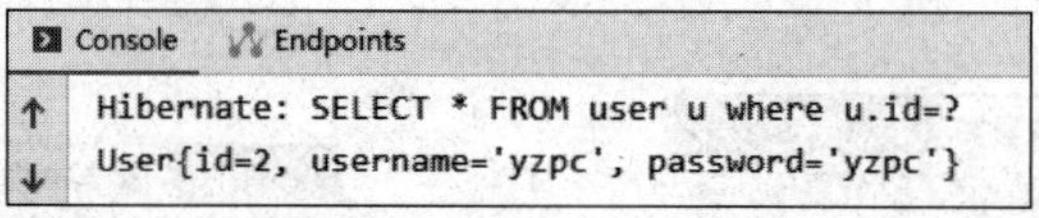
Console
Endpoints
Hibernate: SELECT * FROM user u where u.id=?
User{id=2, username='yzpc', password='yzpc'}

图 5-25　自定义的方法查询 id 为 2 的用户

访问 http://localhost:8080/findUserByName?username=admin,查询 username 为 admin 的用户信息,控制台输出该用户,如图 5-26 所示。

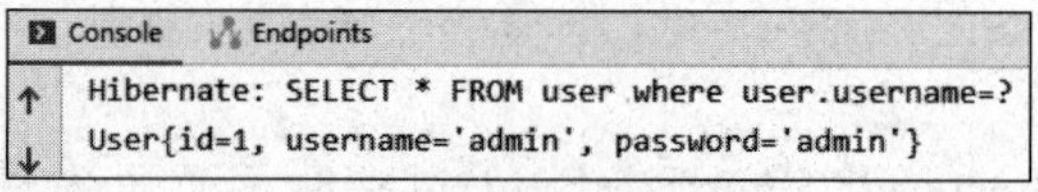
Console
Endpoints
Hibernate: SELECT * FROM user where user.username=?
User{id=1, username='admin', password='admin'}

图 5-26　自定义的方法查询 username 为 admin 的用户

访问 http://localhost:8080/findMaxIdUser,查询最大 id 的用户信息,控制台输出该用户,如图 5-27 所示。

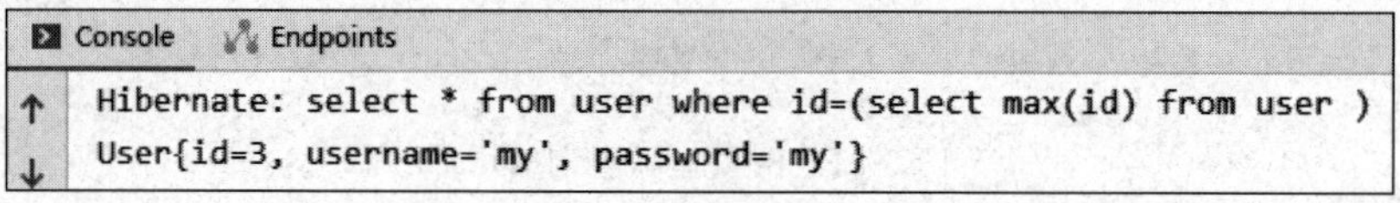
Console
Endpoints
Hibernate: select * from user where id=(select max(id) from user)
User{id=3, username='my', password='my'}

图 5-27　自定义的方法查询最大 id 的用户

如果数据表中没有记录,或者要操作的条件不存在,可能会出现异常,可以参考 5.2 小节中对于相关异常的处理。

5.4　整合 MyBatis

MyBatis 和 JPA 一样,也是一款优秀的持久层框架,它支持定制化 SQL、存储过程以及高级映射,避免了几乎所有的 JDBC 代码和手动设置参数以及获取结果集。MyBatis 可以使用简单的 XML 或注解来配置和映射原生信息,将接口和普通的 Java 对象映射成数据库中的记录。在 Spring Boot 中,MyBatis 官方提供了一套自动化配置方案,自行适配了对应的启动器，可以做到 MyBatis 开箱即用，进一步简化了 MyBatis 对数据的操作。

5.4.1　基于 XML 配置的方式整合 MyBatis

下面通过具体的示例来介绍 Spring Boot 整合 MyBatis，具体步骤如下。

(1) 新建一个 Spring Boot 工程 chapter05mybatis，Group 和 Package name 为 com.yzpc，在 Dependencies 依赖中选择 Web 节点下的 Spring Web 依赖和 SQL 节点下的 MyBatis Framework、MySQL Driver 依赖，单击 Finish 按钮，如图 5-28 所示。

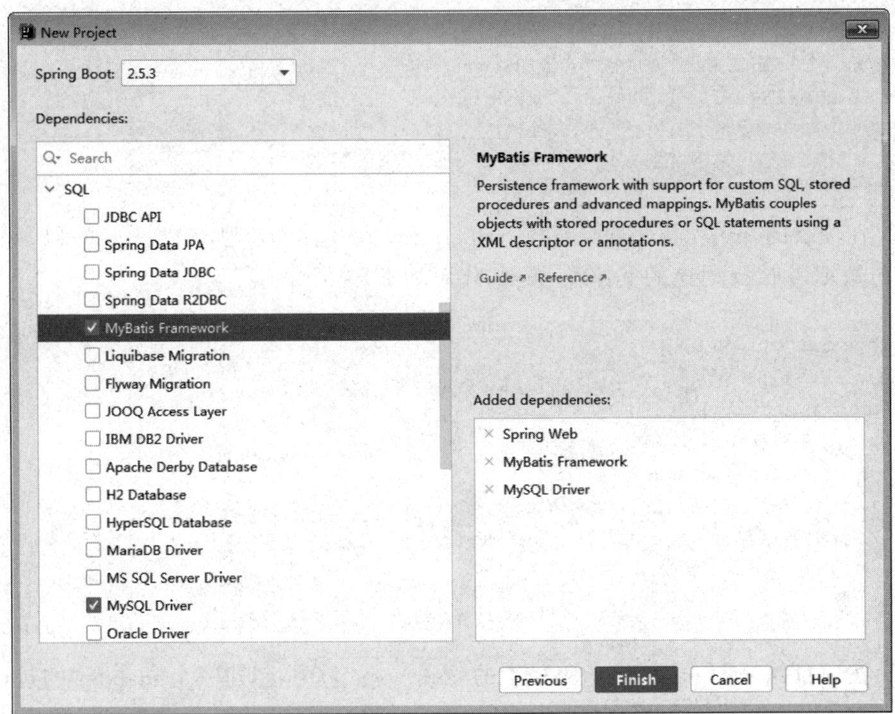

图 5-28　添加 SQL 的多个相关依赖

pom.xml 中自动添加的依赖代码如下。

```
<dependency>
    <groupId>org.mybatis.spring.boot</groupId>
    <artifactId>mybatis-spring-boot-starter</artifactId>
    <version>2.2.0</version>
</dependency>
<!-- 省略Web 中Spring Web 依赖和SQL 中的MySQl Driver 依赖代码 -->
```

在 application.properties 配置文件中加入数据库连接配置，与 5.2.1 小节第(2)步中介绍的一致。

(2) 在 MySQL 中，创建一个名为 chapter05mybatis 的数据库，在该数据库中创建 user 数据表，并预先插入几条测试数据，相应的 SQL 语句如下。

```
# 创建数据库
CREATE DATABASE chapter05mybatis;
# 选择使用数据库
USE chapter05mybatis;
# 创建 user 数据表并插入相关数据
DROP TABLE IF EXISTS USER;
CREATE TABLE USER(
  id INT(10) NOT NULL AUTO_INCREMENT ,
  username VARCHAR(20) DEFAULT NULL ,
  password VARCHAR(20) DEFAULT NULL ,
  PRIMARY KEY (id)
)ENGINE = INNODB DEFAULT CHARSET = utf8;
INSERT INTO USER VALUES(1,'admin','admin');
INSERT INTO USER VALUES(2,'yzpc','yzpc');
INSERT INTO USER VALUES(3,'my','my');
```

(3) 在项目的 src/main/java/路径下的 com. yzpc 包中，新建一个 pojo 的包，并在该包中新建与数据表 user 对应的实体类 User，代码如下。

```
package com.yzpc.pojo;
public class User {
    private int id;
    private String username;
    private String password;
    // 此处省略相应属性的 setter/getter 方法
    // 此处省略构造方法
    // 重写 toString 方法
}
```

(4) 在项目的 src/main/java/路径下的 com. yzpc 包中，新建一个 mapper 的包，在并在该包中新建 UserMapper 的接口，代码如下。

```
package com.yzpc.mapper;
import com.yzpc.pojo.User;
import org.apache.ibatis.annotations.Mapper;
import java.util.List;
@Mapper      // 表明该接口是一个 MyBatis 中的Mapper
public interface UserMapper {
    public int addUser(User user);
    public int updateUser(User user);
    public int deleteUser(int id);
    public User getUserById(int id);
    public List<User> getAllUser();
}
```

在 UserMapper 上添加@Mapper 注解，表明该接口是一个 MyBatis 中的 Mapper，这种方式需要在每个 Mapper 上都要添加；还有一种方式是在主类或配置类上添加@MapperScan("com. yzpc. mapper") 注解，表示扫描 com. yzpc. mapper 包下的所有接口作为 Mapper，这样便不需要在每个接口上配置@Mapper 注解。

（5）在 src/main/resources/路径下新建一个 mapper 目录，并在该目录下，新建 UserMapper.xml 文件，对应前面 UserMapper 接口的方法，#{}用来代替接口中的参数，实体类中的属性可以直接通过#{实体类属性名}获取，代码如下。

```
<?xml version = "1.0" encoding = "UTF - 8" ?>
<!DOCTYPE mapper
        PUBLIC " - //mybatis.org//DTD Mapper 3.0//EN"
        "http://mybatis.org/dtd/mybatis - 3 - mapper.dtd">
<!-- 此处与接口类地址对应 -->
<mapper namespace = "com.yzpc.mapper.UserMapper">
    <!-- 此处与接口方法名对应,指定参数类型与返回结果类型 -->
    <insert id = "addUser" parameterType = "com.yzpc.pojo.User">
        insert into user(username,password) values (#{username}, #{password})
    </insert>
    <update id = "updateUser" parameterType = "com.yzpc.pojo.User">
        update user set username = #{username},password = #{password} where id = #{id}
    </update>
    <delete id = "deleteUser" parameterType = "Integer">
        delete from user where id = #{id}
    </delete>
    <select id = "getUserById" resultType = "com.yzpc.pojo.User">
        select * from user where id = #{id}
    </select>
    <select id = "getAllUser" resultType = "com.yzpc.pojo.User">
        select * from user
    </select>
</mapper>
```

UserMapper.xml 放在 resources 目录下，不用担心打包时被忽略，但是放在 resources 目录下，不能自动被扫描到，此时在 application.properties 中添加 mybatis 扫描 mapper 的路径，需要在 application.properties 配置文件中指定位置，代码如下。

```
#指定映射 XML 文件位置
#classpath 对应resources, *.xml 表示配置mapper 下所有XML 文件
mybatis.mapper - locations = classpath: mapper/*.xml
```

UserMapper.xml 除了放在 resources 目录下，还可以直接放在 UserMapper 所在的 com.yzpc.mapper 包下，该包下的 UserMapper.xml 会被自动扫描到，但是 Maven 会带来 java 目录下的 XML 资源在项目打包时被忽略掉，所以，如果 UserMapper.xml 放在 com.yzpc.mapper 包下，为了避免打包时 java 目录下的 XML 文件被自动忽略掉，需要在 pom.xml 文件中再添加如下配置。

```
<build>
    <resources>
```

```
            <resource>
                <directory>src/main/java</directory>
                <includes>
                    <include>**/*.xml</include>
                </includes>
            </resource>
            <resource>
                <directory>src/main/resources</directory>
            </resource>
        </resources>
    </build>
```

（6）在项目的 src/main/java/路径下的 com.yzpc 包中，新建一个 controller 包，在该包中新建 UserController 类，并在该类中添加增删改查的方法，代码如下。

```
@RestController
public class UserController {
    @Autowired
    UserMapper userMapper;
    @GetMapping("/addUser")
    public String addUser(){
        int result = userMapper.addUser(new User("yzpc1","yzpc1"));
        String message;
        if (result > 0) {
            message = "成功往数据表中插入了 " + result + " 条数据!";
        }else {
            message = "插入数据失败!";
        }
        System.out.println(message);
        return message;
    }
    @GetMapping("/updateUser")
    public String updateUser(){
        int result = userMapper.updateUser(new User(4,"sj","sj"));
        String message;
        if (result > 0) {
            message = "成功在数据表中修改了 " + result + " 条数据!";
        }else {
            message = "修改数据失败!";
        }
        System.out.println(message);
        return message;
    }
    @GetMapping("/deleteUser/{id}")
    public String deleteUser(@PathVariable("id") int id){
        int result = userMapper.deleteUser(id);
        String message;
        if (result > 0) {
            message = "成功在数据表中删除了 " + result + " 条数据!";
```

```
        }else {
            message = "删除数据失败!";
        }
        System.out.println(message);
        return message;
    }
    @GetMapping("/getUserById")
    public User getUserById(int id){
        User user = null;
        try{
            user = userMapper.getUserById(id);
            if (user!= null) {
                System.out.println("查询 id 为" + id + "的 User 对象为: " + user);
            }
        }catch (EmptyResultDataAccessException e){
            System.err.println("未查询到 id 为 " + id + " 的记录!");
            return null;
        }
        return user;
    }
    @GetMapping("/getAllUser")
    public List<User> getAllUser(){
        List<User> users = null;
        try{
             users = userMapper.getAllUser();
             if (users!= null) {
                 System.out.println("所有 User 对象列表为: ");
                 // 循环输出集合中对象
                 for (User user : users) {
                     System.out.println(user);
                 }
             }
        }catch (EmptyResultDataAccessException e){
            System.err.println("数据表没有记录!");
            return null;
        }
        return users;
    }
}
```

(7) 启动项目,在浏览器地址栏中,依次访问增删改查的请求,相应请求如下。

- 增加用户: http://localhost:8080/addUser。
- 修改用户: http://localhost:8080/updateUser。
- 删除用户: http://localhost:8080/deleteUser/4。
- 查询用户: http://localhost:8080/getUserById?id=3。
- 查询所有用户: http://localhost:8080/getAllUser。

浏览器和控制台上会输出相应信息,控制台的运行结果如图 5-29 所示。

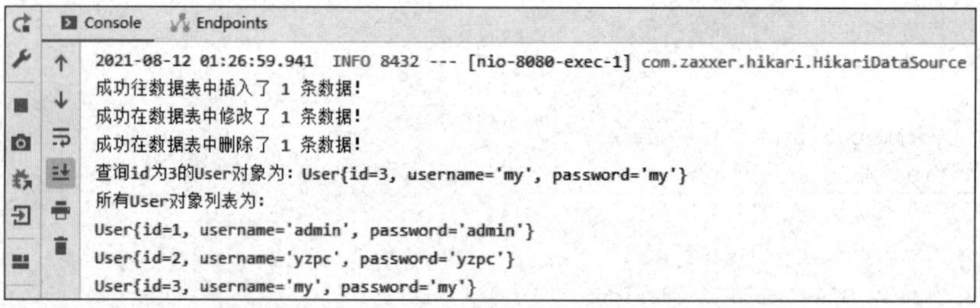

```
Console  Endpoints
2021-08-12 01:26:59.941  INFO 8432 --- [nio-8080-exec-1] com.zaxxer.hikari.HikariDataSource
成功往数据表中插入了 1 条数据!
成功在数据表中修改了 1 条数据!
成功在数据表中删除了 1 条数据!
查询id为3的User对象为: User{id=3, username='my', password='my'}
所有User对象列表为:
User{id=1, username='admin', password='admin'}
User{id=2, username='yzpc', password='yzpc'}
User{id=3, username='my', password='my'}
```

图 5-29　执行增删改查后控制台输出结果

5.4.2　基于注解的方式整合 MyBatis

MyBatis 不仅可以使用基于 XML 的映射文件来操作数据库,还可以使用注解的方式来操作数据库。在前面的示例上做一些修改即可,具体修改如下。

(1) 修改 application.properties 文件。注释或删除配置文件中的如下配置。

```
#mybatis.mapper-locations=classpath:mapper/*.xml
```

(2) 删除 Mapper 映射文件。在 src/main/resources 目录下,删除 mapper 目录。

(3) 修改 UserMapper 接口。将 com.yzpc.mapper 包中的 UserMapper 接口,添加相应的 SQL 注解,代码如下。

```
package com.yzpc.mapper;
import com.yzpc.pojo.User;
import org.apache.ibatis.annotations.*;
import java.util.List;
@Mapper
public interface UserMapper {
    @Insert("insert into user(username,password) values(#{username},
            #{password})")
    public int addUser(User user);
    @Update("update user set username=#{username},password=#{password}
            where id=#{id}")
    public int updateUser(User user);
    @Delete("delete from user where id=#{id}")
    public int deleteUser(int id);
    @Select("select * from user where id=#{id}")
    public User getUserById(int id);
    @Select("select * from user")
    public List<User> getAllUser();
}
```

重新启动项目,分别访问增删改查的请求,访问地址与 5.4.1 节第(7)步中的地址一致,运行结果与使用基于 XML 的映射文件的方式一致。

通过上面的例子可以看到,MyBatis 基本实现开箱即用的特性。自动化配置将开发者从繁杂的配置文件中解脱出来,专注于业务逻辑的开发。

5.5 小结

本章介绍了不同数据库数据源的配置、使用 JdbcTemplate 进行增删改查的操作，重点介绍了 Spring Boot 如何整合 Spring Data JPA 和 MyBatis。通过本章的学习，读者可以掌握 Spring Boot 访问关系型数据库的解决方案。

第6章 Spring Boot使用NoSQL

NoSQL 指非关系型数据库，非关系型数据库和关系型数据库最重要的区别是 NoSQL 不适用 SQL 作为查询语言。NoSQL 数据存储可以不需要固定的表格模式，一般都有水平可扩展的特征。NoSQL 主要有以下几种分类。

- 键值存储数据库：键值数据库就像在传统语言中使用的哈希表。用户可以通过 key 来添加、查询或者删除数据，由于使用主键访问，所以会获得不错的性能及扩展性，这一类的数据库有 Redis、Tokyo Cabinet 等。
- 文档型数据库：将数据以文档的形式储存。每个文档都是自包含的数据单元，是一系列数据项的集合，一般应用在 Web 应用中，这一类的数据库有 MongoDB 等。
- 列存储数据库：功能相对局限，但是查找速度快，容易进行分布式扩展，一般用于分布式文件系统中，这一类的数据库有 HBase、Cassandra 等。
- 图形(Graph)数据库：将数据以图的方式储存，专注于构建关系图谱，例如社交网络、推荐系统等，这一类的数据库有 Neo4J、InfoGrid 等。

近些年来，NoSQL 数据库的发展势头很快，种类繁多，目前已经产生了 100 多个 NoSQL 数据库系统。Spring Boot 对大多数常见的 NoSQL 提供配置支持，本章主要介绍 Spring Boot 整合 Redis 和 MongoDB 两种 NoSQL 数据库。

6.1 整合 Redis

6.1.1 Redis 简介

Redis 是使用 ANSI C 语言编写的基于内存的 NoSQL 数据库，是一个高性能的缓存存储系统，整个数据库加载在内存中进行操作，定期通过异步操作把数据库数据刷新到硬盘上进行保存。Redis 是目前最流行的键值对 Key-Value 存储数据库，以键值对的形式存储数据，没有数据表，直接用键值对的形式存储数据，被称为数据结构服务器。

Redis 完全开源，遵守 BSD 协议，是一个高性能的内存中的数据结构存储系统，主要用于 Java EE 应用的缓存实现，也可作为消息代理(Message Broker)使用，偶尔作为键值对数据库使用。Redis 包括字符串(string)、列表(list)、哈希(hash)、集合(set)和有序集合(sorted set)五大类型。

Redis 与其他键值对缓存产品相比，有以下三个特点。

- Redis 支持数据的持久化，可以将内存中的数据保存在磁盘中，重启的时候可以再次加载进行使用。
- Redis 不仅支持简单的键值对类型的数据，同时还提供列表、集合、哈希等数据结构的存储。
- Redis 支持数据的备份，即 master-slave 模式的数据备份。

6.1.2　Redis 安装

(1) 下载 Redis。Redis 官网(https://redis.io/download)只提供 Linux 版本的下载，官网推荐使用 Linux 版本 Redis，Redis 官方不支持 Windows 系统。若想使用 Windows 版本的 Redis，可从 GitHub 上下载由 Microsoft 开放技术小组开发和维护的针对 64 位 Windows 的版本(https://github.com/tporadowski/redis/releases)，如图 6-1 所示。

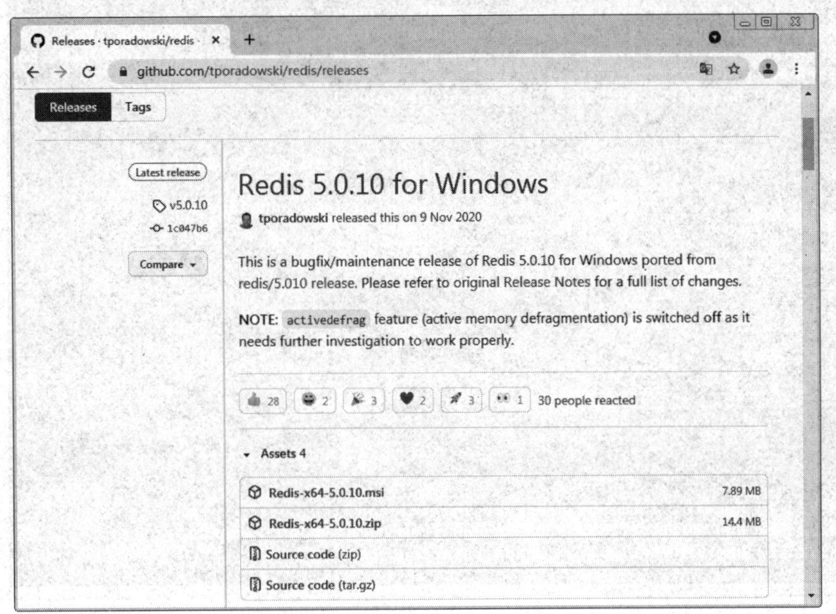

图 6-1　GitHub 上针对 Windows 版本的 Redis

GitHub 提供 zip 或 msi 两种下载格式，这里下载 Redis-x64-5.0.10.zip，将下载的 zip 文件解压缩后，放在 D：盘的根目录下。

(2) 启动 Redis 服务。在“运行”中输入 cmd，然后把目录指向解压缩的 Redis 目录，如图 6-2 所示。

执行 Redis-x64-5.0.10 目录下 redis-server.exe redis.windows.conf 命令行启动 Redis 服务，如图 6-3 所示，表示成功启动 Redis 服务。

也可直接双击打开 Redis 目录下的 redis-server.exe 文件，启动 Redis 服务。

(3) 使用客户端连接 Redis 服务器。Redis 服务器启动成功后，重新打开一个 cmd 命令窗口，把目录指向解压缩的 Redis 目录，使用 redis-cli.exe -h 127.0.0.1 -p 6379 命令，创建一个地址为 127.0.0.1、端口号为 6379 的 Redis 数据库服务，然后输入 ping 命令，就会输出

PONG 的提示，表示连接成功，如图 6-4 所示。

也可直接双击打开 Redis 目录下的 redis-cli. exe 文件，启动 Redis 客户端。

```
Microsoft Windows [版本 6.1.7601]
版权所有 (c) 2009 Microsoft Corporation。保留所有权利。

C:\Users\Administrator>D:

D:\>cd D:\Redis-x64-5.0.10

D:\Redis-x64-5.0.10>
```

图 6-2　将目录指向解压缩的 Redis 目录

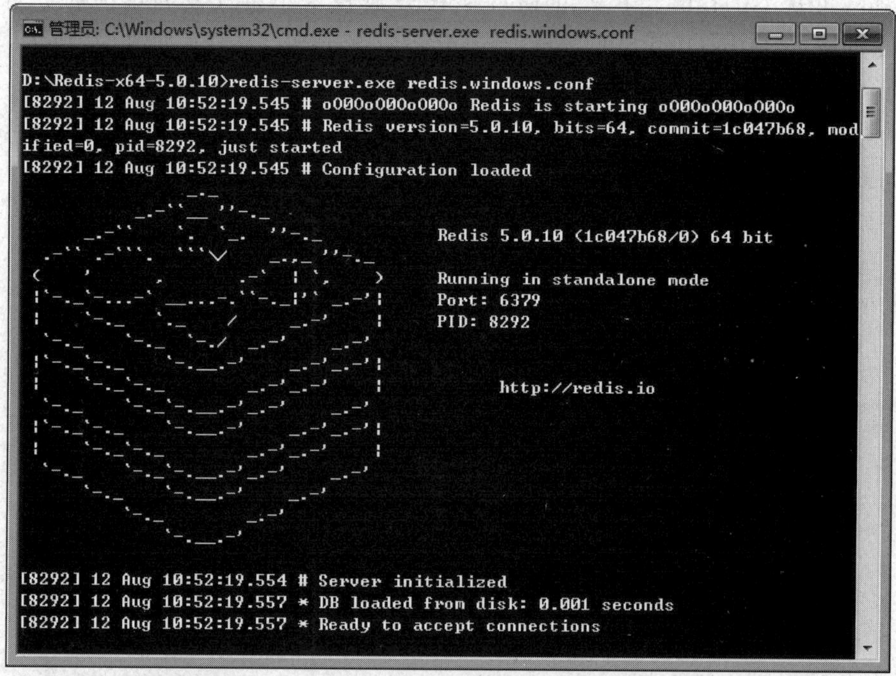

图 6-3　启动 Redis 服务

```
Microsoft Windows [版本 6.1.7601]
版权所有 (c) 2009 Microsoft Corporation。保留所有权利。

C:\Users\Administrator>d:

D:\>cd D:\Redis-x64-5.0.10

D:\Redis-x64-5.0.10>redis-cli.exe -h 127.0.0.1 -p 6379
127.0.0.1:6379> ping
PONG
127.0.0.1:6379>
```

图 6-4　使用 Redis 客户端连接服务器

(4) 操作测试 Redis。使用 set key value 和 get key 命令保存和获得数据。例如,使用 set username yzpc 设置 username 的值为 yzpc,并用 get username 获取 username 的值,如图 6-5 所示。

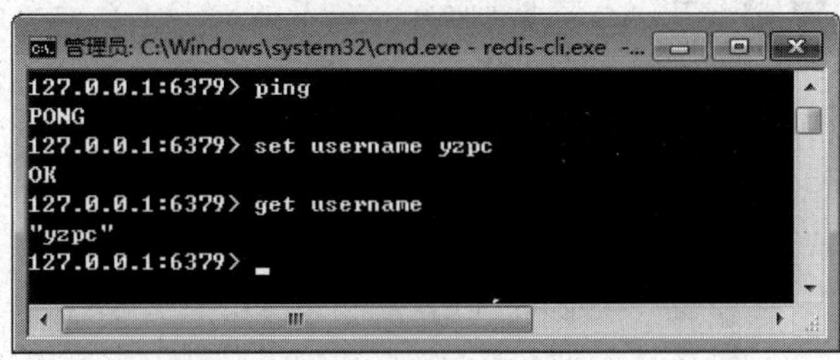

图 6-5 测试 Redis

(5) 使用 Redis 客户端工具查看数据。使用 Redis 自带的 redis-cli. exe 需要通过命令行去操作,可以使用例如 RedisClient、RedisDesktopManager、Redis Studio 等,这里使用 RedisClient 客户端工具,下载该工具后打开 RedisClient,如图 6-6 所示。

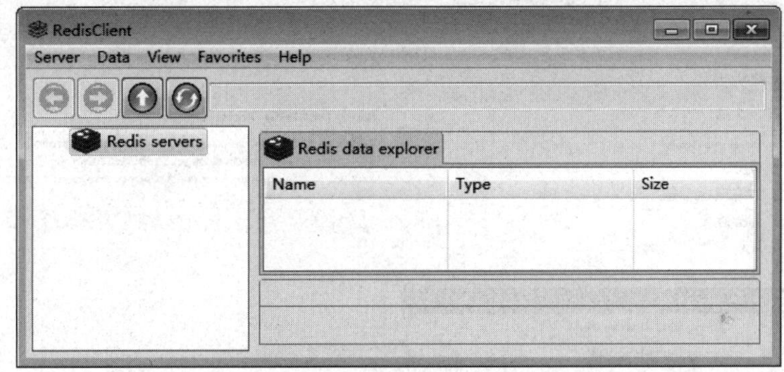

图 6-6 Redis 客户端 RedisClient 界面

在图 6-6 中,选择 Server 菜单下的 Add 命令,或在左窗口中的 Redis servers 上右击并从弹出的快捷菜单中选择 Add server 命令,添加 Redis 服务,填入相关信息,如图 6-7 所示。

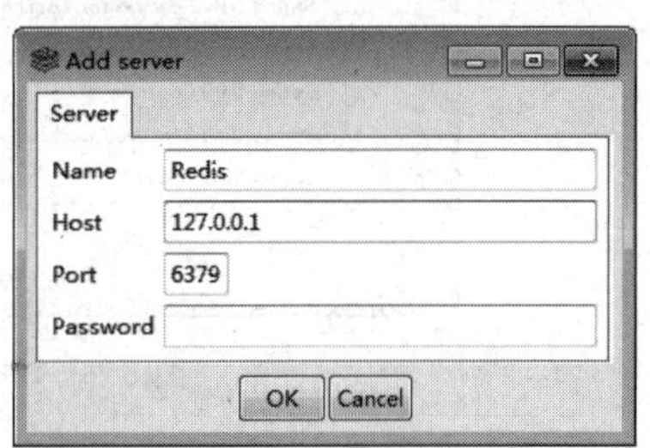

图 6-7 输入 Redis 服务信息界面

在图 6-7 中,单击 OK 按钮后,可以看到 Redis 服务中的数据库,共有 16 个数据库。db0 是默认的数据库名,在第(4)步保存的 username 就在该数据库中,展开 db0 数据库,即可看到 username 的字段数据,如图 6-8 所示。

6.1.3 Spring Boot 整合 Redis

下面通过示例来介绍 Spring Boot 整合 Redis,具体步骤如下。

(1) 新建一个 Spring Boot 工程 chapter06redis,Group 和 Package name 为 com. yzpc,在 Dependencies 依赖中选择 Web 节点下的 Spring Web 依赖和 NoSQL 节点下的 Spring

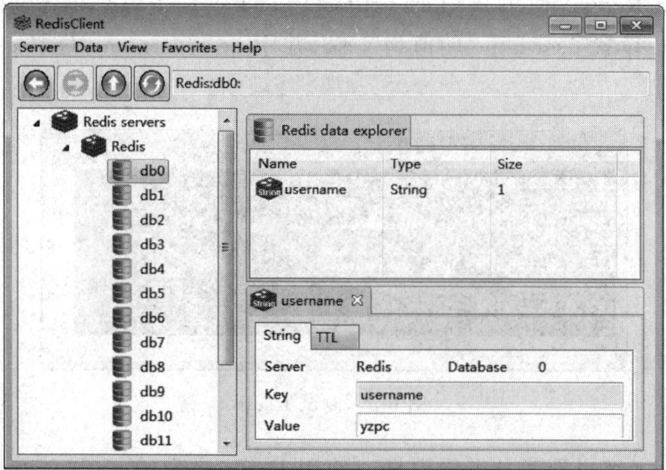

图 6-8 Redis 服务中的数据库

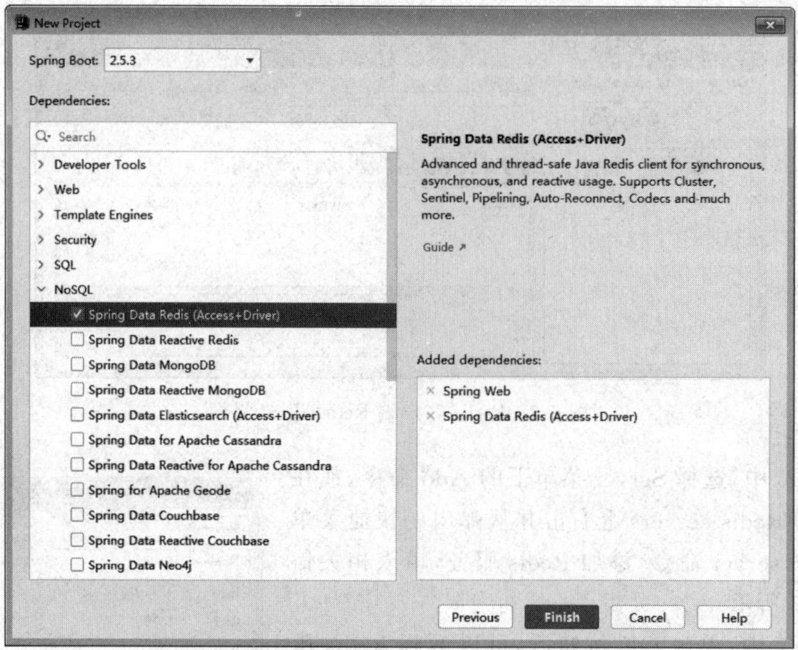

图 6-9 添加 NoSQL 的 Spring Data Redis 依赖

Data Redis(Access+Driver)依赖，单击 Finish 按钮，如图 6-9 所示。

pom.xml 中自动添加的依赖代码如下。

```
<dependency>
    <groupId>org.springframework.boot</groupId>
    <artifactId>spring-boot-starter-data-redis</artifactId>
</dependency>
<!-- 省略Web 中Spring Web 依赖代码 -->
```

(2) 在 application. properties 配置文件中加入有关 redis 的配置,代码如下。

```
spring.redis.database = 0
spring.redis.host = localhost
spring.redis.port = 6379
#如果 redis 有密码,这里就要设置相应的密码
spring.redis.password =
```

(3) 在项目的 src/main/java/路径下的 com. yzpc 包中,新建一个 pojo 包,并在该包中创建与数据表 user 对应的实体类 User,代码如下。

```
package com.yzpc.pojo;
import java.io.Serializable;
public class User implements Serializable {
    private int id;
    private String username;
    private String password;
    //此处省略相应属性的 setter/getter 方法
    //此处省略构造方法
    //省略重写的 toString 方法
}
```

(4) 在项目的 src/main/java/路径下的 com. yzpc 包中,新建一个 controller 包,并在该包中创建 UserController 的类,进行测试,代码如下。

```
package com.yzpc.controller;
import com.yzpc.pojo.User;
import org.springframework.beans.factory.annotation.Autowired;
import org.springframework.data.redis.core.RedisTemplate;
import org.springframework.data.redis.core.StringRedisTemplate;
import org.springframework.data.redis.core.ValueOperations;
import org.springframework.web.bind.annotation.GetMapping;
import org.springframework.web.bind.annotation.RestController;
@RestController
public class UserController {
    @Autowired
    RedisTemplate redisTemplate;
    @Autowired
    StringRedisTemplate stringRedisTemplate;
    @GetMapping("/save")
    public void save(){
        ValueOperations<String,String> ops1 = stringRedisTemplate.opsForValue();
        //添加字符串
        ops1.set("username","张三");
        String username = ops1.get("username");
        System.out.println(username);
        ValueOperations ops2 = redisTemplate.opsForValue();
        User user = new User(1,"admin","admin");
```

```
        //添加实体类
        ops2.set("user",user);
        System.out.println("Redis 是否存在相应的 key: " + redisTemplate.hasKey("user"));
        System.out.println("从 Redis 数据库中获取对象: " + (User)ops2.get("user"));
        //修改
        ops2.getAndSet("user",new User(2,"yzpc","yzpc"));
        //查询
        User u = (User) ops2.get("user");
        System.out.println("从 Redis 数据库中获取对象: " + u.toString());
        //删除
        Boolean b = redisTemplate.delete("user");
        System.out.println("从 Redis 中是否删除相应的键为 user 的对象: " + b);
    }
}
```

StringRedisTemplate 和 RedisTemplate 都是 Spring Data Redis 提供的模板类，用来对数据进行操作，都通过 Spring 提供的 Serializer 序列化到数据库。其中 StringRedisTemplate 是 RedisTemplate 的子类，只针对键值都是字符串的数据进行操作，而 RedisTemplate 可以操作对象。在 Spring Boot 中默认提供这两个模板类，StringRedisTemplate 和 RedisTemplate 都提供了 Redis 的基本操作方法，还提供了下面访问数据库的方法。

- opsForList：操作 List 数据。
- opsForSet：操作 Set 数据。
- opsForZSet：操作 ZSet 数据。
- opsForHash：操作 Hash 数据。

它们都是首先获得一个操作对象，再使用该操作对象完成数据的读写。

(5) 启动项目，在浏览器访问 http://localhost:8080/save，可以看到控制台输出信息，如图 6-10 所示。

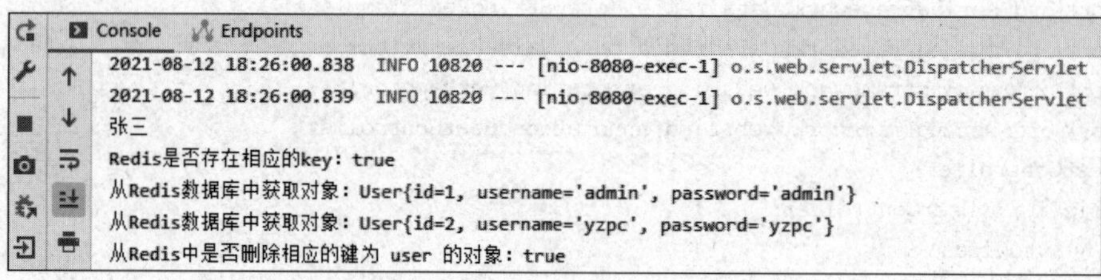

图 6-10 Redis 数据存储操作

6.2 整合 MongoDB

6.2.1 MongoDB 简介

MongoDB 是一个非常成熟的 NoSQL 数据库，与 Redis 采用 key-value 存储机制不同，MongoDB 是基于文档的 NoSQL，由 C++语言编写，旨在为 Web 应用提供可扩展的高性能数据存储解决方案，是当前 NoSQL 数据库产品中最热门的一种。

MongoDB 介于关系数据库和非关系数据库之间，是非关系数据库当中功能最丰富、最像关系数据库的产品。MongoDB 支持的数据结构非常松散，是类似 JSON 的 BSON(Binary JSON，二进制 JSON)格式，因此可以存储比较复杂的数据类型。MongoDB 最大的特点是支持的查询语言非常强大，其语法类似面向对象的查询语言，几乎可以实现类似关系数据库表单查询的绝大部分功能，而且还支持对数据建立索引。

传统的关系数据库一般由数据库(database)、表(table)、记录(record)三个层次概念组成，MongoDB 由数据库(database)、集合(collection)、文档对象(document)三个层次组成。MongoDB 中的集合对应于关系型数据库里的表，但是集合中没有列、行和关系概念，这体现了模式自由的特点。

MongoDB 可以应用于各种规模的企业、各个行业以及各类应用程序的开源数据库，主要功能特性如下。

- 面向集合存储，易存储对象类型的数据。
- 模式自由。
- 支持动态查询。
- 支持完全索引，包含内部对象。
- 支持查询。
- 支持复制和故障恢复。
- 使用高效的二进制数据存储，包括大型对象(如视频等)。
- 自动处理碎片，以支持云计算层次的扩展性。
- 支持 Ruby、Python、Java、C++、PHP 等多种语言。
- 文件存储格式为 BSON(一种 JSON 的扩展)。
- 可通过网络访问。

6.2.2 MongoDB 安装

MongoDB 提供了多种版本操作系统的安装包，这里主要介绍 Windows 操作系统版本的 MongoDB。

(1) 进入 MongoDB 官网 https://www.mongodb.com/try/download/community，下载社区版的安装包，这里选择 Windows 平台下 4.2.12 版本的 msi 安装包，单击 Download 按钮进行下载，如图 6-11 所示。

提示：MongoDB 4.4.x 的版本需要在 Windows 10 或 Windows 2016 Server 以上版本的操作系统上安装，读者可根据自己的操作系统选择相应版本下载。

(2) 下载完成后得到 mongodb-win32-x86_64-2012plus-4.2.12-signed.msi 的安装文件，双击安装文件开始安装，根据提示单击 Next 按钮即可，当进入如图 6-12 所示的界面时，取消选中 Install MongoDB Compass 前面的复选框，否则安装会特别慢。

(3) 如果使用 MongoDB 可视化的图形界面管理工具 MongoDB Compass 操作 MongoDB 数据库，可从官方网站 https://www.mongodb.com/try/download/compass 单独下载相应版本的 MongoDB Compass 进行安装，如图 6-13 所示。

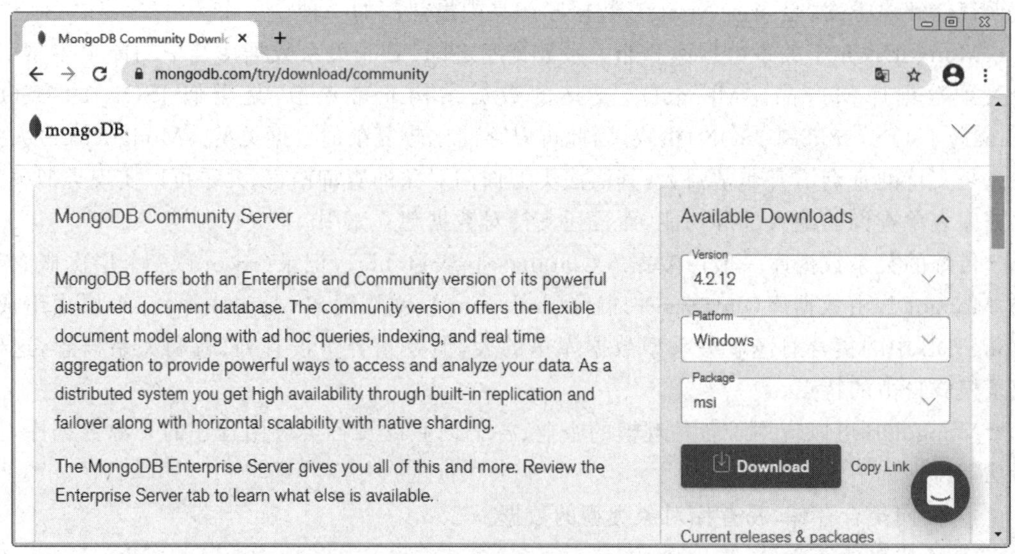

图 6-11　MongoDB 下载页面

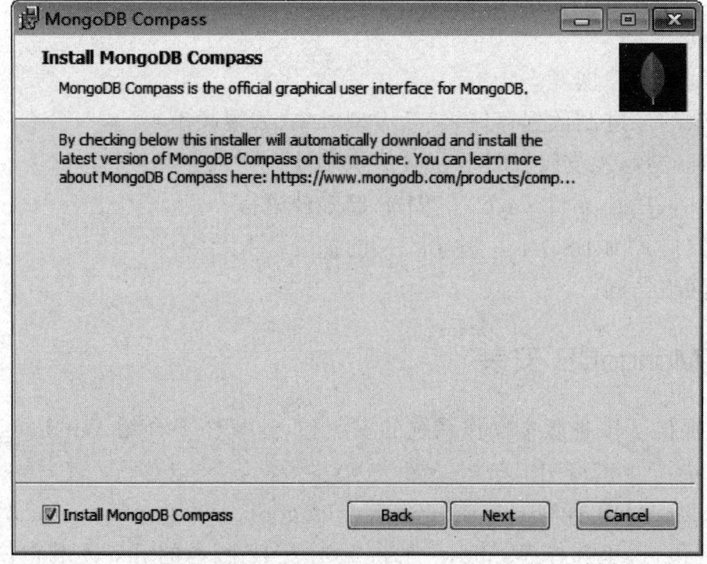

图 6-12　默认选择安装 MongoDB Compass

也可使用 MongoBooster、NoSQL Manager for MongoDB 等可视化客户端来连接 MongoDB 数据库。

(4) 安装 MongoDB Compass 可视化工具，完成后运行 MongoDB Compass 工具，如图 6-14 所示。

(5) 在图 6-14 中，单击 Fill in connection fields individually 链接可填写数据库连接相关配置，这里采用默认值，单击 CONNECT 按钮，连接成功之后，显示出默认自带的三个数据库(admin、config、local)，如图 6-15 所示。

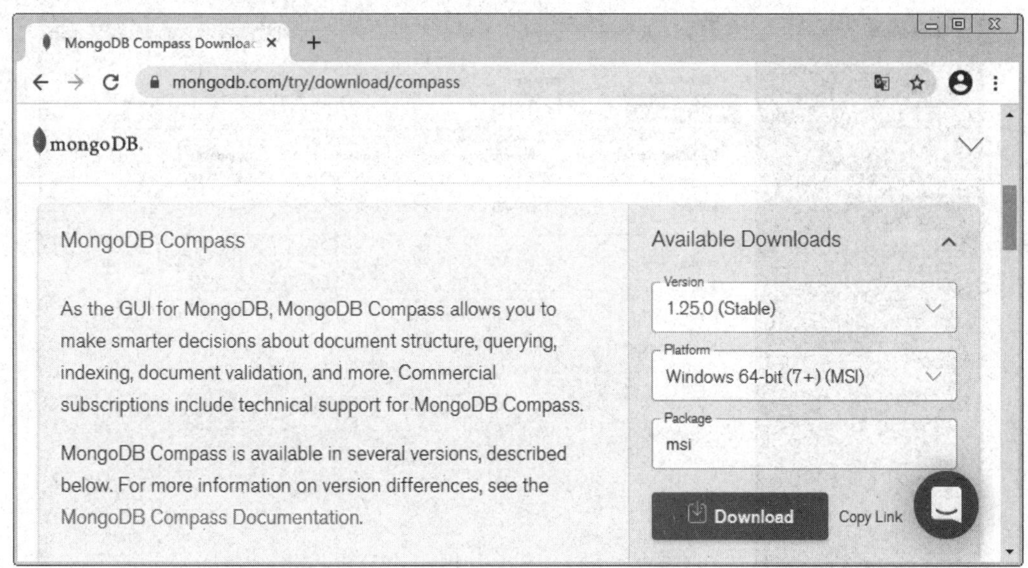

图 6-13 MongoDB Compass 下载页面

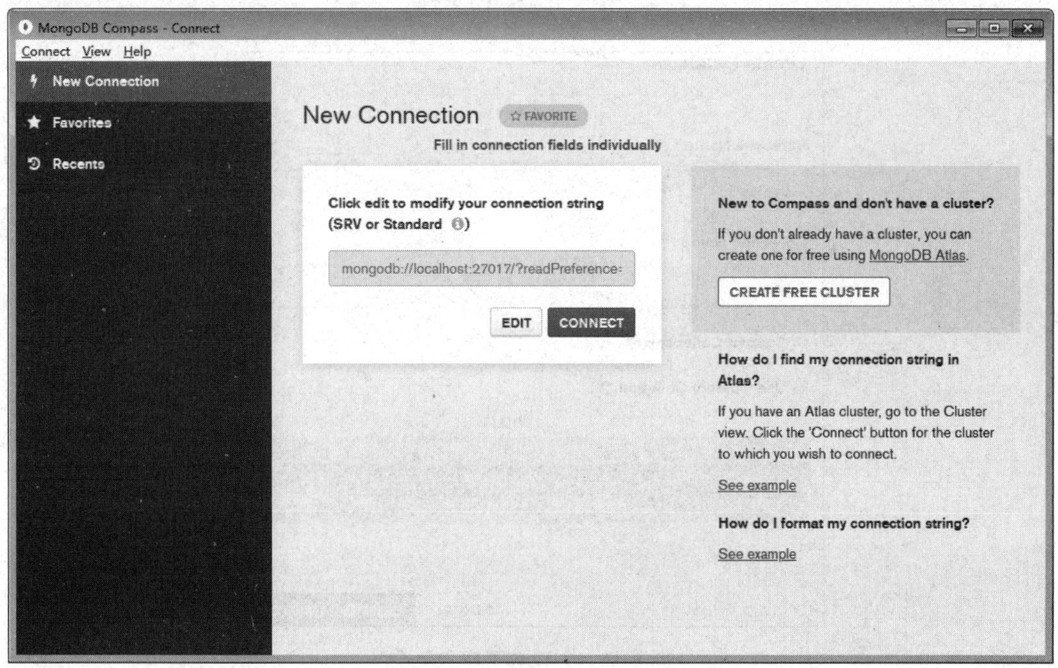

图 6-14 MongoDB Compass 工具

6.2.3 Spring Boot 整合 MongoDB

下面通过示例来介绍 Spring Boot 整合 MongoDB,具体步骤如下。

(1) 使用 MongoDB Compass 工具创建名为 chapter06 的数据库。在图 6-15 中的 MongoDB Compass 工具界面上单击 CREATE DATABASE 按钮或左侧窗口下方的+,弹

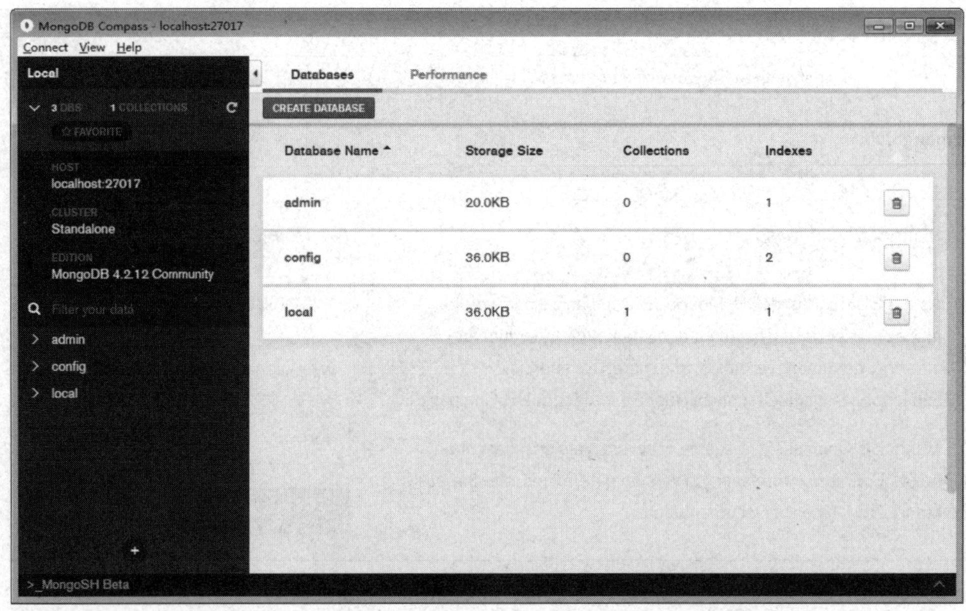

图 6-15　MongoDB Compass 链接 MongoDB 数据库

Create Database

Database Name

chapter06

Collection Name

book

Capped Collection

Use Custom Collation

Before MongoDB can save your new database, a collection name must also be specified at the time of creation. More Information

CANCEL　CREATE DATABASE

图 6-16　创建数据库和集合

出创建数据库的对话框，在 Database Name 处输入 chapter06，Collection Name 处输入 book，如图 6-16 所示。

单击图 6-16 中右下方的 CREATE DATABASE 按钮，即可创建数据库和集合，这里的 book 集合相当于关系型数据库的数据表。

(2) 新建一个 Spring Boot 工程 chapter06mongodb，Group 和 Package name 为 com.yzpc，在 Dependencies 依赖中选择 Web 节点下的 Spring Web 依赖和 NoSQL 节点下的

Spring Data MongoDB 依赖,单击 Finish 按钮,如图 6-17 所示。

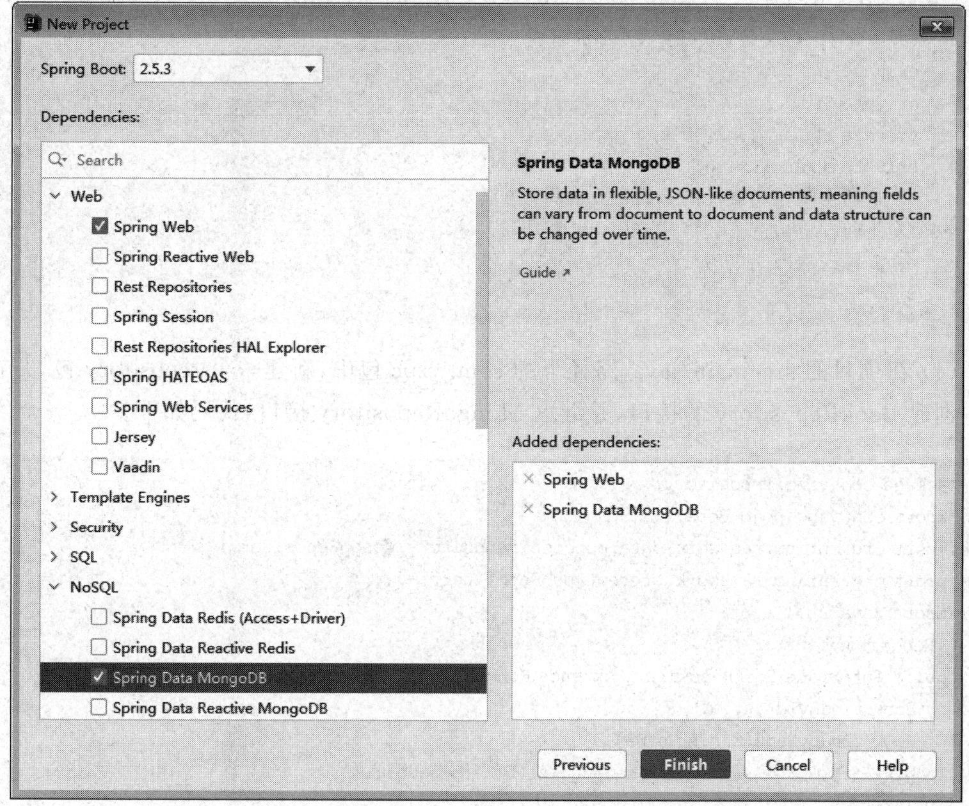

图 6-17 添加 NoSQL 的 Spring Data MongoDB 依赖

pom.xml 文件中自动添加的依赖代码如下。

```
<dependency>
    <groupId>org.springframework.boot</groupId>
    <artifactId>spring-boot-starter-data-mongodb</artifactId>
</dependency>
<!-- 省略 Web 中Spring Web 依赖代码 -->
```

在 application.properties 配置文件中加入 mongodb 数据库连接的有关配置信息如下。

```
spring.data.mongodb.uri=mongodb://localhost:27017/chapter06
```

(3) 在项目的 src/main/java/路径下的 com.yzpc 包中,新建一个 pojo 包,并在该包中新建实体类 Book,使用@Document(collection="book")注解通过 collection 参数指定 Book 类声明为 mongodb 的文档集合 book,代码如下。

```
package com.yzpc.pojo;
import org.springframework.data.annotation.Id;
import org.springframework.data.mongodb.core.mapping.Document;
```

```
@Document(collection = "book")
public class Book {
    @Id        //主键,不可重复,自带索引
    private int id;
    private String name;
    private String publisher;
    private double price;
    // 此处省略相应属性的 setter/getter 方法
    // 此处省略构造方法
    // 重写 toString 方法
}
```

(4) 在项目的 src/main/java/路径下的 com.yzpc 包中,新建一个 repository 包,并在该包中创建 BookRepository 的接口,并继承 MongoRepository 接口,代码如下。

```
package com.yzpc.repository;
import com.yzpc.pojo.Book;
import org.springframework.data.mongodb.repository.MongoRepository;
import org.springframework.stereotype.Repository;
import java.util.List;
@Repository
public interface BookRepository extends MongoRepository<Book,Integer>{
    Book findById(int id);
    Book findByName(String name);
    List<Book> findBooksByNameContains(String name);
}
```

(5) 在项目的 src/main/java/路径下的 com.yzpc 包中,新建一个 controller 包,并在该包中创建 BookController 的类,代码如下。

```
package com.yzpc.controller;
import com.yzpc.pojo.Book;
import com.yzpc.repository.BookRepository;
import org.springframework.beans.factory.annotation.Autowired;
import org.springframework.web.bind.annotation.GetMapping;
import org.springframework.web.bind.annotation.RestController;
import java.util.ArrayList;
import java.util.List;
@RestController
public class BookController {
    @Autowired
    BookRepository bookRepository;
    @GetMapping("/saveBooks")
    public List<Book> saveBooks(){
        List<Book> bookList = new ArrayList<Book>();
        bookList.add(new Book(1,"Spring Boot 编程思想","电子工业出版社",118.00));
        bookList.add(new Book(2,"Spring Boot 开发实战","清华大学出版社",59.90));
        bookList.add(new Book(3,"Spring Boot 2 实战之旅","清华大学出版社",79.00));
```

```
        bookList.add(new Book(4,"Spring Boot 实战派","电子工业出版社",109.00));
        return bookRepository.saveAll(bookList);
    }
    @GetMapping("/addBook")
    public void addBook(){
        Book book = new Book(5,"Spring Boot 技术内幕","机械工业出版社",79.00);
        bookRepository.insert(book);
    }
    @GetMapping("/insertBook")
    public void insertBook(int id,String name,String publisher,double price){
        Book book = new Book(id,name,publisher,price);
        bookRepository.insert(book);
    }
    @GetMapping("/findById")
    public Book findById(int id){
        return bookRepository.findById(id);
    }
    @GetMapping("/findByName")
    public Book findByName(String name){
        return bookRepository.findByName(name);
    }
    @GetMapping("/findByNameContains")
    public List<Book> findByNameContains(String name){
        return bookRepository.findBooksByNameContains(name);
    }
    @GetMapping("/updateBook")
    public Book updateBook(int id,double price){
        Book book = bookRepository.findById(id);
        if (book!= null){
            book.setPrice(price);
        }
        return bookRepository.save(book);
    }
    @GetMapping("/deleteBookByName")
    public void deleteBookByName(String name){
        Book book = bookRepository.findByName(name);
        bookRepository.delete(book);
    }
    @GetMapping("/deleteBookById")
    public void deleteBookById(int id){
        Book book = bookRepository.findById(id);
        bookRepository.delete(book);
    }
}
```

(6) 启动项目，在浏览器中访问 http://localhost:8080/saveBooks，测试向 chapter06 数据库的 book 集合中存入 4 条记录，并将此 4 条记录在页面输出，如图 6-18 所示。

访问 http://localhost:8080/addBook，插入一条固定的记录，如图 6-19 所示。

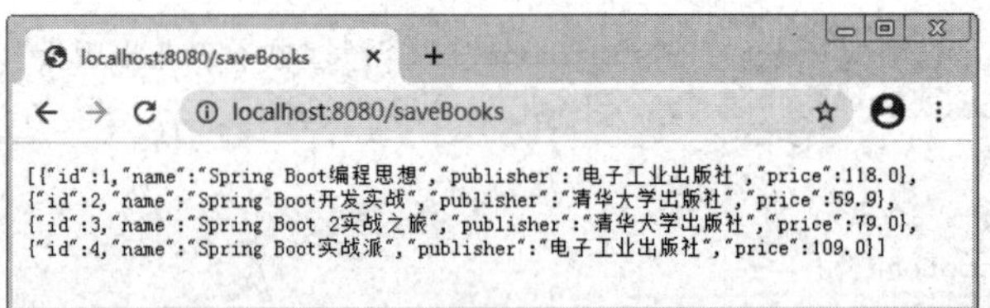

图 6-18　保存多条记录

book

	_id Int32	name String	publisher String	price Double
1	1	"Spring Boot编程思想"	"电子工业出版社"	118
2	2	"Spring Boot开发实战"	"清华大学出版社"	59.9
3	3	"Spring Boot 2实战之旅"	"清华大学出版社"	79
4	4	"Spring Boot实战派"	"电子工业出版社"	109
5	5	"Spring Boot技术内幕"	"机械工业出版社"	79

图 6-19　插入一条固定纪录

访问 http://localhost:8080/insertBook?id=6&name=Spring 编程 &publisher=人民邮电出版社 &price=99.00，通过参数传递的方法，插入一条记录，在 MongoDB Compass 工具中查看 book 集合中的记录，如图 6-20 所示。

chapter06.book　DOCUMENTS 0　TOTAL SIZE 0B　AVG. SIZE 0B　INDEXES 1　TOTAL SIZE 4.0KB　AVG. SIZE 4.0KB

Documents　Aggregations　Schema　Explain Plan　Indexes　Validation

FILTER　OPTIONS　FIND　RESET　…

ADD DATA　VIEW　Displaying documents 1 - 6 of 6　REFRESH

book

	_id Int32	name String	publisher String	price Double
1	1	"Spring Boot编程思想"	"电子工业出版社"	118.0
2	2	"Spring Boot开发实战"	"清华大学出版社"	59.9
3	3	"Spring Boot 2实战之旅"	"清华大学出版社"	79.0
4	4	"Spring Boot实战派"	"电子工业出版社"	109.0
5	5	"Spring Boot技术内幕"	"机械工业出版社"	79.0
6	6	"Spring编程"	"人民邮电出版社"	99.0

图 6-20　MongoDB Compass 工具查看 book 集合

访问 http://localhost:8080/findById?id=4，查询 id 为 4 的记录，如图 6-21 所示。

图 6-21　通过 id 查询记录

访问 http://localhost:8080/findByName?name=Spring Boot 编程思想，查询 name 为“Spring Boot 编程思想”的记录，如图 6-22 所示。

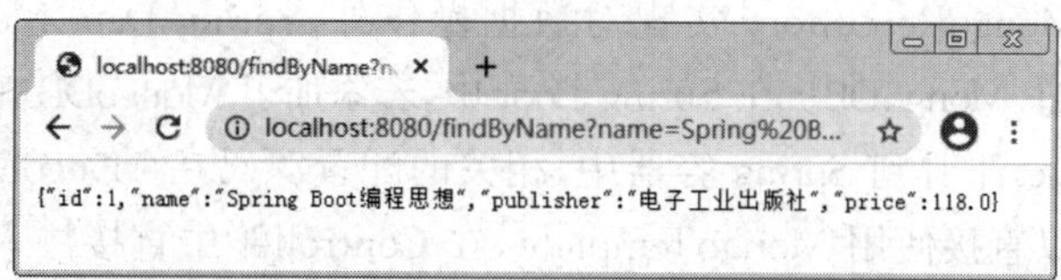

图 6-22　通过 name 查询记录

访问 http://localhost:8080/findByNameContains?name=编程,查询 name 中包含有“编程”的记录,如图 6-23 所示。

图 6-23　查询 name 中包含关键字“编程”的记录

访问 http://localhost:8080/updateBook?id=5&price=69.90,修改 id 为 5 的记录价格,如图 6-24 所示。

图 6-24　修改记录

访问 http://localhost:8080/deleteBookByName?name=Spring 编程,删除 name 为“Spring 编程”的记录,如图 6-25 所示。

book

	_id Int32	name String	publisher String	price Double
1	1	"Spring Boot编程思想"	"电子工业出版社"	118
2	2	"Spring Boot开发实战"	"清华大学出版社"	59.9
3	3	"Spring Boot 2实战之旅"	"清华大学出版社"	79
4	4	"Spring Boot实战派"	"电子工业出版社"	109
5	5	"Spring Boot技术内幕"	"机械工业出版社"	79

图 6-25　删除 name 为“Spring 编程”后的 book 集合

访问 http://localhost:8080/deleteBookById?id=5,删除 id 为 5 的记录,如图 6-26 所示。

book

	_id Int32	name String	publisher String	price Double
1	1	"Spring Boot编程思想"	"电子工业出版社"	118
2	2	"Spring Boot开发实战"	"清华大学出版社"	59.9
3	3	"Spring Boot 2实战之旅"	"清华大学出版社"	79
4	4	"Spring Boot实战派"	"电子工业出版社"	109

图 6-26　删除 id=5 的纪录后的 book 集合

(7) 除了继承 MongoRepository 实现对数据操作外，Spring Data MongoDB 还通过使用 MongoTemplate 来操作 MongoDB。在 Spring Boot 中，若添加了 MongoDB 相关的依赖，则默认会有一个 MongoTemplate 注册到 Spring 容器中，相关的配置源码在 MongoDataAutoConfiguration 类中。因此，用户可以直接使用 MongoTemplate，在 Controller 中直接注入 MongoTemplate 就可以使用，添加代码到 BookController 中，这里以添加记录和查询记录为例，代码如下。

```
@Autowired
MongoTemplate mongoTemplate;
@GetMapping("/saveBook")
public void saveBook(){
    List<Book> books = new ArrayList<Book>();
    books.add(new Book(7,"Java 编程思想(第 4 版)","机械工业出版社",108.00));
    books.add(new Book(8,"Java Web 应用开发技术","清华大学出版社",69.00));
    mongoTemplate.insertAll(books);
    Book book = new Book(9,"MySQL 基础教程","人民邮电出版社",129.00);
    mongoTemplate.insert(book);
}
@GetMapping("/findBook")
public void findBook(){
    List<Book> books = mongoTemplate.findAll(Book.class);
    if (books!= null) {
        System.out.println("所有 Book 对象列表为: ");
        for (Book book : books) {      //循环输出集合中对象
            System.out.println(book);
        }
    }
    Book book = mongoTemplate.findById(4,Book.class);
    System.out.println("根据 Id 查询单条记录");
    System.out.println(book);
}
```

重新启动项目，访问 http://localhost:8080/saveBook，测试向 chapter06 数据库的 book 集合中插入记录，使用 MongoDB Compass 工具查看集合中的记录，查看是否插入成功，如图 6-27 所示。

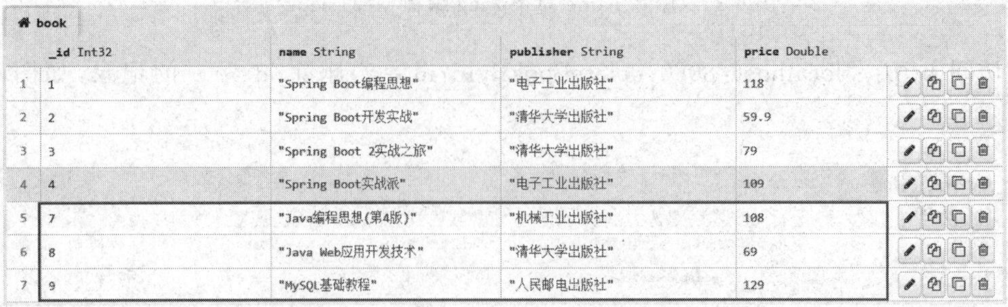

book

	_id Int32	name String	publisher String	price Double
1	1	"Spring Boot编程思想"	"电子工业出版社"	118
2	2	"Spring Boot开发实战"	"清华大学出版社"	59.9
3	3	"Spring Boot 2实战之旅"	"清华大学出版社"	79
4	4	"Spring Boot实战派"	"电子工业出版社"	109
5	7	"Java编程思想(第4版)"	"机械工业出版社"	108
6	8	"Java Web应用开发技术"	"清华大学出版社"	69
7	9	"MySQL基础教程"	"人民邮电出版社"	129

图 6-27 使用 MongoTemplate 添加数据

访问 http://localhost:8080/findBook,查询 id 为 4 的记录,控制台输出结果如图 6-28 所示。

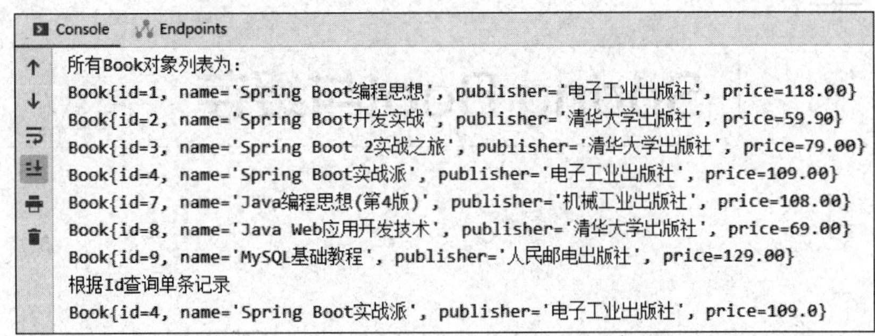

```
Console    Endpoints
所有Book对象列表为:
Book{id=1, name='Spring Boot编程思想', publisher='电子工业出版社', price=118.00}
Book{id=2, name='Spring Boot开发实战', publisher='清华大学出版社', price=59.90}
Book{id=3, name='Spring Boot 2实战之旅', publisher='清华大学出版社', price=79.00}
Book{id=4, name='Spring Boot实战派', publisher='电子工业出版社', price=109.00}
Book{id=7, name='Java编程思想(第4版)', publisher='机械工业出版社', price=108.00}
Book{id=8, name='Java Web应用开发技术', publisher='清华大学出版社', price=69.00}
Book{id=9, name='MySQL基础教程', publisher='人民邮电出版社', price=129.00}
根据Id查询单条记录
Book{id=4, name='Spring Boot实战派', publisher='电子工业出版社', price=109.0}
```

图 6-28　使用 MongoTemplate 查询数据

6.3　小结

本章主要介绍 NoSQL 数据库的分类,分别介绍了 Redis 和 MongoDB 数据库的安装,重点介绍了如何使用 Spring Boot 整合 Redis 和 MongoDB。

第7章 Spring Boot与缓存

缓存是分布式系统中的重要组件，主要解决数据库数据的高并发访问问题，用来加速系统访问，提升系统性能。为了提高服务器访问性能，减少数据库的压力、提高用户体验，使用缓存尤为重要。缓存的工作机制是先从缓存中读取数据，如果缓存中没有此数据，再从慢速设备上读取实际数据(读取时数据也会存入缓存)。Spring Boot 对缓存提供了良好的支持，可以使用自带的缓存，也可以使用第三方缓存。本章将对 JSR-107 规范、缓存抽象与缓存注解、EhCache 2. x 缓存和 Redis 缓存进行介绍，让读者对 Spring Boot 缓存技术有初步的了解。

7.1 JCache (JSR-107)规范

缓存在数据访问量比较大的系统中使用非常多，为了统一缓存的开发规范，提升系统的扩展性，J2EE 发布了 JSR-107 规范。

1. JSR-107 规范

JSR 是 Java Specification Requests 的缩写，意思是 Java 规范提案，2012 年 10 月 26 日 JSR 规范委员会发布了 JSR 107(JCache API)。JCache 规范定义了一种对 Java 对象临时在内存中进行缓存的方法，包括对象的创建、共享访问、假脱机(spooling)、失效、各 JVM 的一致性等，可被用户缓存最经常访问的数据。

Java Caching 定义了 5 个核心接口，分别是 CachingProvider(缓存提供者)、CacheManager(缓存管理器)、Cache、Entry 和 Expiry。

(1) CachingProvider：定义了创建、配置、获取、管理和控制多个 CacheManager。一个应用可以在运行期间访问多个 CachingProvider。

(2) CacheManager：定义了创建、配置、获取、管理和控制多个唯一命名的 Cache，这些 Cache 存在于 CacheManager 的上下文中。一个 CacheManager 仅被一个 CachingProvider 所拥有。

(3) Cache：是一个类似 Map 的数据结构并临时存储以 key 为索引的值。一个 Cache 仅被一个 CacheManager 所拥有。

(4) Entry：是一个存储在 Cache 中的 key-value 对。

(5) Expiry：每一个存储在 Cache 中的条目有一个定义的有效期。一旦超过这个时间，条目为过期的状态。一旦过期，条目将不可访问、更新和删除。缓存有效期可以通过

ExpiryPolicy 设置。

2. 应用调用缓存

应用首先会调用 CachingProvider，Caching Provider 管理了多个 CacheManager，Cache Manager 中才是真正的 Cache 缓存。Cache Manager 中可以管理不同类型的缓存，比如 Redis、EhCache 等。

在具体缓存组件中，可以设置不同模块的缓存。例如，Redis 中可以缓存商品信息、热点数据等不同模块数据，每个缓存都是 Entry<K，V>键值对类型，并且可以对缓存设置 Expiry 过期时间，指定缓存存活的时间，如图 7-1 所示。

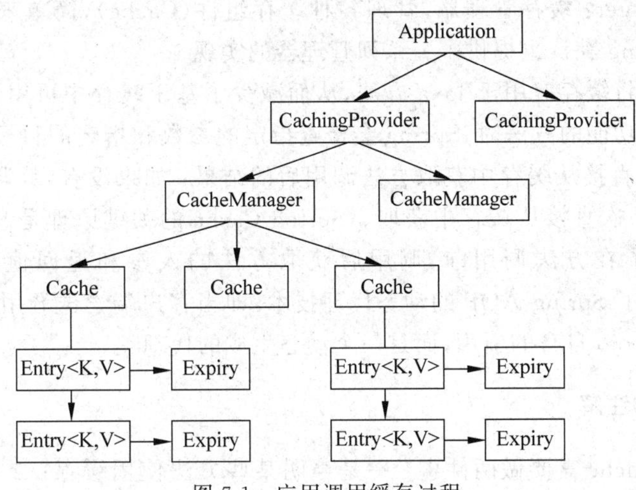

图 7-1 应用调用缓存过程

JSR-107 作为一个 Java 规范，定义的都是一些接口，类似 JDBC 规范。JSR-107 的好处是用户直接面向接口编程，需要用到哪种缓存的实现，直接引入该缓存实现，系统就能运行起来，使用 JSR-107 时需引入如下依赖。

```
<dependency>
    <groupId>javax.cache</groupId>
    <artifactId>cache-api</artifactId>
</dependency>
```

然而，并不是市面上所有的缓存组件都提供了 JSR-107 规范的实现。如果用户选择的缓存中间件没有实现 JSR-107 规范接口，那么就需要用户自己去实现。所以，JSR-107 规范在实际开发中使用得不多。为了简化开发，Spring 提供了自己的缓存抽象，也定义了一些类似的注解，在实际开发中一般使用 Spring 的缓存抽象。

7.2 缓存抽象与缓存注解

1. Spring 缓存抽象

Spring 框架从 3.1 开始定义了 org. springframework. cache. Cache 和 org. springframework.

cache.CacheManager 接口，提供对缓存功能的声明，能够与多种流行的缓存实现集成。Spring Cache 不是具体的缓存技术，而是基于具体的缓存产品（如 EhCache、Redis 等）的共性进行了一层抽象，可以通过简单的配置切换底层使用的缓存。

Spring 定义了自己缓存抽象用于统一缓存的操作，仅仅提供抽象，而不是具体的实现，只要实现 Cache 和 CacheManager 接口，就可以接入 Spring 通过缓存注解的方式使用缓存（一些主流的缓存都提供该抽象的实现）。

缓存抽象中定义的 Cache 和 CacheManager 两个接口如下。

- Cache：缓存接口，组件规范定义，包含缓存的各种操作集合。Spring 提供了各种 xxxCache 的实现，如 RedisCache，EhCacheCache，ConcurrentMapCache 等。
- CacheManager：缓存管理器，管理各种缓存组件（Cache），简单来说就是用于存放 Cache，Spring 默认也提供了一系列管理器的实现。

抽象的核心是将缓存应用于 Java 方法，从而减少了基于缓存中可用信息的执行次数。每次调用需要缓存功能的方法时，Spring 会检查指定的参数和指定的目标方法是否已经被调用过，如果有，就直接从缓存中获取方法调用后的结果；如果没有，就调用方法并将缓存结果返回给用户，下次直接从缓存中获取。Spring Cache 的关键原理是 Spring AOP，通过 Spring AOP 实现了在方法调用前、调用后获取方法的入参和返回值，进而实现缓存。Spring Cache 利用了 Spring AOP 的动态代理技术，即当客户端尝试调用 pojo 某个方法的时候，给它的不是 pojo 自身的引用，而是一个动态生成的代理类。

2. Spring 缓存注解

使用 Spring Cache 需要做两件事：一是声明某些方法使用缓存；二是配置 Spring 对 Cache 的支持。Spring 对 Cache 的支持有基于注解和基于 XML 配置两种方式，下面针对 Spring Boot 中的缓存注解及相关属性进行介绍。

（1）@Cacheable：该注解由 Spring 框架提供，可作用于类或方法（一般标注在 service 层，通常用在数据查询方法上），用于对方法的查询结果进行缓存存储。@Cacheable 注解的执行顺序是，先进行缓存查询，如果为空，则进行方法查询，并将结果进行缓存；如果缓存中有数据，不进行方法查询，而是直接使用缓存数据。@Cacheable 注解提供了多个属性，用于对缓存存储进行相关配置，具体属性及说明如表 7-1 所示。

表 7-1 @Cacheable 注解属性及说明

属 性 名	说 明
value/cacheNames	指定缓存空间的名称，必配属性，这两个属性二选一使用
key	指定缓存数据的 key，默认使用方法参数值，可用 SpEL 表达式
keyGenerator	指定缓存数据的 key 的生成器，与 key 属性二选一使用
cacheManager	指定缓存管理器
cacheResolver	指定缓存解析器，与 cacheManager 属性二选一使用
condition	指定在符合某条件下，进行数据缓存
unless	指定在符合某条件下，不进行数据缓存
Sync	指定是否使用异步缓存，默认值为 false

（2）@CachePut：该注解由 Spring 框架提供，可以作用于类或方法（一般标注在 service

层,通常用在数据更新方法上),作用是更新缓存数据。@CachePut 注解的执行顺序是,先进行方法调用,然后将方法结果更新到缓存中。@CachePut 注解也提供了多个属性,这些属性与@Cacheable 注解的属性完全相同。

(3) @CacheEvict:该注解由 Spring 框架提供,可以作用于类或方法(一般标注在 service 层,通常用在数据删除方法上),作用是删除数据缓存。@CacheEvict 注解的执行顺序是,先进行方法调用,然后清除缓存。@CacheEvict 注解提供了多个属性,这些属性与@Cacheable 注解的属性基本相同。除此之外,还提供了两个特殊属性。

- allEntries 属性:表示是否清除指定缓存空间中的所有缓存数据,默认值为 false。
- beforeinvocation 属性:表示是否在方法执行之前进行缓存清除,默认值为 false。

(4) @Caching:该注解用于同时添加多个缓存注解,定义复杂规则的数据缓存。该注解作用于类或方法。@Caching 注解包含 cacheable、put 和 evict 三个属性,作用相当于@Cacheable、@CachePut 和@CacheEvict。

(5) @CacheConfig:该注解作用于类,用于统筹管理类中所有使用@Cacheable、@CachePut 和 @CacheEvict 注解标注的方法中的公共属性,这些公共属性包括 cacheNames、keyGenerator、cacheManager 和 cacheResolver。

(6) @EnableCaching:该注解由 Spring 框架提供,Spring Boot 框架对该注解进行了继承,用于开启基于注解的缓存支持,该注解需要配置类上(在 Spring Boot 中,通常配置在项目的启动类上)。

3. Spring Boot 支持的缓存组件

在 Spring Boot 中,数据的管理存储依赖于 Spring 框架中 cache 相关的 org. springframework. cache. Cache 和 org. springframework. cache. CacheManager 缓存管理接口。如果程序中没有定义类型为 CacheManager 的 Bean 组件或者名为 cacheResolver 的 CacheResolver 缓存解析器,Spring Boot 将根据以下指定的顺序尝试并启用以下缓存组件。

- Generic。
- JCache (JSR-107) (EhCache 3、Hazelcast、Infinispan 等)。
- EhCache 2. x。
- Hazelcast。
- Infinispan。
- Couchbase。
- Redis。
- Caffeine。
- Simple。

上面按照 Spring Boot 缓存组件的加载顺序列举了支持的 9 种缓存组件,在项目中添加某个缓存管理组件(例如 EhCache 2. x)后,Spring Boot 项目会选择并启用对应的缓存管理器。如果项目中同时添加了多个缓存组件,且没有指定缓存管理器或者缓存解析器(CacheManager 或者 CacheResolver),那么 Spring Boot 会按照上述顺序在添加的多个缓存中优先启用指定的缓存组件进行缓存管理。

在 Spring Boot 默认缓存管理中,没有添加任何缓存管理组件也能实现缓存管理。这是因为开启缓存管理后，Spring Boot 会按照上述列表顺序查找有效的缓存组件进行缓存管理，如果没有任何缓存组件，会默认使用最后一个 Simple 缓存组件进行管理。Simple 缓存组件是 Spring Boot 默认的缓存管理组件，它默认使用内存中的 ConcurrentHashMap 进行缓存存储，所以在没有添加任何第三方缓存组件的情况下，也可以实现内存中的缓存管理。在默认情况下使用的是 Simple 简单缓存，不建议在正式环境中使用这种缓存管理方式。

4. Sprig Boot 默认缓存管理

使用缓存的主要目的是减小数据库的访问压力、提高用户体验。Spring 框架支持透明地向应用程序添加缓存并对缓存进行管理，其管理缓存的核心是将缓存应用于操作数据的方法中，从而减少操作数据的次数，同时不会对程序本身造成任何干扰。Spring Boot 继承了 Spring 的缓存管理功能，通过@EnableCaching 注解开启基于注解的缓存支持，自动化配置合适的缓存管理器(CacheManager)。

下面将针对 Spring Boot 支持的默认缓存管理，结合数据库的访问操作对 Spring Boot 的缓存管理进行讲解，步骤如下。

(1) 使用 MySQL 数据库创建一个名为 chapter07 的数据库，并新建 user 数据表，包含 id、username、password 属性，预先插入几条测试记录。

(2) 新建一个 Spring Boot 工程 chapter07cache，Group 和 Package name 为 com. yzpc，在 Dependencies 依赖中选择 Web 节点下的 Spring Web 依赖，SQL 节点下的 Spring Data JPA 依赖、MySQL Driver 依赖和 I/O 节点下的 Spring cache abstraction 依赖，单击 Finish 按钮，如图 7-2 所示。

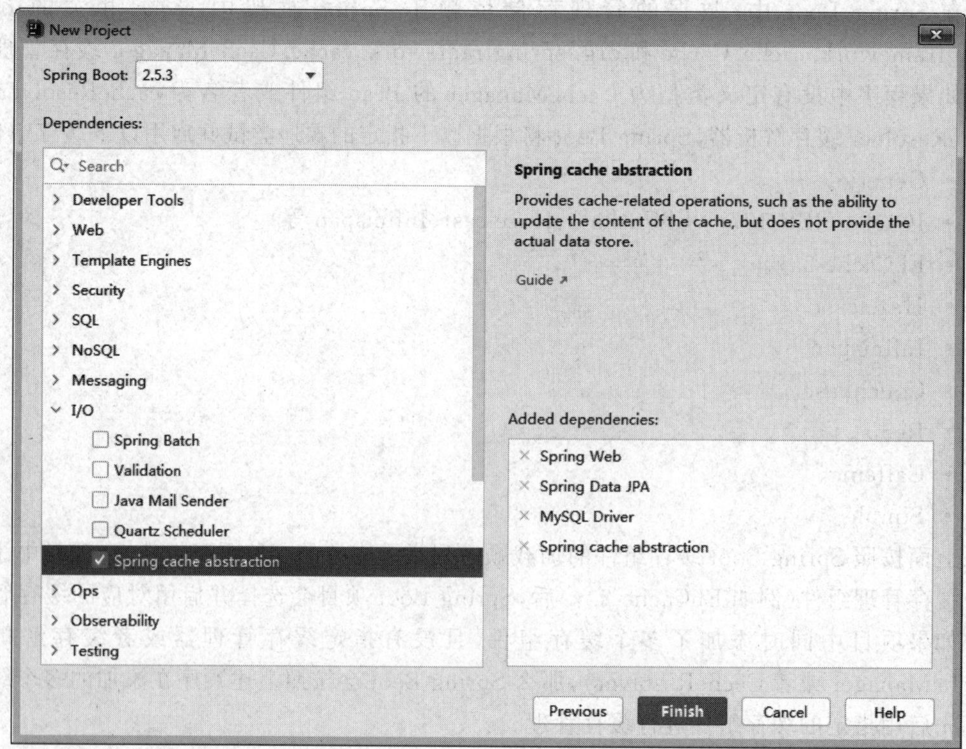

图 7-2 添加 Web 等相关依赖

pom.xml 中自动添加的依赖代码如下。

```
<!-- 开启 cache 缓存 -->
<dependency>
    <groupId>org.springframework.boot</groupId>
    <artifactId>spring-boot-starter-cache</artifactId>
</dependency>
<!-- 省略 Spring Web、Spring Data JPA、MySQL Driver 依赖代码 -->
```

(3) 在项目的 src/main/resources/ 路径下的 application.properties 配置文件中加入数据库连接配置信息(与 5.2.1 小节中介绍的一致),设置 JPA 的基本配置以及 Spring cache 的目标缓存管理器配置如下。

```
spring.datasource.driver-class-name=com.mysql.cj.jdbc.Driver
spring.datasource.url=jdbc:mysql://localhost:3306/chapter07?serverTimezone=UTC
spring.datasource.username=root
spring.datasource.password=123456
spring.jpa.database=mysql
spring.jpa.show-sql=true
spring.jpa.hibernate.ddl-auto=update
# Spring cache 目标缓存管理器默认为simple,如果不加,会按顺序查找到该缓存组件
spring.cache.type=simple
```

(4) 在项目的 src/main/java/ 路径下的 com.yzpc 包中,新建一个 pojo 包,并在该包中新建与数据表 user 对应的实体类 User,添加 JPA 对应的注解进行映射配置,代码如下。

```
package com.yzpc.pojo;
import javax.persistence.*;
@Entity(name = "user")                                  // 设置 ORM 实体类,并指定映射的表名
public class User implements Serializable{
    @Id                                                 // 表示映射对应的主键
    @GeneratedValue(strategy = GenerationType.AUTO) //设置主键自增策略
    private int id;
    private String username;
    private String password;
    // 此处省略构造方法
    // 此处省略相应属性的 setter/getter 方法
    // 重写 toString 方法
}
```

(5) 在项目的 src/main/java/ 路径下的 com.yzpc 包中,新建一个 repository 包,并在该包中新建一个用于操作 User 实体的接口 UserRepository,该接口继承自 JpaRepository,也可在该接口中自定义相应的方法,代码如下。

```
public interface UserRepository extends JpaRepository<User, Integer>{
}
```

(6) 在项目的 src/main/java/ 路径下的 com.yzpc 包中,新建一个 service 包,并在该包

中新建一个用于 User 相关业务操作的 UserService 类，代码如下。

```
package com.yzpc.service;
import com.yzpc.pojo.User;
import com.yzpc.repository.UserRepository;
import org.springframework.beans.factory.annotation.Autowired;
import org.springframework.cache.annotation.Cacheable;
import org.springframework.stereotype.Service;
import java.util.Optional;
@Service
public class UserService {
    @Autowired
    UserRepository userRepository;
    //根据 id 查询用户信息
    @Cacheable(cacheNames = "user",unless = "#result == null")
    public User findUserById(int id){
        Optional<User> user = userRepository.findById(id);
        if (user.isPresent()){
            return user.get();
        }
        return null;
    }
}
```

上述自定义了一个业务操作类 UserService，使用注入的 UserRepository 实例对象完成对 User 的操作。在 UserService 类中的 findUserById(int id)方法上添加了查询缓存@Cacheable，该注解的作用是将查询结果 User 存放在 Spring Boot 默认缓存中名称为 user 的名称空间中，对应缓存的唯一标识(即缓存数据对应的主键 key)默认为方法参数 id 的值。

(7) 在项目的 src/main/java/ 路径下的 com.yzpc 包中，新建一个 controller 包，并在该包中新建一个访问控制类 UserController，代码如下。

```
package com.yzpc.controller;
import com.yzpc.pojo.User;
import com.yzpc.service.UserService;
import org.springframework.beans.factory.annotation.Autowired;
import org.springframework.web.bind.annotation.GetMapping;
import org.springframework.web.bind.annotation.RestController;
@RestController
public class UserController {
    @Autowired
    UserService userService;
    @GetMapping("/findUserById")
    public User findUserById(int id){
        User user = userService.findUserById(id);
        return user;
    }
}
```

在 UserController 用户管理控制类，使用注入 UserService 实例对象完成对 User 的操作。

(8) 在项目的 Chapter07cacheApplication 启动类上使用@EnableCaching 注解开启基于注解的缓存支持，代码如下。

```
@SpringBootApplication
@EnableCaching      //开启 Spring Boot 基于注解的缓存管理支持
public class Chapter07cacheApplication {
    public static void main(String[] args) {
        SpringApplication.run(Chapter07cacheApplication.class, args);
    }
}
```

(9) 启动项目，访问 http://localhost:8080/findUserById?id=3，查询 id 为 3 的用户信息，调用方法查询数据库，并将查询结果的数据存储到缓存user中，页面的查询结果如图 7-3 所示。

图 7-3　第一次访问页面显示结果

此时控制台输出 Hibernate 的查询语句，如图 7-4 所示。

```
Console   Endpoints
2021-08-13 12:55:25.775  INFO 6304 --- [nio-8080-exec-1] o.a.c.c.C.[Tomcat].[localhost].[/]       : Initializing Spring DispatcherServlet 'dispatcherServlet'
2021-08-13 12:55:25.775  INFO 6304 --- [nio-8080-exec-1] o.s.web.servlet.DispatcherServlet        : Initializing Servlet 'dispatcherServlet'
2021-08-13 12:55:25.777  INFO 6304 --- [nio-8080-exec-1] o.s.web.servlet.DispatcherServlet        : Completed initialization in 2 ms
Hibernate: select user0_.id as id1_0_0_, user0_.password as password2_0_0_, user0_.username as username3_0_0_ from user user0_ where user0_.id=?
```

图 7-4　第一次访问控制台输出结果

再次访问 http://localhost:8080/findUserById?id=3，此时控制台没有任何输出内容，同图 7-4，但页面还是输出图 7-3 的查询结果，这表明没有调用查询方法，页面的数据直接从数据缓存中获得。

7.3　EhCache 2.x 缓存

EhCache 是一种广泛使用的开源的高性能 Java 分布式缓存框架，它具有内存和磁盘存储、缓存加载器、缓存扩展、缓存异常处理、GZIP 缓存、Servlet 过滤器，以及支持 REST 等特点。EhCache 有很高的拓展性和伸缩性，广泛使用在各种 Java 项目中(如 Hibernate 默认使用 Ehcache 作为二级缓存)。在 Spring Boot 中，只要配置一个配置文件就可以将 EhCache 集成到项目中。EhCache 2.x 缓存的使用步骤如下。

(1) 复制 7.2 小节的 chapter07cache 项目并重命名为 chapter07ehcache。用 IDEA 打开该项目，选择 chapter07ehcache 项目，右击，选择 Refactor→Rename 命令，然后输入新的

Module 名称为 chapter07ehcache，单击 OK 按钮，如图 7-5 所示。打开 pom. xml 文件，修改 < artifactId >和< name >标签下的名称为 chapter07ehcache。

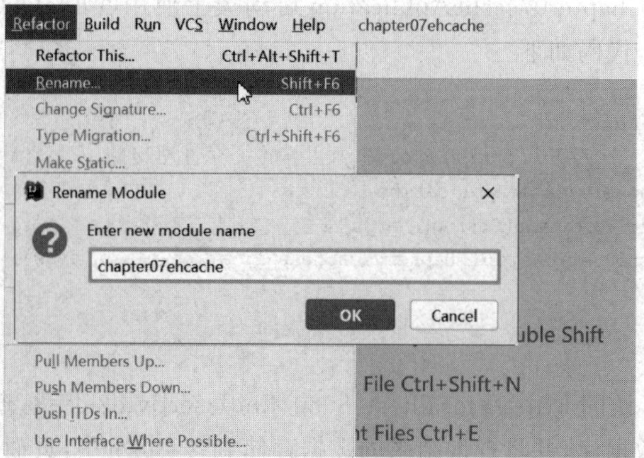

图 7-5 重命名项目的 Module

右击启动类 Chapter07cacheApplication，选择 Refactor→Rename 命令，修改新名称为 Chapter07ehcacheApplication，如图 7-6 所示。在图 7-6 中，单击 Refactor 按钮，在弹出的 Rename Tests 对话框中，选中 class 前面的复选框，通知修改测试类，单击 OK 按钮，如图 7-7 所示。

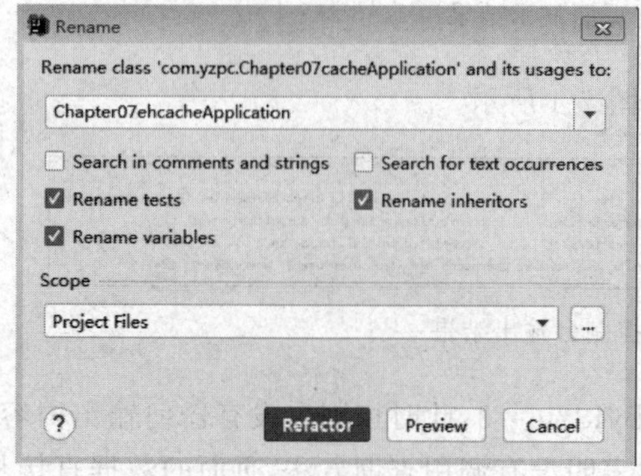

图 7-6 重命名启动类的名称

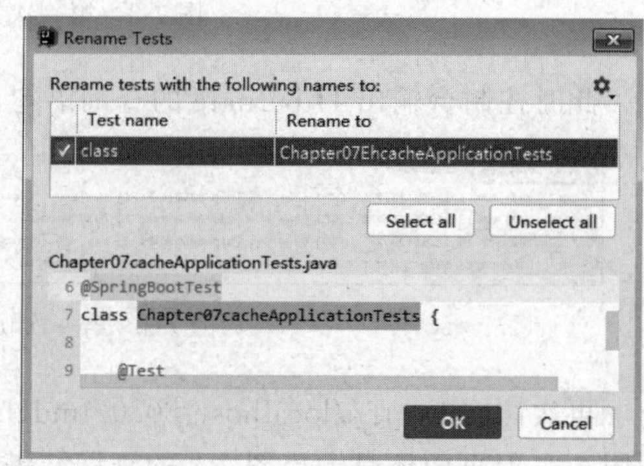

图 7-7 重命名关联的测试的名称

选择 Run 菜单下 Edit Configurations...命令，在弹出的对话框中，将 Name 后的名称修改为 Chapter07ehcacheApplication，单击 OK 按钮，如图 7-8 所示。

(2) 在 pom. xml 文件中，添加 Ehcache2. x 缓存依赖坐标，代码如下。

```
<!-- 此处省略其他依赖代码 -->
<!-- ehcache 的缓存坐标 -->
<dependency>
    <groupId>net.sf.ehcache</groupId>
    <artifactId>ehcache</artifactId>
</dependency>
```

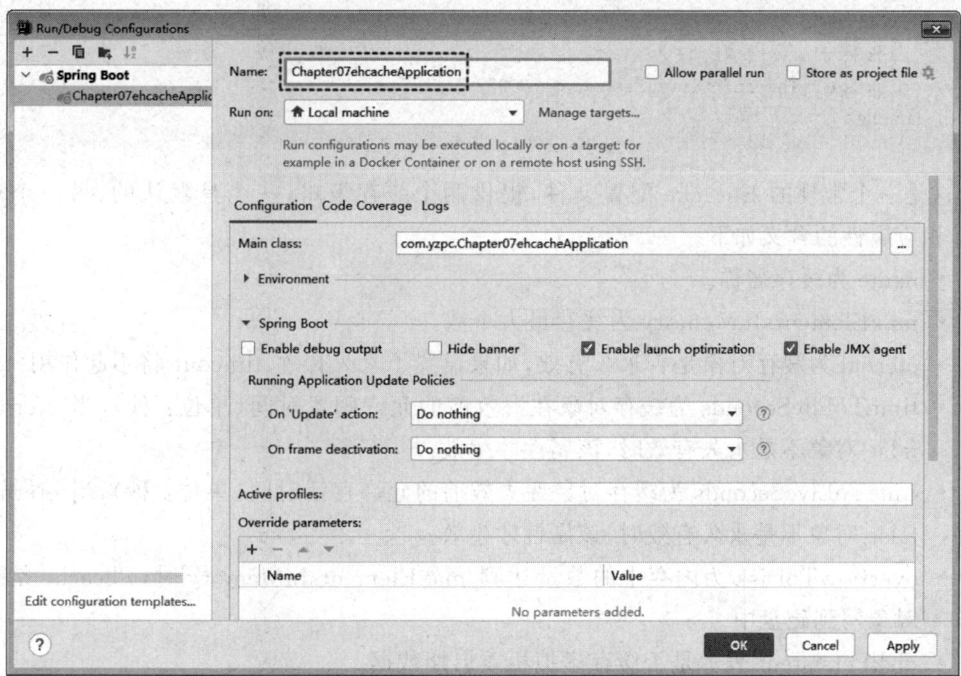

图 7-8 Edit Configurations...下的名称修改

通过配置属性 spring. cache. type 来指定缓存管理器，修改 application. propertes 配置文件，添加代码如下。

```
# Spring cache 目标缓存管理器修改为ehcache
spring.cache.type = ehcache
```

（3）在项目的 src/main/resources/ 目录下创建 ehcache. xml 文件，作为 Ehcache 缓存的配置文件，配置如下。

```
<?xml version = "1.0" encoding = "UTF - 8"?>
< ehcache >
    < diskStore path = "java.io.tmpdir/cache"/>
    < defaultCache
        maxElementsInMemory = "10000"
        eternal = "false"
        timeToIdleSeconds = "120"
        timeToLiveSeconds = "120"
        overflowToDisk = "false"
        diskPersistent = "false"
        diskExpiryThreadIntervalSeconds = "120"/>
    < cache name = "user"
        maxElementsInMemory = "10000"
        eternal = "true"
        timeToIdleSeconds = "120"
        timeToLiveSeconds = "120"
```

```
        overflowToDisk = "true"
        diskPersistent = "true"
        diskExpiryThreadIntervalSeconds = "600"/>
</ehcache>
```

这是一个常规的 Ehcache 配置文件，提供两个缓存策略，一个是默认的，另一个名为 user，相应属性的含义如下。

- name 为缓存名称。
- maxElementsInMemory 为缓存最大个数。
- eternal 为缓存对象是否永久有效，如果设置了永久有效，timeout 将不起作用。
- timeToIdleSeconds 为缓存对象在失效前的允许闲置时间（单位：秒），当 eternal=false 对象不是永久有效时，该属性才生效。
- timeToLiveSeconds 为缓存对象在失效前的允许存活时间（单位：秒），当 eternal=false 对象不是永久有效时，该属性才生效。
- overflowToDisk 为内存中的数量达到 maxElementsInMemory 时，Ehcache 是否将对象写到磁盘中。
- diskPersistent 表示是否缓存虚拟机重启期数据。
- diskExpiryThreadIntervalSeconds 为磁盘失效线程运行时间间隔。

默认情况下，这个文件名是固定的，必须为 ehcache.xml，如果一定要使用其他名称，则需要在 application.properties 中明确指定配置文件名，配置内容如下。

```
spring.cache.ehcache.config = classpath: another - config.xml
```

（4）修改 UserRepository 接口，添加一个修改的方法 update()，并在方法上方添加@Query 注解，代码如下。

```
package com.yzpc.repository;
import com.yzpc.pojo.User;
import org.springframework.data.jpa.repository.JpaRepository;
import org.springframework.data.jpa.repository.Modifying;
import org.springframework.data.jpa.repository.Query;
import org.springframework.transaction.annotation.Transactional;
public interface UserRepository extends JpaRepository < User, Integer >{
    @Transactional
    @Modifying
    @Query("update user u set u.username = ?1, u.password = ?2 where u.id = ?3")
    public void update(String username, String password, int id);
}
```

@Transactional 注解一般用于 Spring 中的事务管理，在 JPA 中提供了@Query 注解用于使用 JPQL 执行数据库操作，如果数据库操作是修改数据而非查询数据，则需要再额外使用@Modifying 注解提示 JPA 该操作是修改操作。

（5）修改 UserService 类，在该类中添加增删改查的方法，分别为 save()、delete()、

update()、findOne()方法，在相应的方法上加上缓存注解，代码如下。

```
@Cacheable(cacheNames = "user", unless = "#reult == null")
public User findOne(int id) {
    Optional<User> u = userRepository.findById(id);
    System.out.println("为 Key = " + u.get().getId() + "的数据做了缓存");
    return u.get();
}
@CachePut(value = "user",key = "#result.id")
public User save(User user) {
    User u = userRepository.save(user);
    System.out.println("为 Key = " + u.getId() + "的数据做了缓存");
    return u;
}
@CachePut(value = "user",key = "#result.id")
public User update(User user) {
    userRepository.update(user.getUsername(), user.getPassword(), user.getId());
    System.out.println("为 key = " + user.getId() + "数据记录做了缓存");
    return user;
}
@CacheEvict(value = "user")
public void delete(int id) {
    System.out.println("删除了 Key = " + id + "的数据缓存");
}
```

使用@CachePut 注解将新增的或修改的数据保存到缓存，其中缓存名为 user，数据的 key 是 user 的 id；使用@CacheEvict 注解从缓存 user 中删除 key为 user 的 id 的数据；使用@Cacheable 注解中没有标记 key 值，将会使用默认参数值 user_id 作为 key 进行数据保存，在进行缓存更新时必须使用同样的 key，同时在查询缓存@Cache 注解中，定义了unless＝"＃reult＝＝null"表示查询结果为空，不进行缓存。

(6) 修改 UserController 类，在该类中添加相应增删改查方法并添加访问映射，代码如下。

```
@GetMapping("/findOne")
public User findOne(int id){
    return userService.findOne(id);
}
@GetMapping("/save")
public User save(User user){
    return userService.save(user);
}
@GetMapping("/update")
public User update(User user){
    User u = userService.update(user);
    return u;
}
@GetMapping("/delete")
public String delete(int id){
    userService.delete(id);
    return "删除 key = " + id + "的数据缓存";
}
```

(7) 启动项目,第一次访问 http://localhost: 8080/findOne?id=4,将调用方法查询数据库。并将数据存储到缓存中,此时控制台输出如图 7-9 所示,同时网页显示数据如图 7-10 所示。

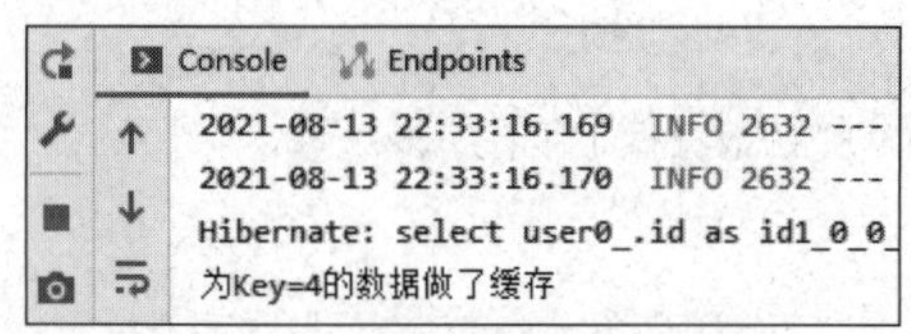

图 7-9 第一次访问控制台输出结果

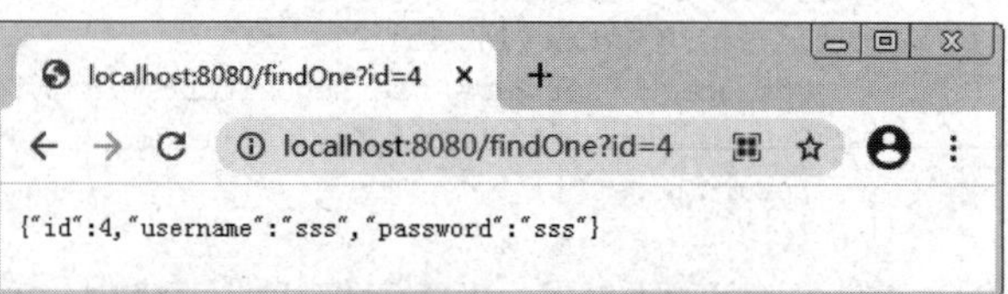

图 7-10 第一次访问页面显示效果

再次访问 http://localhost:8080/findOne?id=4,此时控制台没有任何输出,这表明没有调用查询方法,页面显示的数据是直接从缓存中获取的。

访问 http://localhost:8080/save?username=lisi&password=lisi,此时控制台输出如图 7-11 所示,页面显示数据如图 7-12 所示。

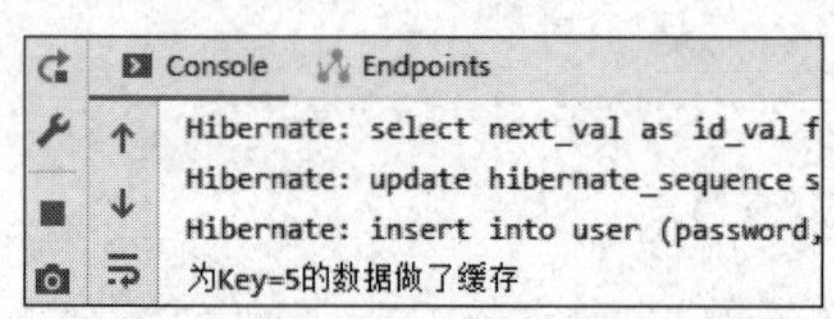

图 7-11 测试插入数据时控制台输出

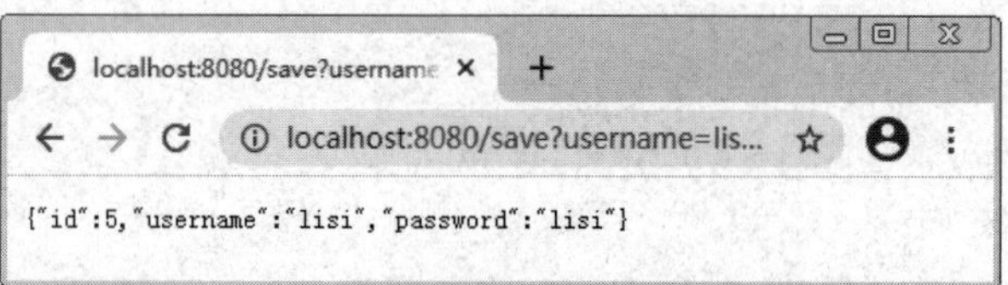

图 7-12 测试插入数据时页面输出

插入数据后,访问 http://localhost:8080/findOne?id=5,控制台无任何输出,从缓存直接获得 id 为 5 的数据,页面显示结果与图 7-12 相同。

访问 http://localhost:8080/update?id=5&username=李四 &password=123456,此时控制台输出如图 7-13 所示,页面显示数据如图 7-14 所示。

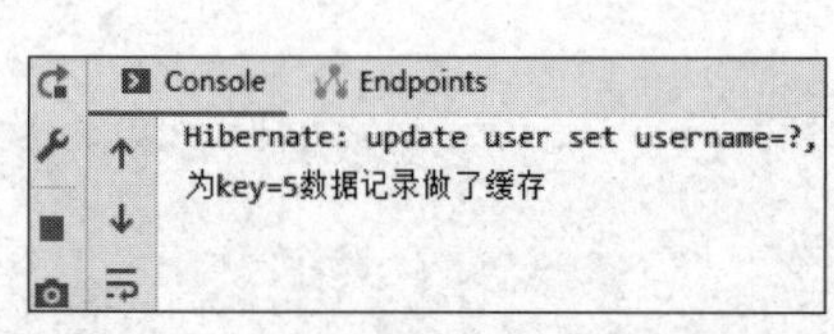

图 7-13 测试修改数据时控制台输出

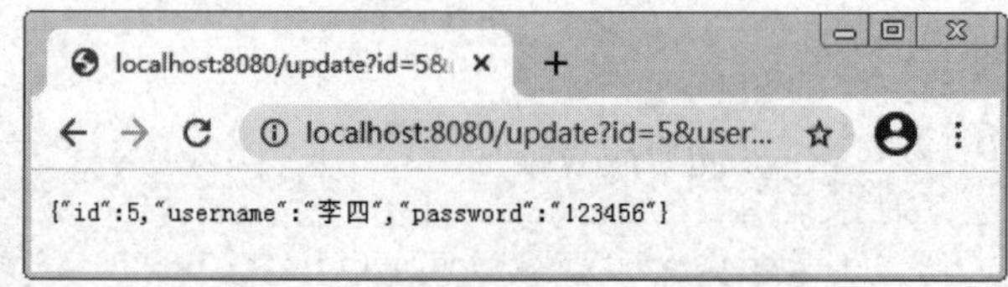

图 7-14 测试修改数据时页面输出

修改数据后,访问 http://localhost:8080/findOne?id=5,控制台无任何输出,从缓存直接获得 id 为 5 的数据,页面显示结果与图 7-14 相同。

访问 http://localhost:8080/delete?id=5,删除 key 为 5 的缓存,控制台和浏览器都输出"删除 key=5 的数据缓存"。再次访问 http://localhost: 8080/findOne?id=5,此时重新做了缓存,控制台输出结果如图 7-15 所示。

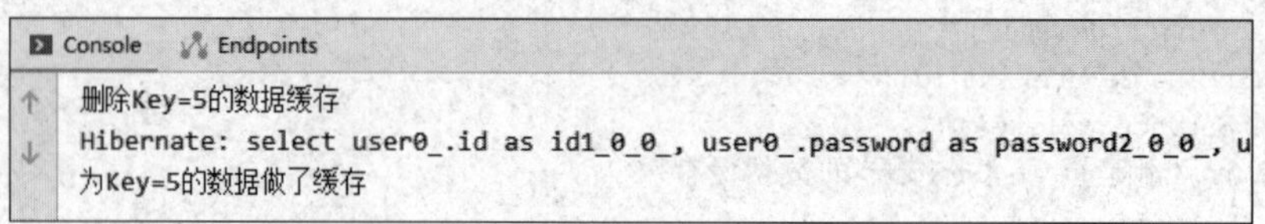

图 7-15 测试删除缓存后重做缓存数据

7.4 Redis缓存

Redis是目前使用最广泛的内存数据存储系统之一。它支持更丰富的数据结构，支持数据持久化、事务、双机集群系统、主从库等。Redis是key-value存储系统，它支持的value类型包括String、List、Set、Hash和ZSet(有序集合)。Redis会周期性地把更新后的数据写入磁盘，或把修改操作写入追加的记录文件中，并且在此基础上实现了master-slave(主从)同步。机器重启后，能通过持久化数据建立自动重建内存。如果使用Redis作为Cache，则机器宕机后热点数据不会丢失。丰富的数据结构加上Redis兼具缓存系统和数据库的特性，使得Redis适用于高并发的读写、计数器、排行榜、分布式会话、互动场景、最新列表等应用场景。

和Ehcache缓存一样，如果在classpath下存在Redis并且已经配置好，此时默认就会使用RedisCacheManager作为缓存提供者。下面通过复制修改7.3小节的示例来介绍Spring Boot使用Redis缓存，步骤如下。

(1) 复制7.3小节的chapter07ehcache项目并重命名为chapter07redis。用IDEA打开该项目，按照7.3小节的第(1)步的修改方法，将其他相关名称修改为chapter07redis。

(2) 在pom.xml文件中，删除或注释Ehcache2.x缓存依赖，添加Spring Data Redis依赖启动器，代码如下。

```
<!-- 此处省略其他依赖代码 -->
<!-- Sprign Data Redis 依赖启动器 -->
<dependency>
    <groupId>org.springframework.boot</groupId>
    <artifactId>spring-boot-starter-data-redis</artifactId>
</dependency>
```

(3) 修改src/main/resources/路径下的application.propertes配置文件，通过配置属性spring.cache.type来指定redis，并配置Redis服务连接配置，代码如下。

```
# 修改目标缓存管理器为Redis
spring.cache.type = redis
spring.redis.host = localhost
spring.redis.port = 6379
# Redis 服务器连接密码(默认为空)
spring.redis.password =
```

使用类似Redis的第三方缓存组件进行缓存管理时，缓存数据并不是像Spring Boot默认缓存管理那样存储在内存中，而是需要预先搭建类似Redis服务的数据仓库进行缓存存储。所以，这里首先需要安装并开启Redis服务(具体可参考第6章中6.1.2小节的介绍)。

(4) 删除resources下的ehcache.xml配置文件，User实体类、UserRepository接口、UserService服务类和UserController控制类不做修改，运行测试步骤与7.3小节的第(7)步一致，这里不再说明。

（5）通过 Redis Desktop Manager 可视化工具查看 Redis 缓存，如图 7-16 所示。

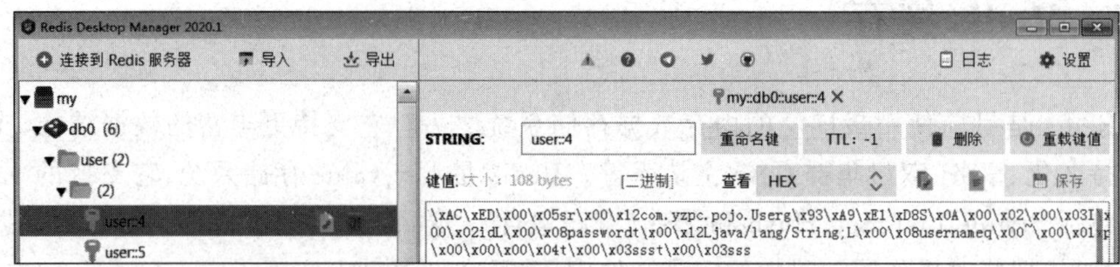

图 7-16　可视化工具查看 Redis 缓存

7.5　小结

本章介绍了 JSR-107 规范、缓存抽象与缓存注解，分别通过示例介绍了默认 Simple 缓存、EhCache 2.x 缓存和 Redis 缓存，让读者对 Spring Boot 缓存技术有初步了解。

第8章 Spring Boot消息服务

在项目开发中，有时候需要与其他系统进行集成，这时需要发送消息给其他系统，让其完成对应功能。这种情况，可以在程序内部相互调用，还可以使用消息服务中间件进行业务处理，能够提升系统的异步通信和扩展解耦能力。Spring Boot 对消息服务管理提供了非常好的支持。本章将针对 Spring Boot 消息服务的原理和整合使用进行介绍。

8.1 消息服务概述

消息（Message）是指软件对象之间进行交互作用和通信利用的一种方式。中间件（Middleware）是处于操作系统和应用程序之间的软件，也有人认为它应该属于操作系统中的一部分。MQ 全称 Message Queue（消息队列），是在消息的传输过程中保存消息的容器，多用于分布式系统之间进行通信。

消息服务是将软件与软件之间的交互消息、交互方式进行存储和管理的一种技术。在多数应用尤其是分布式系统中，消息服务是不可或缺的重要部分，它使用起来比较简单，同时也解决了不少难题，例如异步处理、应用解耦、流量削峰、分布式事务管理等，使用消息服务可以实现一个高性能、高可用、高扩展的系统。下面通过实际开发中的若干场景来分析和说明为什么要使用消息服务，以及使用消息服务的好处。

1. 异步处理

应用场景说明：用户注册后，系统需要将信息写入数据库，并发送注册邮件和注册短信通知，下面介绍不同的处理方式如何处理注册业务需求。

（1）串行处理方式：用户发送注册请求后，服务会先将注册信息写入数据库，依次发送注册邮件和注册短信，服务器只有在消息处理完毕后才会将处理结果返回给客户端，如图 8-1 所示。

串行处理方式非常耗时，用户体验不友好。

（2）并行处理方式：用户发送注册请求后，将注册信息写入数据库，可以使用多线程并发操作同时发送注册邮件和注册短信，最后返回给客户端，如图 8-2 所示。

并行处理方式在一定程度上提高了后台业务处理的效率，但如果遇到较为耗时的业务处理，仍然显得不够完善。

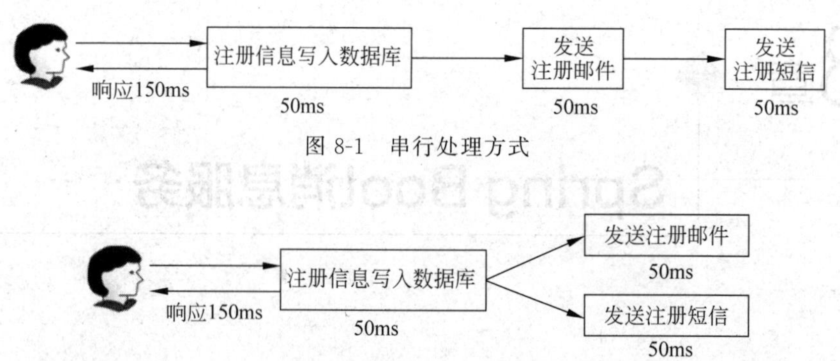

图 8-1 串行处理方式

图 8-2 并行处理方式

(3) 消息服务处理方式：可以在业务中嵌入消息服务进行业务处理，先将注册信息写入数据库，在非常短的时间内将注册信息写入消息队列后即可返回响应信息。此时，前端业务不需要理会后台业务处理，而发送注册邮件和短信可通过异步读取的方式自动读取消息队列中的相关消息进行后续业务处理，如图 8-3 所示。

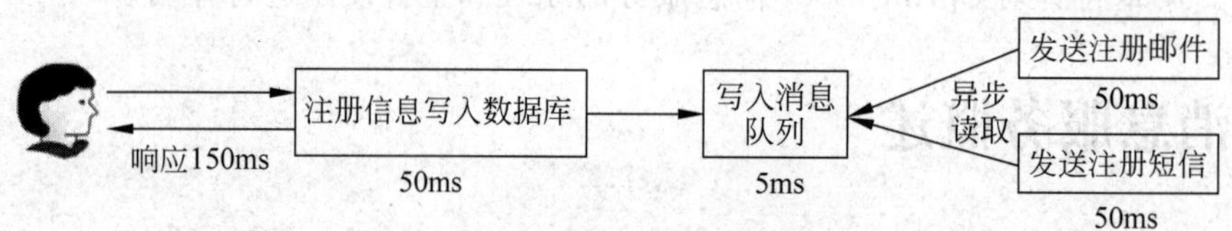

图 8-3 使用消息服务处理

由于消息队列服务器处理速度快于数据库(消息队列也比数据库有更好的伸缩性)，因此响应速度得到大幅改善。

2. 应用解耦

应用场景说明：订单系统→库存系统，用户下单后，订单服务需要通知库存服务。如果采用传统方式处理订单业务，下单后，订单服务会直接调用库存服务接口进行库存更新，如图 8-4 所示。

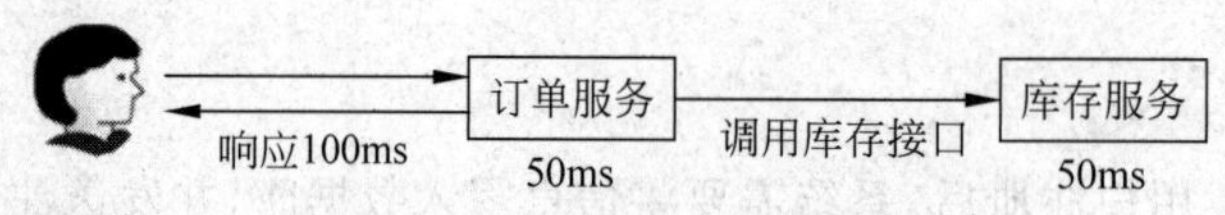

图 8-4 传统处理方式应用解耦

传统方式应用解耦一旦库存系统出现异常，订单服务会失败导致订单丢失。

如果使用消息服务模式应用解耦，订单服务的下单会快速写入消息队列，库存服务会监听并读取到订单，从而修改库存，如图 8-5 所示。

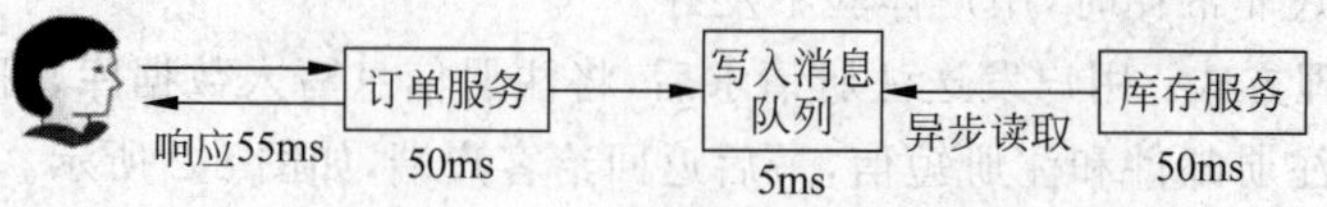

图 8-5 消息服务模式应用解耦

相较于传统方式，消息服务模式显得更加高效、可靠。

3. 流量削峰

应用场景说明：在大型购物节时，购物网站开展秒杀活动，是流量削峰的一种应用场景。秒杀活动一般由于瞬时访问量过大，服务器接收过大，会导致流量暴增，相关系统无法处理请求甚至崩溃。为了解决这个问题，通常会采用消息队列缓冲瞬时高峰流量，系统可以从消息队列中读取数据，对请求进行分层过滤，从而过滤掉一些请求，如图 8-6 所示。

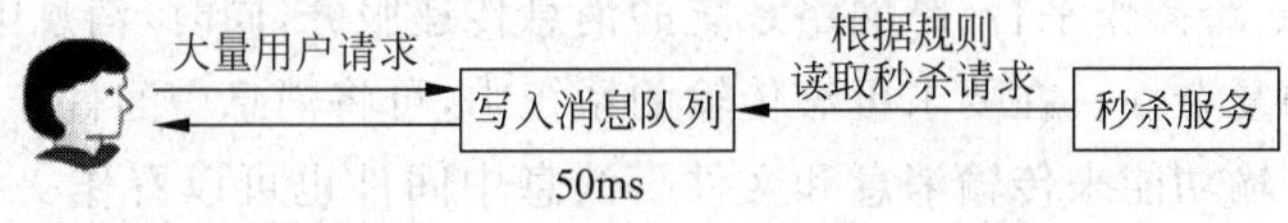

图 8-6　流量削峰场景说明

针对秒杀业务的需求，如果专门增设服务器来应对秒杀活动期间的请求瞬时高峰，在非秒杀活动期间，这些多余的服务器和配置则有些浪费；如果不进行处理的话，秒杀活动瞬时高峰流量请求有可能压垮服务。因此，在秒杀活动中加入消息服务是较为理想的解决方案。在秒杀活动中，当瞬时高并发请求到来时，服务器接收所有请求，但并不处理这些请求（也可能来不及处理），只是将它们写入消息队列中。当消息队列中的请求数量达到最大值时，用户请求将被转发到错误页面，这样可控制参加活动的人数。当消息队列缓存所有用户请求之后，系统中多个秒杀业务处理系统将会根据秒杀规则读取这些请求消息，并对消息逐一进行处理，通过这种方式即实现了负载均衡，也缓解了瞬时高流量对秒杀业务处理系统的压力。

4. 分布式事务管理

应用场景说明：在分布式系统中，分布式事务是开发中必须要面对的技术难题，分布式系统的请求业务处理的数据一致性通常是重点考虑的问题，较为可靠的处理方式是基于消息队列的二次提交。在失败的情况下可以进行多次尝试，或者基于队列数据进行回滚操作。因此，在分布式系统中加入消息服务是一个既能保证性能不变，又能保证业务一致性的方案。针对这种分布式事务处理的需求，使用消息服务的处理机制，如图 8-7 所示。

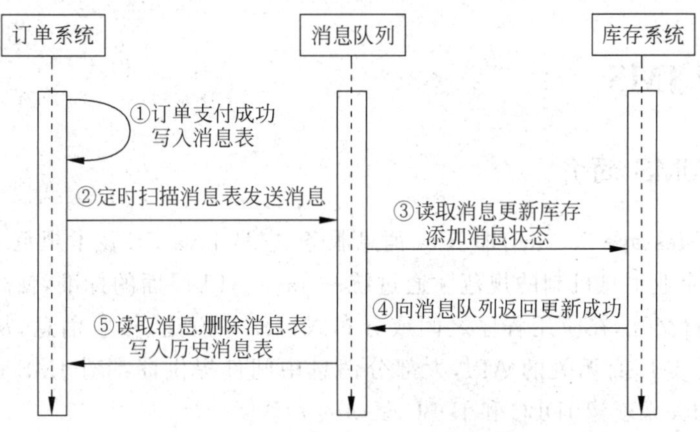

图 8-7　分布式事务管理说明

如果使用传统方式在订单系统中写入订单支付成功信息后，再调用库存系统进行库存更新，一旦库存系统异常，很有可能导致库存更新失败而订单支付成功的情况，从而导致数据不一致。针对这种分布式系统的事务管理，通常会在分布式系统之间加入消息服务进行管理。

消息队列中间件，是一种异步通信的中间件，利用高效可靠的异步消息传递机制进行与平台无关的数据交流，并基于数据通信来进行分布式系统的集成。消息中间件的发展非常迅速，在分布式事务处理环境中，它往往能够充当通信资源管理器的角色，为分布式应用提供实时、高效、可靠、跨操作平台、跨网络系统的消息传递服务，同时，消息中间件降低了开发跨平台应用程序的复杂性。在要求可靠传输的系统中，可将消息中间件作为通信平台，向应用程序提供可靠传输功能来传输消息和文件。消息中间件也可以看作一种容器，可以理解为邮局，发送者将消息投递到邮局，然后邮局负责发送给具体的接收者，数据的发送和接收由邮局来完成。

当前使用较多的消息队列中间件有 ActiveMQ、RabbitMQ、Kafka、RocketMQ 等。目前市面上消息中间件各有侧重点，读者可根据需要进行选择，下面通过对比来了解各消息队列中间件，如表 8-1 所示。

表 8-1　常用消息队列中间件

项　目	ActiveMQ	RabbitMQ	Kafka	RocketMQ
公司/社区	Apache	Rabbit	Apache	Alibaba
开发语言	Java	Erlang	Scala&Java	Java
协议支持	AMQP，OpenWire，STOMP，REST	AMQP，XMPP，SMTP，STOMP	自定义协议，社区封装 http 协议支持	自定义
客户端支持语言	Java、C、C++、Python、PHP、Perl、.NET 等	官方支持 Erlang、Java、Ruby 等	官方支持 Java，社区产出多种 API，如 PHP、Python 等	Java、C++
单击吞吐量	万级(最差)	万级(其次)	十万级(次之)	十万级(最好)
消息延迟	毫秒级	微秒级	毫秒以内	毫秒级
功能特性	老牌产品，成熟度高，文档较多	并发能力强，延时低，社区活跃，管理界面丰富	只支持主要的 MQ 功能，为大数据领域服务	MQ 功能比较完善，扩展性佳

8.2　整合 JMS

8.2.1　JMS 简介

JMS(Java Message Service)即 Java 消息服务，它是 Java EE 技术规范中的一个重要组成部分，是一种企业消息机制的规范。通过统一 Java API 层面的标准，使得多个客户端可以通过 JMS 进行交互，在应用程序之间或分布式系统中发送、接收消息，从而进行异步通信。JMS 是一个与厂商无关的 API，大部分消息中间件提供商都对 JMS 提供支持。JMS 和 ActiveMQ 的关系就像 JDBC 和 JDBC 驱动的关系。

JMS 具有以下优势。

(1) 异步：JMS本身是异步的，客户端获取消息的时候，不需要主动发送请求，消息会自动发送给可用的客户端。

(2) 可靠：JMS保证消息只会传递一次。

JMS消息机制模型主要分为2类。

(1) P2P模型：即点对点(Point-to-Point)消息传递模型，生产者生产了一个消息，只能由一个消费者进行消费，点对点消息传递域的特点如下。

- 每个消息只能有一个消费者，类似1对1的关系，就像个人快递只能由自己领取。
- 消息的生产者和消费者之间没有时间上的相关性。
- 消息被消费后，队列中不会再存储，所以消费者不会消费到已经被消费掉的消息。

(2) Pub-Sub模型：即基于发布/订阅(Publish/Subscribe)消息传递模型，生产者生产了一个消息，可以由多个消费者进行消费，基于发布/订阅模式的传输特点如下。

- 生产者将消息发布到主题中，每个消息可以有多个消费者，属于1：N的关系。
- 生产者和消费者之间有时间上的相关性，订阅某一个主题的消费者只能消费自它订阅之后发布的消息。
- 生产者生产时，主题不保存消息，这时它是无状态的。如果无人订阅时生产，那就是一条废消息。

在实际工作中，实现JMS服务的规范有很多，其中比较常用的有传统的ActiveMQ和分布式的Kafka。为了更加可靠和安全，还存在AMQP协议，实现AMQP比较常用的有RabbitMQ等。

8.2.2 Spring Boot 整合 JMS

由于JMS是一套标准规范，因此Spring Boot整合JMS就是整合JMS的某一实现，而底层JMS的实现则可选择ActiveMQ、ActiveMQ Artemis或JORAM等。这里以ActiveMQ作为JMS实现来介绍Spring Boot对JMS的支持。

1. ActiveMQ 安装

访问http://activemq.apache.org/，页面中存在ActiveMQ Classic和ActiveMQ Artemis两个项目，其中，ActiveMQ Classic代表传统的ActiveMQ，ActiveMQ Artemis则通常简称为Artemis。这里下载ActiveMQ Classic的Windows版本apache-activemq-5.16.3-bin.zip，该版本的ActiveMQ解压缩后即可使用，双击apache-activemq-5.16.3\bin\win64目录下的activemq.bat或wrapper.exe启动ActiveMQ，如图8-8所示。

通过浏览器访问http://localhost:8161/admin/，提示输入用户名和密码，默认为admin和admin，进入后能看到如图8-9所示的界面，表示ActiveMQ已经启动成功。

在图8-9的管理界面上，有Queues、Topics、Subscribers、Connections、Network、Scheduled、Send链接，通过这些链接可以管理ActiveMQ上的消息队列、消息主题、订阅者、连接、网络桥接等内容。

在ActiveMQ中，61616为消息代理的端口，采用TCP协议发送；8161为管理界面的端口，采用内嵌jetty服务器。如果要修改ActiveMQ管理界面的一些参数，例如修改服务端口(默认为8161)，可以打开解压路径下的conf/jetty.xml文件进行修改。

```
ActiveMQ
jvm 1    | INFO | Connector amqp started
jvm 1    | INFO | Listening for connections at: stomp://SK-20210124RNGK:61613?maximumConnections=1000&wireFormat.maxFrameSize=104857600
jvm 1    | INFO | Connector stomp started
jvm 1    | INFO | Listening for connections at: mqtt://SK-20210124RNGK:1883?maximumConnections=1000&wireFormat.maxFrameSize=104857600
jvm 1    | INFO | Connector mqtt started
jvm 1    | INFO | Starting Jetty server
jvm 1    | INFO | Creating Jetty connector
jvm 1    | WARN | ServletContext@o.e.j.s.ServletContextHandler@1abb3e9d{/,null,STARTING} has uncovered http methods for path: /
jvm 1    | INFO | Listening for connections at ws://SK-20210124RNGK:61614?maximumConnections=1000&wireFormat.maxFrameSize=104857600
jvm 1    | INFO | Connector ws started
jvm 1    | INFO | Apache ActiveMQ 5.16.3 (localhost, ID:SK-20210124RNGK-51375-1629395501024-0:1) started
jvm 1    | INFO | For help or more information please see: http://activemq.apache.org
jvm 1    | WARN | Store limit is 102400 mb (current store usage is 0 mb). The data directory: D:\apache-activemq-5.16.3\bin\win64\..\..\data\kahadb only has 12163 mb of usable space. - resetting to maximum available disk space: 12163 mb
jvm 1    | WARN | Temporary Store limit is 51200 mb (current store usage is 0 mb). The data directory: D:\apache-activemq-5.16.3\bin\win64\..\..\data only has 12163 mb of usable space. - resetting to maximum available disk space: 12163 mb
jvm 1    | INFO | ActiveMQ WebConsole available at http://127.0.0.1:8161/
jvm 1    | INFO | ActiveMQ Jolokia REST API available at http://127.0.0.1:8161/api/jolokia/
```

图 8-8　启动 ActiveMQ

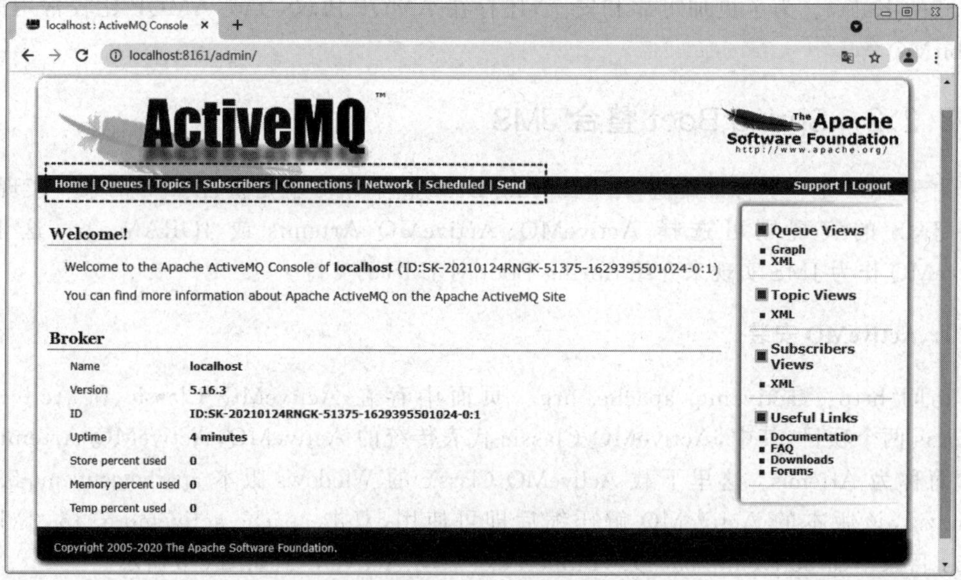

图 8-9　进入 ActiveMQ 管理界面

2. 整合 Spring Boot

（1）创建基于 ActiveMQ 的 Spring Boot 项目。新建一个 Spring Boot 工程 chapter08activemq，Group 和 Package name 为 com. yzpc，在 Dependencies 依赖中选择 Web 节点下的 Spring Web 依赖，Messaging 节点下的 Spring for Apache ActiveMQ 5 依赖，单击 Finish 按钮，如图 8-10 所示。

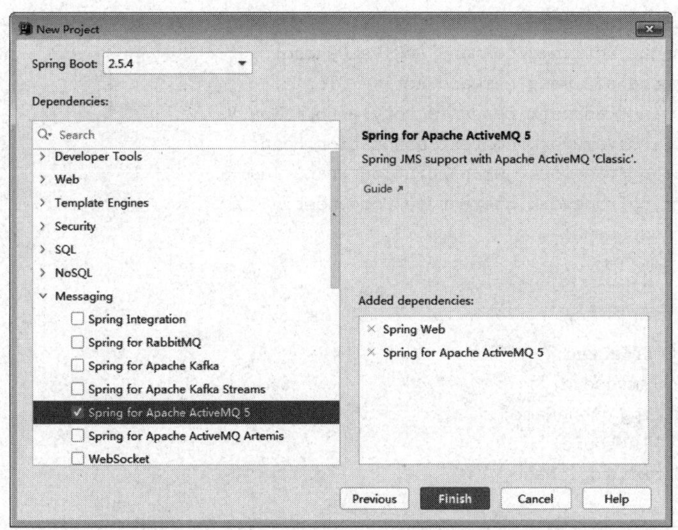

图 8-10 创建基于 Apache ActiveMQ5 的项目

pom.xml 中自动添加的依赖代码如下。

```
<dependency>
    <groupId>org.springframework.boot</groupId>
    <artifactId>spring-boot-starter-activemq</artifactId>
</dependency>
<!-- 省略Web 中Spring Web 依赖代码 -->
```

(2) 配置 ActiveMQ 的消息代理地址。在项目的配置文件 application.properties 中，配置 ActiveMQ 的消息代理地址等信息，配置代码如下。

```
# ActiveMQ 服务器地址
spring.activemq.broker-url=tcp://localhost:61616
spring.activemq.user=admin
spring.activemq.password=admin
# 设置是Queue 队列还是Topic,false 为Queue,true 为Topic ,默认 false
spring.jms.pub-sub-domain=false
# spring.jms.pub-sub-domain=true
# 变量,定义队列和topic 的名称
myqueue: activemq-queue
mytopic: activemq-topic
```

首先配置程序连接 activemq 的通讯地址，默认端口是 61616，然后配置信任所有的包，这个配置是为了支持发送对象消息，最后配置 ActiveMQ 的用户名、密码。

(3) 新建 ActiveMQ 配置类 ConfigBean。在项目的 src/main/java/ 路径下的 com.yzpc 包中，新建一个 config 包，并在该包中新建 ActiveMQ 配置类 ConfigBean，配置 Queue 队列和 topic 两种模式，代码如下。

```
package com.yzpc.config;
import org.apache.activemq.command.ActiveMQQueue;
import org.apache.activemq.command.ActiveMQTopic;
import org.springframework.beans.factory.annotation.Value;
import org.springframework.context.annotation.Bean;
import org.springframework.jms.annotation.EnableJms;
import org.springframework.stereotype.Component;
import javax.jms.Queue;
import javax.jms.Topic;
@Component
@EnableJms
public class ConfigBean {
    @Value("${myqueue}")
    private String queueName;
    @Value("${mytopic}")
    private String topicName;
    @Bean                          //队列
    public Queue queue(){
        return new ActiveMQQueue(queueName);
    }
    @Bean                  //主题
    public Topic topic(){
        return new ActiveMQTopic(topicName);
    }
}
```

(4) 队列模式下创建队列生产者和队列消费者。在项目的 src/main/java/ 路径下的 com.yzpc 包中，新建一个 controller 包，并在该包中新建 QueueProducerController 类为队列生产者控制器，主要向消息队列中发送消息，代码如下。

```
package com.yzpc.controller;
import org.springframework.beans.factory.annotation.Autowired;
import org.springframework.jms.core.JmsMessagingTemplate;
import org.springframework.web.bind.annotation.RequestMapping;
import org.springframework.web.bind.annotation.RestController;
import javax.jms.Queue;
//队列Queue 消息生产者
@RestController
public class QueueProducerController {
    @Autowired
    private JmsMessagingTemplate jmsMessagingTemplate;
    @Autowired
    private Queue queue;
    //消息生产者
    @RequestMapping("/sendmsg")
    public void sendmsg(String msg) {
        System.out.println("发送消息到队列：" + msg);
        // 指定消息发送的目的地及内容
        this.jmsMessagingTemplate.convertAndSend(this.queue, msg);
    }
}
```

在 com. yzpc. controller 的包中，新建 QueueConsumerController 类为队列消费者控制器，代码如下。

```
package com.yzpc.controller;
import org.springframework.jms.annotation.JmsListener;
import org.springframework.web.bind.annotation.RestController;
//队列 Queue 消费者控制器
@RestController
public class QueueConsumerController {
    // 消费者接收消息
    @JmsListener(destination = "${myqueue}")
    public void readActiveQueue(String message) {
        System.out.println("接收到: " + message);
    }
}
```

(5) 运行测试项目。启动项目，访问 http://localhost:8080/sendmsg?msg = Hello ActiveMQ Queue!，向消息队列中发送消息 Hello ActiveMQ Queue!，控制台的输出效果如图 8-11 所示。

```
Console   Endpoints
2021-08-20 03:03:41.138  INFO 8216 --- [nio-8080-exec-1]
2021-08-20 03:03:41.138  INFO 8216 --- [nio-8080-exec-1]
2021-08-20 03:03:41.139  INFO 8216 --- [nio-8080-exec-1]
发送消息到队列: Hello ActiveMQ Queue!
接收到: Hello ActiveMQ Queue!
```

图 8-11　队列模式下发送和接收消息控制台输出

在 ActiveMQ 管理界面上，单击 Queues 链接，则显示在 application. properties 配置文件定义的 activemq-queue 的队列名称，其效果如图 8-12 所示。

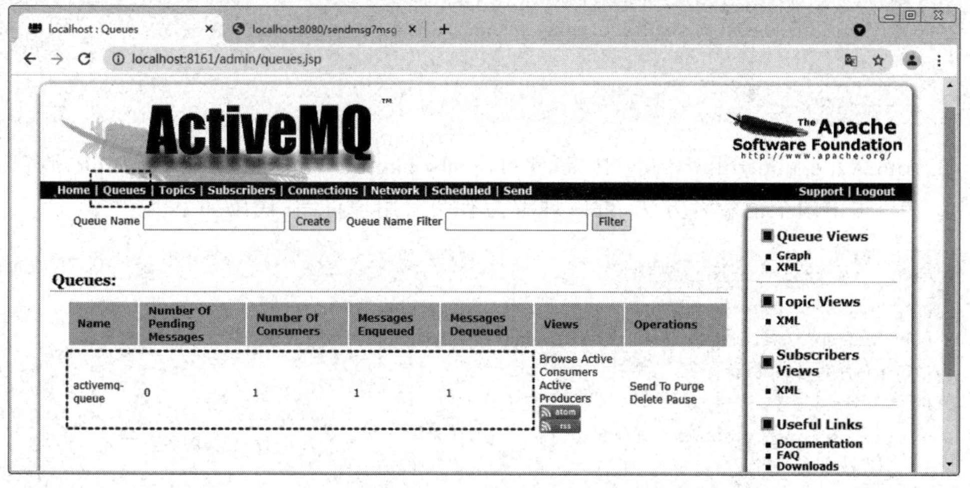

图 8-12　ActiveMQ 管理界面 Queues 模式

ActiveMQ 管理界面的 Queues 模式中的属性含义如下。

- Number Of Pending Messages：消息队列中待处理的消息。

- Number Of Consumers：消费者的数量。
- Messages Enqueued：累计进入过消息队列的总量。
- Messages Dequeued：累计消费过的消息总量。

注意：队列模式时，配置文件 application.properties 中 spring.jms.pub-sub-domain 属性必须设置为 false；当设为 true 时，为主题模式。

（6）Topic 模式下创建队列生产者和队列消费者。在 application.properties 配置文件中将 spring.jms.pub-sub-domain 属性的值设置为 true，即为主题模式。

在 com.yzpc.controller 的包下，新建 TopicProducerController 类为 Topic 生产者控制器，主要向消息队列中发送消息，代码如下。

```
package com.yzpc.controller;
import org.springframework.beans.factory.annotation.Autowired;
import org.springframework.jms.core.JmsMessagingTemplate;
import org.springframework.web.bind.annotation.RequestMapping;
import org.springframework.web.bind.annotation.RestController;
import javax.jms.Topic;
// Topic 消息生产者
@RestController
public class TopicProducerController {
    @Autowired
    private JmsMessagingTemplate jmsMessagingTemplate;
    @Autowired
    private Topic topic;
    // 消息生产者
    @RequestMapping("/topicsendmsg")
    public void sendmsg(String msg) {
        System.out.println("发送消息到 MQ: " + msg);
        // 指定消息发送的目的地及内容
        this.jmsMessagingTemplate.convertAndSend(this.topic, msg);
    }
}
```

在 com.yzpc.controller 的包中，新建 TopicConsumerController 类为 Topic 消费者控制器，其中，写了两个消费者方法，可以理解为有两个用户订阅，代码如下。

```
package com.yzpc.controller;
import org.springframework.jms.annotation.JmsListener;
import org.springframework.web.bind.annotation.RestController;
// Topic 消费者控制器
@RestController
public class TopicConsumerController {
    // 消费者接收消息
    @JmsListener(destination = "${mytopic}")
    public void readActiveTopic(String message) {
        System.out.println("接收到: " + message);
    }
    @JmsListener(destination = "${mytopic}")
```

```
    public void readActiveTopic1(String message) {
        System.out.println("接收到: " + message);
    }
}
```

重新启动项目，访问 http://localhost:8080/topicsendmsg?msg = Hello ActiveMQ Topic!，向消息队列中发送消息 Hello ActiveMQ Topic!，控制台输出效果（有两个消费者方法），如图 8-13 所示。

```
Console    Endpoints
2021-08-20 03:14:18.981  INFO 7472 --- [nio-8080-exec-1]
2021-08-20 03:14:18.982  INFO 7472 --- [nio-8080-exec-1]
发送消息到MQ: Hello ActiveMQ Topic!
接收到: Hello ActiveMQ Topic!
接收到: Hello ActiveMQ Topic!
```

图 8-13 主题模式下发送和接收消息控制台输出

在 ActiveMQ 管理界面上，单击 Topics 链接，则显示在 application. properties 配置文件定义的 activemq-topic 的 Topics 名称，如图 8-14 所示。

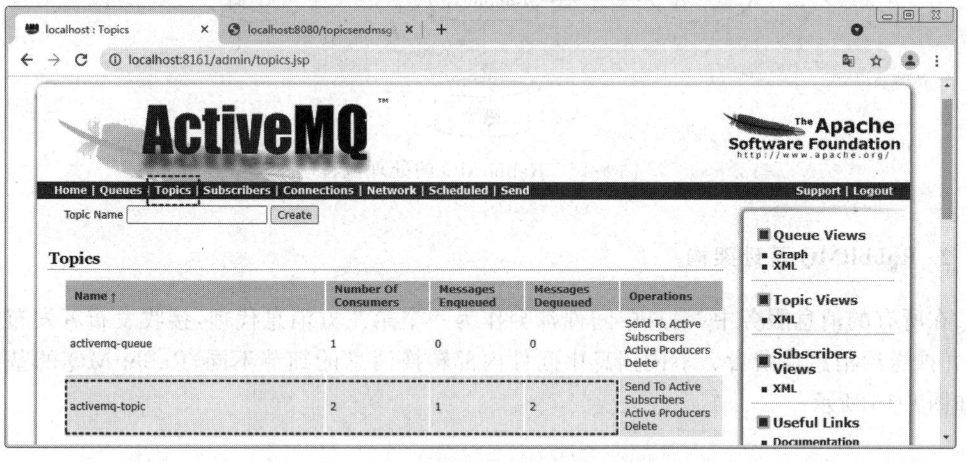

图 8-14 ActiveMQ 管理界面 Topics 模式

ActiveMQ 管理界面的 Topics 模式中的属性含义如下。

- Number Of Consumers：消费者的数量。
- Messages Enqueued：累计进入过消息队列的总量。
- Messages Dequeued：累计消费过的消息总量。

8.3 整合 AMQP

AMQP（Advanced Message Queuing Protocol）为高级消息队列协议，是一个提供统一消息服务的应用层标准高级消息队列协议，是应用层协议的一个开放标准，为面向消息的中间件设计。基于此协议的客户端与消息中间件可传递消息，并不受客户端、中间件的不同产品、不同开发语言等条件的限制。只要按照规范的格式发送数据，任何平台都可以通过

AMQP 进行消息交互，目前流行的 RabbitMQ、StormMQ 等都实现了 AMQP。

8.3.1 RabbitMQ

1. RabbitMQ 简介

RabbitMQ 是基于 AMQP 协议的轻量级、可靠、可伸缩和可移植的消息代理，是 Rabbit 技术公司于 2007 年基于 AMQP 标准开发并提供对其的支持。Rabbit 科技是 LSHIFT 和 CohesiveFT 在 2007 年成立的合资企业，2010 年 4 月被 VMware 旗下的 SpringSource 收购，在 2013 年 5 月并入 Pivotal。RabbitMQ 采用 Erlang 语言开发，支持多种客户端。Erlang 语言由 Ericson 设计，专门为开发高并发和分布式系统的一种语言，在电信领域使用广泛。

RabbitMQ 是目前应用非常广泛的消息中间件。在企业级应用、微服务应用中，RabbitMQ 担当着十分重要的角色，RabbitMQ 对消息的处理流程如图 8-15 所示。

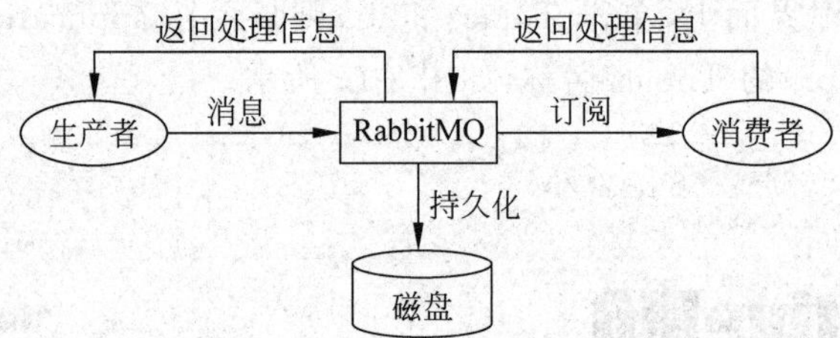

图 8-15　RabbitMQ 的处理流程

2. RabbitMQ 基础架构

在所有的消息服务中，消息中间件都会作为一个第三方消息代理，接收发布者发布的消息，并推送给消息消费者。不同消息中间件内部转换消息的细节不同，RabbitMQ 的基础架构如图 8-16 所示。

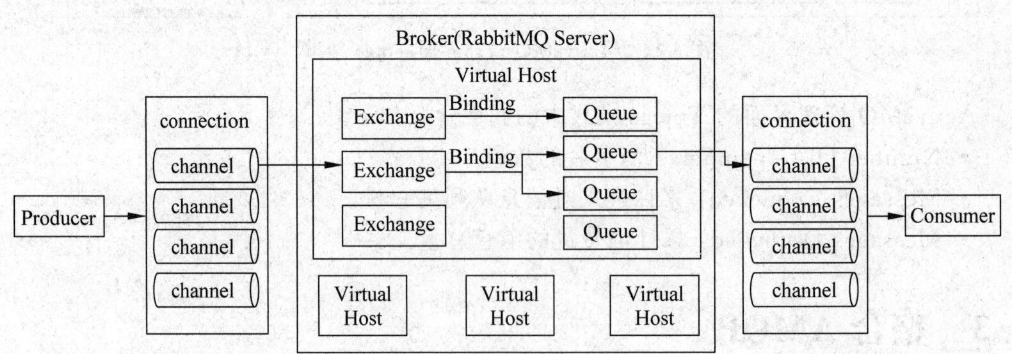

图 8-16　RabbitMQ 的基础架构

RabbitMQ 做了一层抽象，在发消息者和队列之间加入了交换器(Exchange)。这样，发消息者和队列之间就没有直接联系，转而变成发消息者把消息交给交换器，交换器根据调度策略再把消息发给队列。RabbitMQ 的基础架构中有很多细节内容和内部组件，部分说明如下。

- Broker：消息队列服务进程，此进程包括 Exchange 和 Queue 两个部分。
- Exchange：消息队列交换器，按一定的规则将消息路由转发到某个队列，对消息进行过滤。
- Queue：消息队列，存储消息的队列，消息到达队列并转发给指定的消费方。
- Producer：消息生产者，即生产方客户端，生产方客户端将消费发送到 MQ。
- Consumer：消息消费者，即消费方客户端，接收 MQ 转发的消息。

使用 RabbitMQ 进行消息发布和接收的流程。

(1) 发送消息。

- 生产者和 Broker 建立 TCP 连接。
- 生产者和 Broker 建立通道。
- 生产者通过通道消息发送 Broker，由 Exchange 将消息进行转发。
- Exchange 将消息转发到指定的 Queue(队列)。

(2) 接收消息。

- 消费者和 Broker 建立 TCP 连接。
- 消费者和 Broker 建立通道。
- 消费者监听指定的 Queue(队列)。
- 当有消息到达 Queue 时，Broker 默认将消息推送给消费者。
- 消费者接收到消息。

3. RabbitMQ 工作模式

RabbitMQ 消息中间件针对不同的服务需求，提供了多种工作模式，下面对 RabbitMQ 支持的工作模式进行简要说明。

(1) 简单模式。生产者把消息放入队列，消费者获得消息，这个模式只有一个消费者、一个生产者、一个队列，只需要配置主机参数，其他参数使用默认值即可通信，如图 8-17 所示。

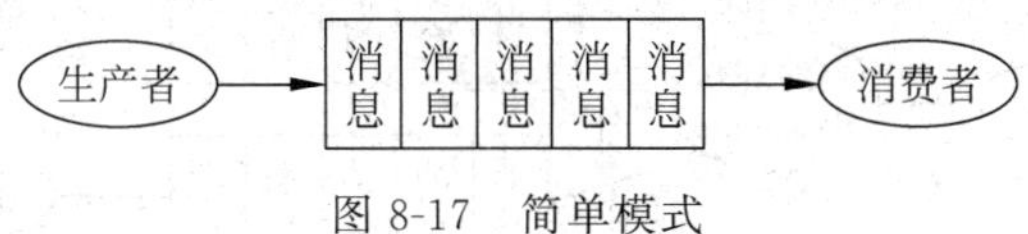

图 8-17 简单模式

(2) 工作队列模式。这种模式出现了多个消费者，为了保证消费者之间负载均衡和同步，需要在消息队列之间加上同步功能，如图 8-18 所示。

图 8-18 工作队列模式

这种模式下，多个消费者通过轮询的方式依次接收消息队列中存储的消息，一旦消息被某一个消费者接收，消息队列会将该消息移除，而接收并处理消息的消费者必须在消费完一条消息后再准备接收下一条消息。

(3) 发布订阅模式。发布订阅模式也称为交换器模式,在该模式中,必须先配置一个 fanout 类型的交换器,不需要指定的路由键值,同时会将消息路由到每一个消息队列上,然后每个消息队列都可对相同的消息进行接收存储,进而由各自消息队列关联的消费者进行消费,如图 8-19 所示。

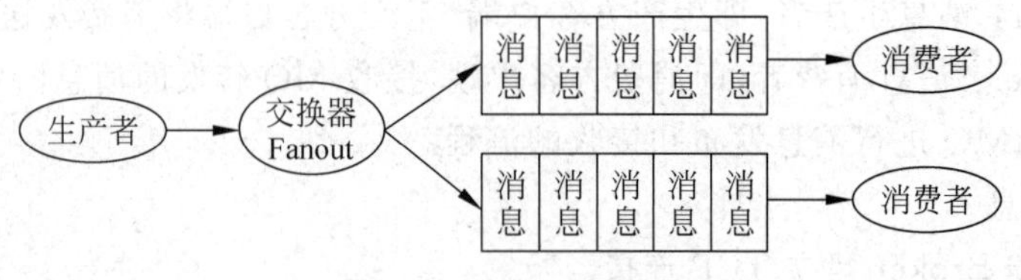

图 8-19 发布订阅模式

这种类型的 Exchange 会将消息广播到所有与它绑定的消息队列,大致相当于 JMS 中的 Pub-Sub 消息模型。

(4) 路由模式。在路由工作模式中,必须先配置一个 direct 类型的交换器,并指定不同的路由键值(Routing Key)将对应的消息从交换器路由到不同的消息队列中进行存储,由消费者进行各自消费,如图 8-20 所示。

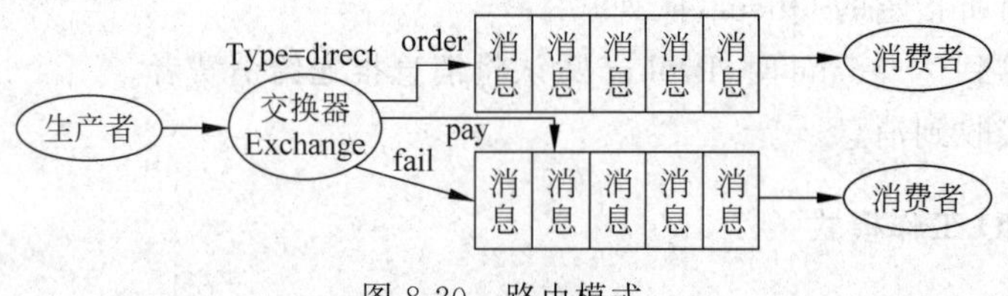

图 8-20 路由模式

(5) 主题模式。在主题转发工作模式中,必须先配置一个 topic 类型的交换器,并指定不同的路由键值将对应的消息从交换器路由到不同的消息队列进行存储,然后由消费者进行各自消费,如图 8-21 所示。

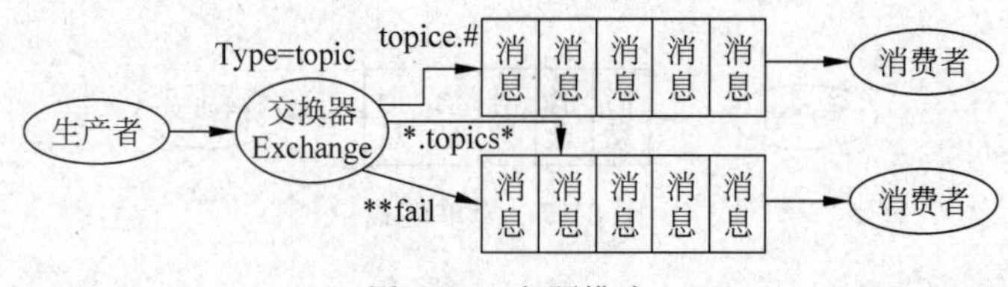

图 8-21 主题模式

主题模式与路由模式的主要不同在于发送到主题交换器的信息,路由键值包含通配符。匹配规则为: * 代替一个字符; # 代替零个或多个字符,然后与其他字符一起使用. 通配符进行连接,从而组成动态路由键值,在发送消息时可以根据需求设置不同的路由键,从而将消息路由到不同的消息队列。

(6) RPC 模式。RPC 模式与工作队列模式主体流程相似,都不需要设置交换器,需要指定唯一的消息队列进行消费。RPC 模式是一个回环结构,主要针对分布式架构的消息传递业务,客户端先发送消息到消息队列,远程服务端获取消息,然后再写入另一个消息队列,向原始客户端相应消息处理结果,如图 8-22 所示。

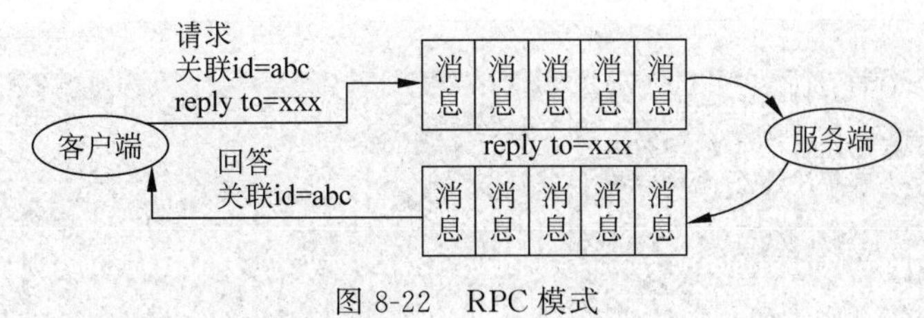

图 8-22 RPC 模式

RPC 工作模式适用于远程服务调用的业务处理场合，例如，在分布式架构中考虑的分布式事务管理问题。

8.3.2 安装 RabbitMQ 以及整合环境搭建

1. 下载安装 RabbitMQ

因为 RabbitMQ 是基于 erlang 语言开发的，所以安装 RabbitMQ 之前，需先下载安装 Erlang。Erlang 语言的下载地址为 https://www.erlang.org/downloads，RabbitMQ 的下载地址为 https://github.com/rabbitmq/rabbitmq-server/releases/，有安装版和解压版。此处下载的 erlang 语言的版本为 otp_win64_24.0.exe，下载的 RabbitMQ 版本是 rabbitmq-server-3.9.4.exe，此版本需要 Erlang 23.2 或更新版本，并支持 Erlang 24.0。

RabbitMQ 安装包依赖于 Erlang 语言包的支持，需要先安装 Erlang 语言包，再安装 RabbitMQ 安装包。运行 Erlang 语言安装包 otp_win64_24.0.exe，按照提示操作即可完成 Erlang 语言包的安装。安装 Erlang 后需要配置环境变量 ERLANG_HOME，变量值为 Erlang 选择安装的具体路径，并在 path 中新增 %ERLANG_HOME%\bin，如图 8-23 和图 8-24 所示。

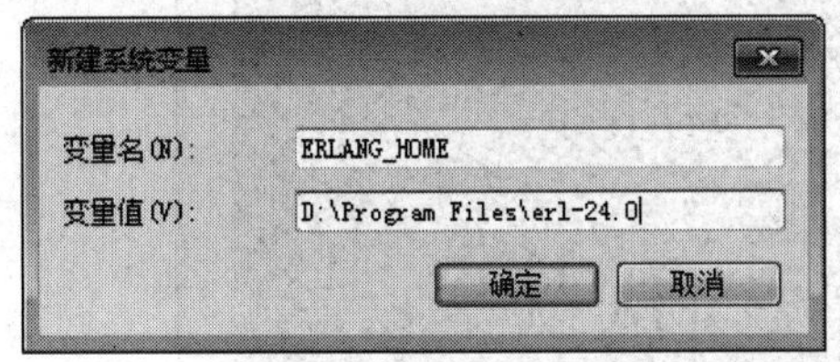

图 8-23 新建 ERLANG_HOME

图 8-24 在 Path 中新增值

运行 RabbitMQ 安装包 rabbitmq-server-3.9.4.exe，按照提示操作即可完成 RabbitMQ 的安装。然后将安装路径下的 sbin 子目录添加到环境变量的 Path 中，值为 D:\Program Files\RabbitMQ Server\rabbitmq_server-3.9.4\sbin（根据自己的实际安装路径）。设置环境变量是为了方便操作系统能找到子目录下的 rabbitmqctl.bat 等命令，如果未配置环境变量，就需要在命令窗口中进入相应文件所在目录，执行相应命令。

在 cmd 命令窗口，进入 RabbitMQ 的 sbin 目录下，运行 rabbitmq-plugins enable rabbitmq_management 命令，开启 rabbitmq_management 插件，为 RabbitMQ 的管理界面提供支持，如图 8-25 所示。

如果运行完命令后，最后一行出现 Plugin Configuration unchanged 的提示，表明插件

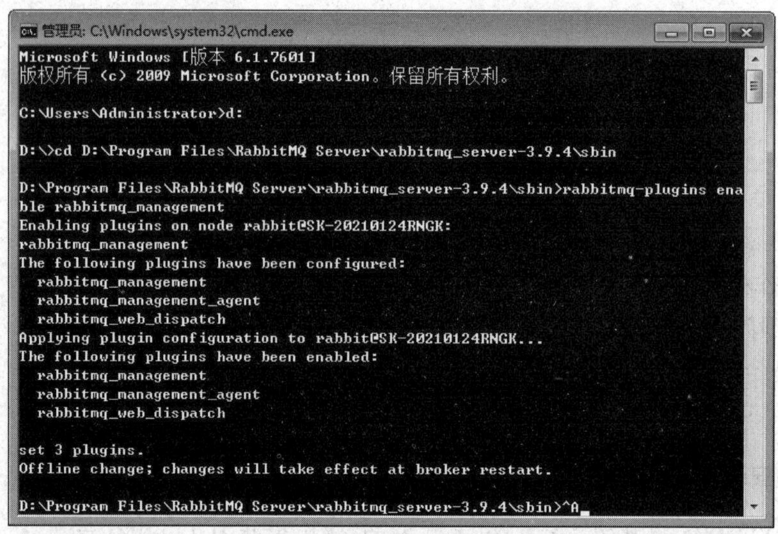

图 8-25　RabbitMQ 启用插件

启用失败,可删除用户目录(C:\Users\<用户名>)下 AppData\Roaming\RabbitMQ 子目录里的内容,重新运行上述命令。

运行 rabbitmq-server.bat 命令启动 RabbitMQ 服务器,RabbitMQ 服务器启动完成后,信息提示如图 8-26 所示。

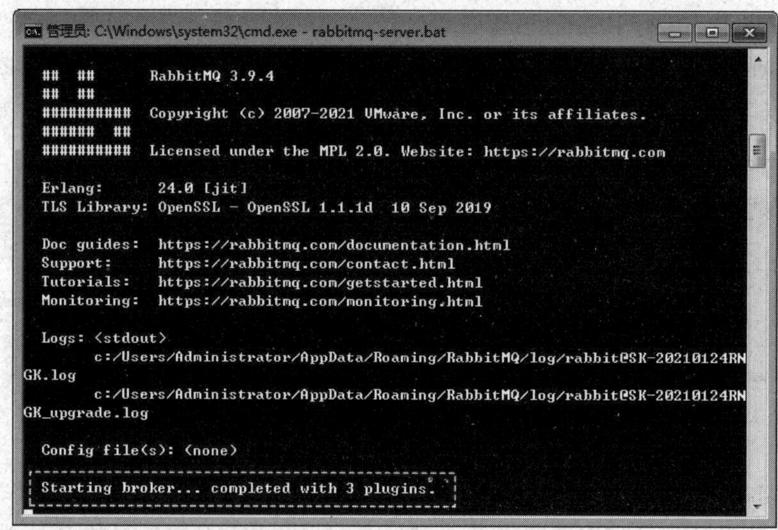

图 8-26　启动 RabbitMQ 服务器

RabbitMQ 默认提供两个端口号 5672 和 15762,其中 5672 用作服务端口号,15762 用作可视化管理端口号。打开浏览器访问 http://localhost:15672/,进入登录页面,如图 8-27 所示。

在该界面中输入默认的账号和密码(guest 和 guest),进入 RabbitMQ 的 Web 管理界面,通过可视化方式查看 RabbitMQ,如图 8-28 所示。

在图 8-28 的可视化管理页面中,显示了 RabbitMQ 版本,用户信息等内容,界面上方有

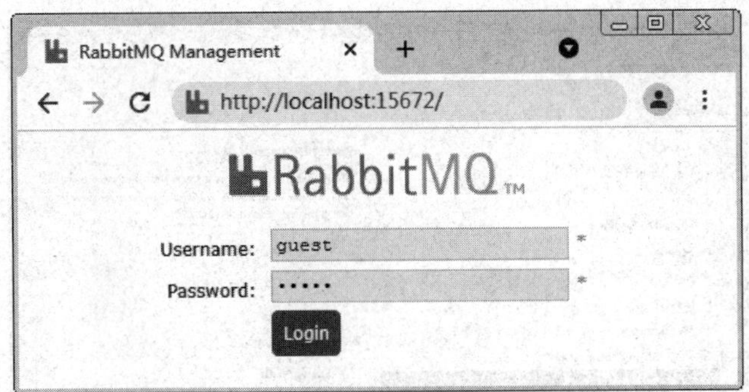

图 8-27　RabbitMQ 可视化登录页面

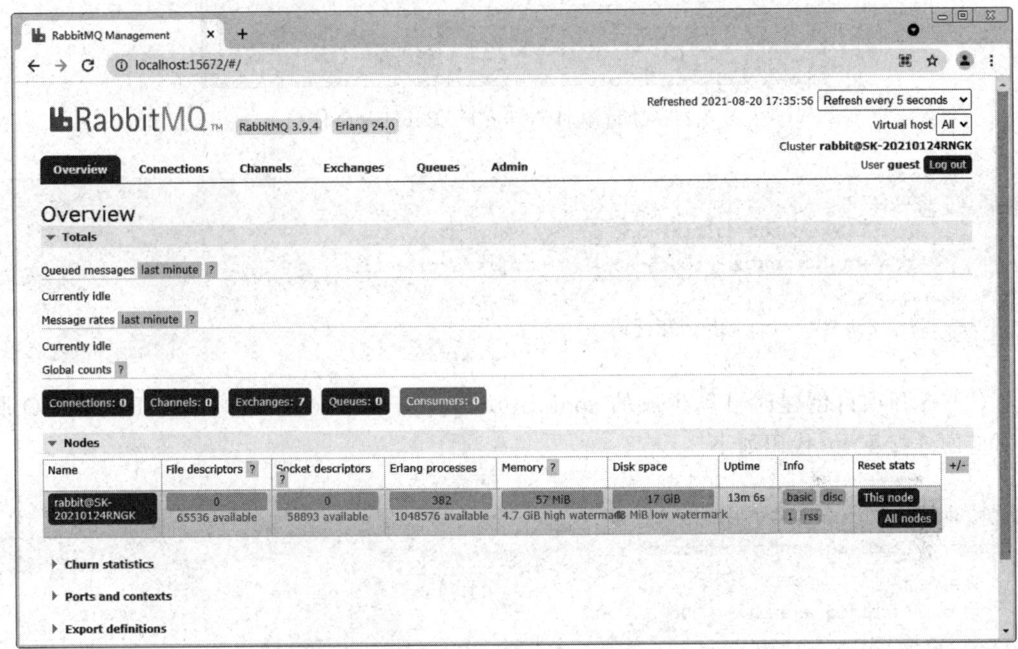

图 8-28　RabbitMQ 可视化 Web 管理页面

Connections、Channels、Exchanges、Queues、Admin 等标签，分别代表查看或管理 RabbitMQ 的连接、内存通道、交换机、消息队列、系统管理。

2. Spring Boot 整合 RabbitMQ 环境搭建

下面对 Spring Boot 整合 RabbitMQ 实现消息服务需要的整合环境进行搭建，步骤如下。

(1) 使用 Spring Initializr 方式创建一个名为 chapter08rabbitmq 的项目，Group 和 Package name 为 com. yzpc，在 Dependencies 依赖中选择 Web 节点下的 Spring Web 依赖，Messaging 节点下的 Spring for RabbitMQ 依赖，单击 Finish 按钮，如图 8-29 所示。

pom. xml 中自动添加的依赖代码如下。

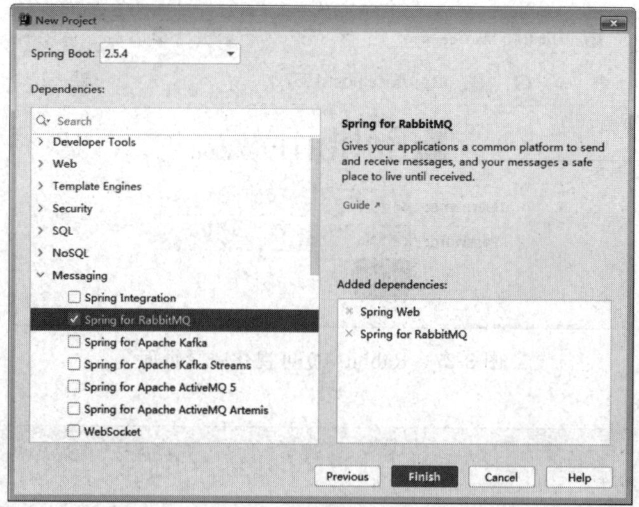

图 8-29 项目添加 Spring for RabbitMQ 依赖

```
<dependency>
    <groupId>org.springframework.boot</groupId>
    <artifactId>spring-boot-starter-amqp</artifactId>
</dependency>
<!-- 省略Web 中Spring Web 依赖代码 -->
```

(2) 打开项目创建时自动生成的 application.properties 配置文件，添加 RabbitMQ 服务对应的连接配置，代码如下。

```
# 配置RabbitMQ 消息中间件连接配置
spring.rabbitmq.host=localhost
spring.rabbitmq.port=5672
spring.rabbitmq.username=guest
spring.rabbitmq.password=guest
#配置RabbitMQ 的虚拟主机路径/,默认可以省略
spring.rabbitmq.virtual-host=/
```

连接的 RabbitMQ 服务端口号为 5672，使用默认用户 guest。

8.3.3 Spring Boot 整合 RabbitMQ 实现

在 RabbitMQ 的后几种工作模式中，消息由 Exchange 进行再分配，Exchange 会根据不同的策略将消息分发到不同 Queue 中。下面选取常用的发布订阅模式、路由模式和主题模式完成在 Spring Boot 项目中的消息服务整合实现。

1. 发布订阅模式(Publisher/Subscribe)

该模式策略把所有到达交换器的消息转发给所有与它绑定的队列。Spring Boot 整合 RabbitMQ 中间件实现消息服务，主要围绕定制中间件、生产者发送消息、消费者接收消息

这 3 个部分进行，具体步骤如下。

(1) 基于配置类方式实现消息发送组件。在项目的 src/main/java/ 目录下的 com.yzpc 包中，新建一个 config 包，并在该包中新建 RabbitMQConfig 的消息配置类，用来定制消息发送相关组件，代码如下。

```
package com.yzpc.config;
import org.springframework.amqp.core.*;
import org.springframework.context.annotation.Bean;
import org.springframework.context.annotation.Configuration;
@Configuration
public class RabbitMQConfig {
    //定义 fanout 类型的转换器
    @Bean
    public FanoutExchange fanoutExchange(){
        return new FanoutExchange("fanout_exchange");
    }
    //定义两个不同的消息队列
    @Bean
    public Queue fanoutQueue1(){
        return new Queue("fanout_queue1");
    }
    @Bean
    public Queue fanoutQueue2(){
        return new Queue("fanout_queue2");
    }
    //将两个不同名称的消息队列与交换器进行绑定
    @Bean
    public Binding fanoutBinding1(){
        return BindingBuilder.bind(fanoutQueue1()).to(fanoutExchange());
    }
    @Bean
    public Binding fanoutBinding2(){
        return BindingBuilder.bind(fanoutQueue2()).to(fanoutExchange());
    }
}
```

上述使用@Bean 注解定制了交换器、队列和队列与交换器的绑定这 3 种类型组件。

(2) 消息生产者发送消息。在项目的测试类 Chapter08rabbitmqApplicationTests 中，创建消息发送者发送消息队列，这里使用 Spring 框架提供的 AmqpTemplate 模板类实现消息发送，代码如下。

```
@Autowired
private AmqpTemplate amqpTemplate;
private String msg = "Hello Fanout!";
@Test
public void fanoutTest(){
    amqpTemplate.convertAndSend("fanout_exchange",null,msg);
    System.out.println("向 fanout_exchange 交换器发送消息: " + msg);
}
```

消息发送由 convertAndSend()方法实现，第一个参数为发送消息的交换器；第二个参数为路由键，因为 Fanout 类型的交换器不处理路由键，这里设置为空，只是简单地将队列绑定到交换器上，每个送到交换器的消息都会被转发到与该交换器绑定的所有队列上；第三个参数为发送的消息内容，接收 Object 类型。

运行单元测试方法 fanoutTest()，在控制台可以看到相应的“向 fanout_exchange 交换器发送消息：Hello Fanout!”的输出语句。验证 RabbitMQ 消息组件的效果，通过 RabbitMQ 可视化页面 Exchanges 选项卡，可以看到一个类型为 fanout 的交换器，名称为 fanout_exchange，其余 7 个交换器是 RabbitMQ 自带的，如图 8-30 所示。单击 fanout_exchange 交换器进入详情页，在 Bindings 节点下，可以看到绑定的两个消息队列，如图 8-31 所示。

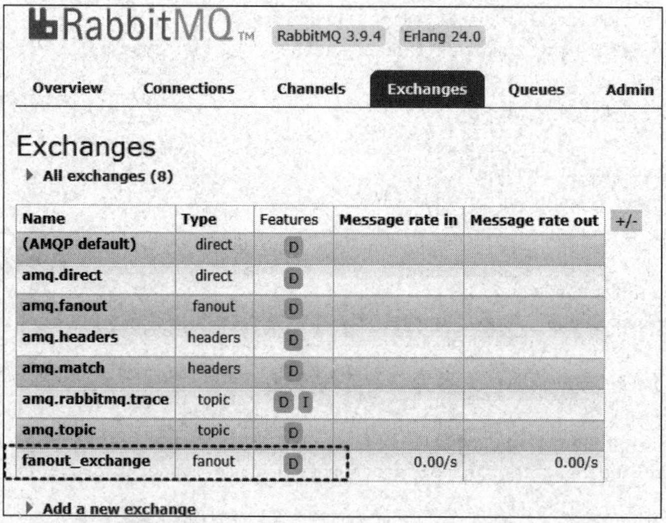

RabbitMQ™ RabbitMQ 3.9.4 Erlang 24.0

Overview Connections Channels Exchanges Queues Admin

Exchanges

▸ All exchanges (8)

Name	Type	Features	Message rate in	Message rate out	+/-
(AMQP default)	direct	D			
amq.direct	direct	D			
amq.fanout	fanout	D			
amq.headers	headers	D			
amq.match	headers	D			
amq.rabbitmq.trace	topic	D I			
amq.topic	topic	D			
fanout_exchange	fanout	D	0.00/s	0.00/s	

▸ Add a new exchange

图 8-30 Exchanges 选项卡效果

在可视化管理页面，单击 Queues 选项卡，查看定制生成的两个消息队列信息，由于目前尚未提供消费者，在发布订阅模式下绑定的两个消息队列中都拥有一条待接收的消息，所以测试类发送的消息会暂存在队列。在图 8-31 中，单击 fanout_queue1 消息队列进入队列详情页，如图 8-32 所示。在 Get Messages 节点下，单击 Get Message(s) 按钮，可以看到消息信息，如图 8-33 所示。

(3) 消息消费者接收消息。在项目 src/main/java/ 目录下的 com.yzpc 包中，新建一个 receiver 包，并在该包中新建 RabbitMQReceiver 业务类，用来实现消息的接收和处理，代码如下。

```
package com.yzpc.receiver;
import org.springframework.amqp.rabbit.annotation.RabbitListener;
import org.springframework.stereotype.Component;
@Component
public class RabbitMQReceiver {
    // 发布订阅工作模式下接收、处理业务
    @RabbitListener(queues = "fanout_queue1")
```

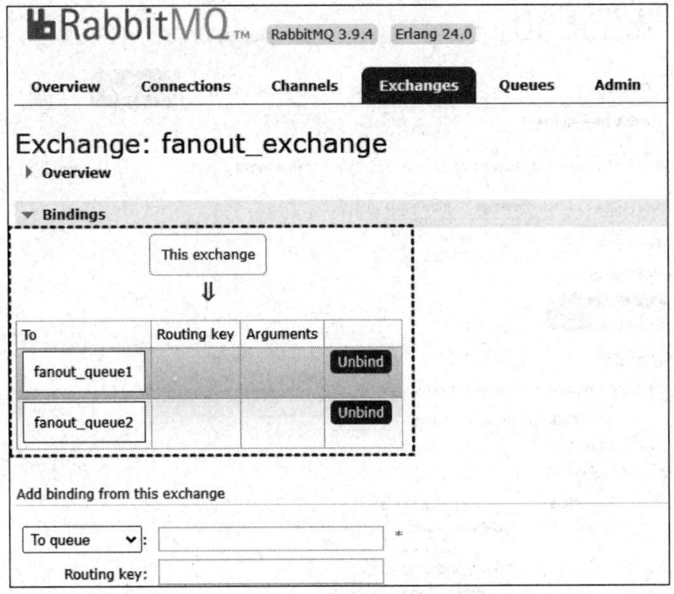

图 8-31 交换器详情页面

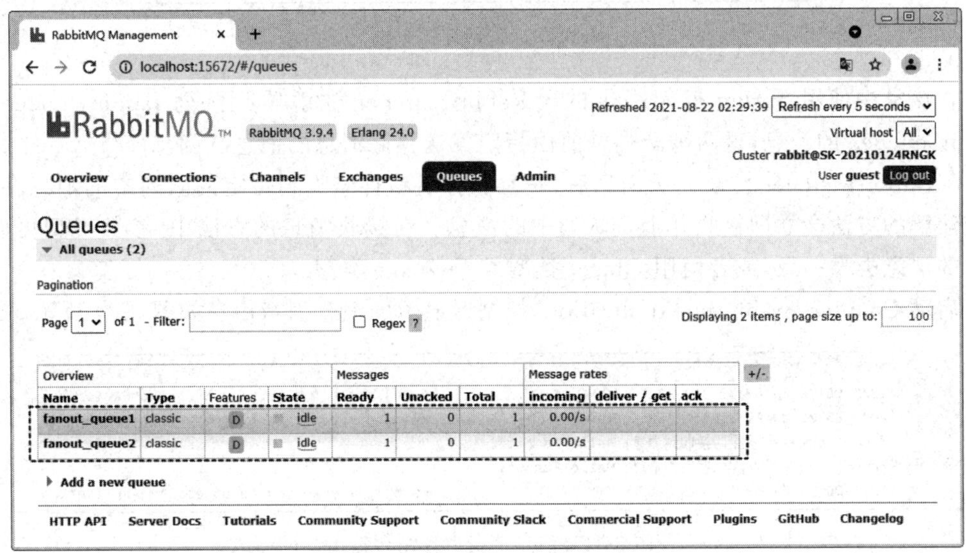

图 8-32 Queues 队列面板信息

```
    public void receiveFanout1(String msg){
        System.out.println("fanout_queue1 队列监听接收到消息: " + msg);
    }
    @RabbitListener(queues = "fanout_queue2")
    public void receiveFanout2(String msg){
        System.out.println("fanout_queue2 队列监听接收到消息: " + msg);
    }
}
```

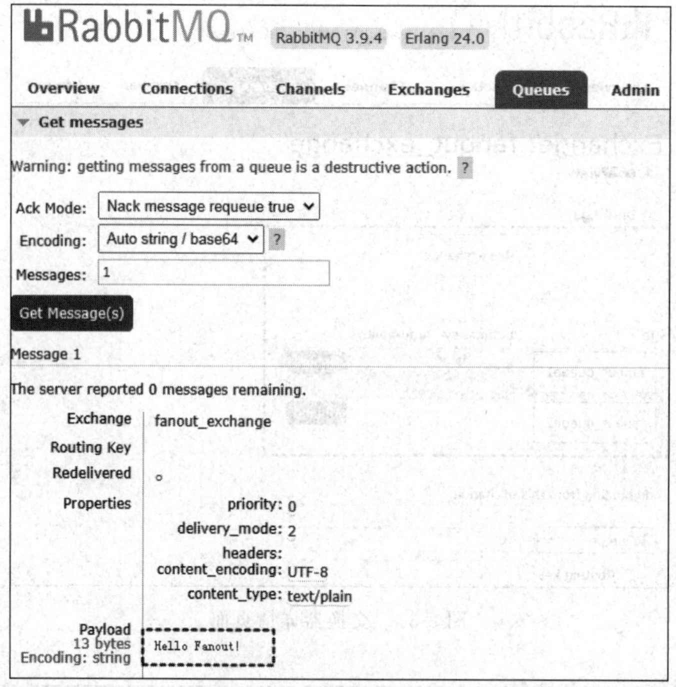

图 8-33 队列详情页面

在该类中使用 Spring 框架提供的@RabbitListener 注解监听队列 fanout_queue1 和 fanout_queue2 的消息，这两个队列是前面指定发送存储消息的消息队列。

@RabbitListener 注解监听消息后，一旦服务启动且监听到指定的队列中的消息存在，对应注解的方法会立即接收并消费队列中的消息。在接收消息的方法中，参数类型可以与发送的消息类型一致，或者使用 Object 类型或 Message 类型。

启动 Chapter08rabbitmqApplication，控制台显示的消息消费信息如图 8-34 所示。

```
Console    Endpoints
2021-08-22 03:09:33.340  INFO 7004 --- [           main] o.s.a.r.c.CachingConnectionFactory
2021-08-22 03:09:33.440  INFO 7004 --- [           main] o.s.a.r.c.CachingConnectionFactory
fanout_queue2队列监听接收到消息: Hello Fanout!
fanout_queue2队列监听接收到消息: Hello Fanout!
2021-08-22 03:09:33.502  INFO 7004 --- [           main] com.yzpc.Chapter08rabbitmqApplication
```

图 8-34 消息消费后控制台输出

项目启动后，消息接收者监听到消息队列中存在的两条消息，并进行了各自的消费。同时，在 RabbitMQ 的可视化管理页面，单击 Queues 选项卡，查看两个消息队列信息，发现表格 Messages 字段下面两个队列中的消息已经被消费，如图 8-35 所示。

提示：在 RabbitMQReceiver 类的代码中，常用的@RabbitListener 注解监听指定队列的消费情况，这种方式会在监听到消息后立即进行消费处理。除此之外，还可以使用 RabbitTemplate 模板类的 receiveAndConvert(String queueName)方法手动消费指定队列中的消息。

图 8-35 测试方法执行结果 1

2. 路由模式(Routing)

路由模式策略是将消息队列绑定到一个 direct 类型的交换器上,当消息到达该交换器时会被转发到与该条消息 routing key 相同的队列上。这里使用基于注解的方式实现路由模式的整合进行讲解。

(1) 定制消息组件和消息消费者。打开 com. yzpc. receiver 包中的 RabbitMQReceiver 业务类,使用@RabbitListener 注解及其相关属性定制路由模式的消息组件,并编写消息消费者接收消息的方法,代码如下。

```
import org.springframework.amqp.rabbit.annotation.Exchange;
import org.springframework.amqp.rabbit.annotation.Queue;
import org.springframework.amqp.rabbit.annotation.QueueBinding;
…
// 2.1 路由模式消息接收,处理业务 1
@RabbitListener(bindings = @QueueBinding(value = @Queue("routing_queue1"),
        exchange = @Exchange(value = "routing_exchange",type = "direct"),
        key = "routing_key1"))
public void receiveRouting1(String message) {
    System.out.println("routing_queue1 队列监听接收消息: " + message);
}
// 2.2 路由模式消息接收,处理业务 2
@RabbitListener(bindings = @QueueBinding(value = @Queue("routing_queue2"),
        exchange = @Exchange(value = "routing_exchange",type = "direct"),
        key = {"routing_key1","routing_key2","routing_key3"}))
public void receiveRouting2(String message) {
    System.out.println("routing_queue2 队列监听接收消息: " + message);
}
```

在消息业务处理类中,新增 2 个用来处理 Routing 路由模式的消息接收者方法,并使用@RabbitListener 注解及其相应属性定制了路由模式下的消息服务组件。与发布订阅模式下的注解相比,路由模式下的交换器类型 type 属性是 direct,必须指定 key 属性,每个消息队列可以映射多个路由键。

(2) 消息生产者发送消息。在项目测试类 Chapter08rabbitmqApplicationTests 中,使用 RabbitTemplate 模板类实现路由模式下的消息发送,代码如下。

```
@Autowired
private RabbitTemplate rabbitTemplate;
private String routing_msg = "routing send routing_key1 message!";
```

```
@Test
public void routingTest(){
    rabbitTemplate.convertAndSend("routing_exchange","routing_key1",
routing_msg);
    System.out.println("向 routing_exchange 交换器发送消息：" + routing_msg);
}
```

在测试类中，调用 RabbitTemplate 的 convertAndSend()方法发送消息，在路由工作模式下发送消息时，必须指定第二个路由键参数，并且要与消息队列映射中的路由键保持一致，否则发送的消息将会丢失。这里使用的是 routing_key1 路由键，两个消息接收者方法都可以正常接收并消费发送端发送的消息。

执行上述测试方法 routingTest()，控制台输出显示如图 8-35 所示。

从图 8-35 可以看到两个消费者都对 routing_key1 路由键的消息进行了消费。

修改 routingTest() 方法中的消息传递参数，调整发送 routing_key2 级别的消息(同时修改路由键和消息内容)，再次运行 routingTest() 方法，控制台输出显示如图 8-36 所示。

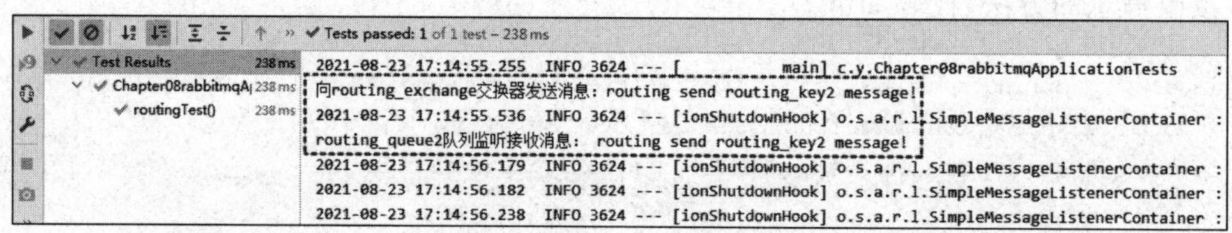

图 8-36　测试方法执行结果 2

在图 8-36 中，控制台输出使用 routing_key2 路由键的信息，说明只有配置映射 routing_key2 路由键的消息消费者的方法接收了消息。

打开 RabbitMQ 可视化管理页面查看定制的 Routing 模式的消息组件，使用注解方式同样自动生成路由模式下的消息组件，并进行了自动绑定，如图 8-37 所示。

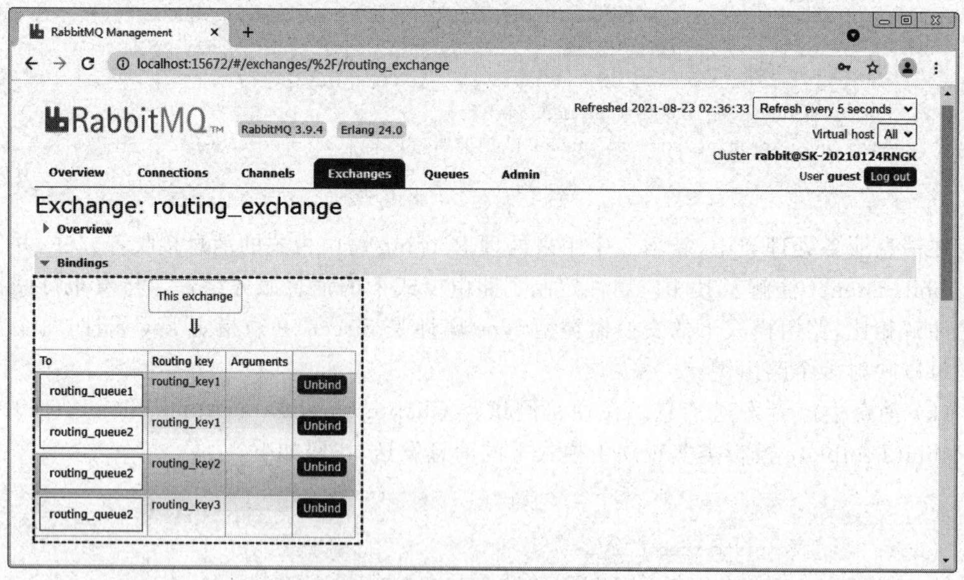

图 8-37　路由模式下的定制组件

8.4 小结

本章主要介绍了消息服务的相关内容。通过 Spring Boot 整合 JMS，介绍了 JMS 的概念，ActiveMQ 消息中间件，以及具体的整合 JMS 过程。通过 Spring Boot 整合 AMPQ，介绍了 RabbitMQ 消息中间件的工作模式，RabbitMQ 的安装，以及几种不同工作模式下的整合过程。

Spring Boot任务管理

在开发 Web 应用时,多数应用都具备任务调度的功能,常见的任务包括异步任务、定时任务和邮件任务。下面以数据库报表为例介绍任务调度如何帮助改善系统设计。报表可能是错综复杂的,用户可能需要花费很长时间才能找到所需的报表数据,此时,可以在报表应用中添加异步任务以减少用户等待时间,从而提升用户体验;除此之外,还可以在报表应用中添加定时任务和邮件任务,以便用户可以安排在任何他们需要的时间定时生成报表,并在 Email 中发送。下面将介绍如何使用 Spring Boot 开发这些常见的任务。

9.1 异步任务

在 Web 应用开发中,大多数情况通过同步方式完成数据交互处理,但是,当处理与第三方系统交互时,容易造成响应迟缓的情况,之前大部分使用多线程完成此类任务,除此之外,还可以使用异步调用的方式。根据异步处理方式的不同,可以将异步任务的调用分为无返回值异步任务调用和有返回值异步任务调用,接下来在 Spring Boot 项目中分别针对这两种方式进行介绍。

1. 无返回值异步任务调用

在实际开发中,项目可能会向新注册用户发送短信验证码,这时,可以使用异步任务调用的方式实现,一方面是因为用户对这个时效性要求不是特别高;另一方面,在特定时间范围内如果没有收到验证码,用户可以选择再次发送验证码。下面使用 Spring Boot 框架演示这种场景需求,进一步说明无返回值的异步任务调用。

(1) 新建一个 Spring Boot 工程 chapter09,Group 和 Package name 为 com.yzpc,在 Dependencies 依赖中选择 Web 节点下的 Spring Web,单击 Finish 按钮。

(2) 在项目的 src/main/java/ 路径下的 com.yzpc 包中,新建一个 service 包,并在该包中新建一个业务实现类 AsyncService,添加 asyncMethod()的方法,代码如下。

```
package com.yzpc.service;
package com.yzpc.service;
import org.springframework.stereotype.Service;
```

```
@Service
public class AsyncService {
    public void asyncMethod(){
        System.out.println("无返回值异步方法执行……");
        try {
            Thread.sleep(6000);
        } catch (InterruptedException e) {
            e.printStackTrace();
        }
        System.out.println("无返回值异步方法执行结束!");
    }
}
```

(3) 在项目的 src/main/java/路径下的 com.yzpc 包中,新建一个 controller 包,并在该包中新建一个 AsyncController 类,添加 async()的方法,代码如下。

```
package com.yzpc.controller;
import com.yzpc.service.AsyncService;
import org.springframework.beans.factory.annotation.Autowired;
import org.springframework.web.bind.annotation.GetMapping;
import org.springframework.web.bind.annotation.RestController;
@RestController
public class AsyncController {
    @Autowired
    private AsyncService asyncService;
    @GetMapping("/async")
    public String async(){
        System.out.println("请求接收到……");
        asyncService.asyncMethod();
        System.out.println("完成响应!");
        return "无返回值异步请求方法调用结束!";
    }
}
```

(4) 启动 chapter09 的项目,在浏览器访问 http://localhost:8080/async,此时发现浏览器无显示,如图 9-1 所示,控制台输出内容如图 9-2 所示。

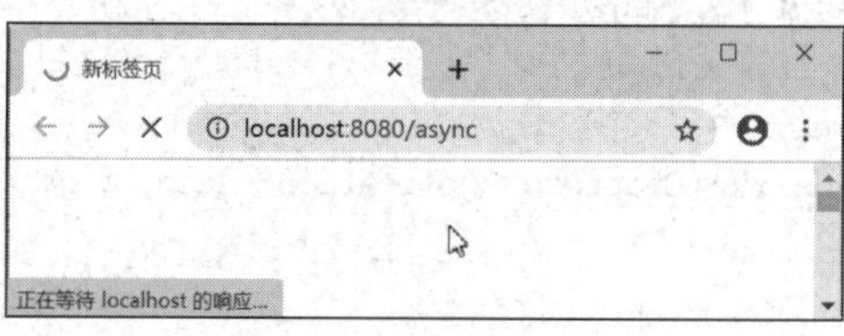

图 9-1 同步调用浏览器无显示

大约 6 秒钟后,浏览器显示返回的字符串,如图 9-3 所示,控制台输出内容如图 9-4 所示。控制台输出内容和浏览器显示内容分别为单独的图片。

(5) 为了实现异步任务,需要在 AsyncService 类的 asyncMethod()方法上添加@Async 注解,实现将 asyncMethod()方法标注为异步方法,代码如下。

```
Console  Endpoints
2021-08-26 17:27:41.952  INFO
2021-08-26 17:27:45.895  INFO
2021-08-26 17:27:45.897  INFO
2021-08-26 17:27:45.900  INFO
请求接收到……
无返回值异步方法执行……
```

图 9-2　同步调用控制台输出效果

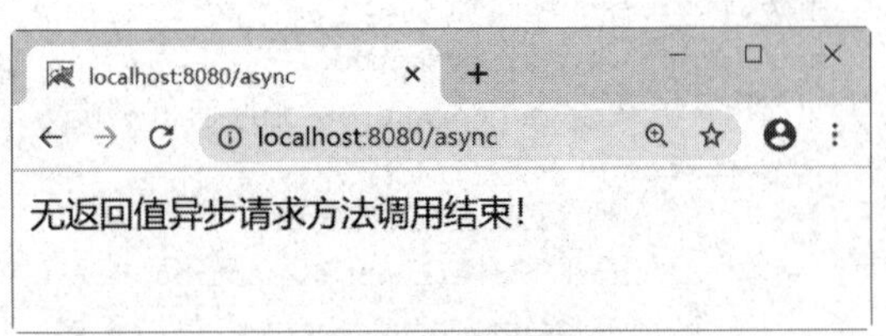

图 9-3　同步调用浏览器显示效果

```
Console  Endpoints
2021-08-26 17:27:45.897  INFO
2021-08-26 17:27:45.900  INFO
请求接收到……
无返回值异步方法执行……
无返回值异步方法执行结束!
完成响应!
```

图 9-4　同步调用控制台输出效果

```
@Service
public class AsyncService {
    // 无返回值的异步方法
    @Async
    public void asyncMethod(){
        // 省略已实现的代码,实现代码与前面一致
    }
}
```

在 Spring Boot 中希望异步方法生效，还需在项目启动类上使用@EnableAsync 注解开启基于注解的异步任务支持，代码如下。

```
package com.yzpc;
import org.springframework.boot.SpringApplication;
import org.springframework.boot.autoconfigure.SpringBootApplication;
import org.springframework.scheduling.annotation.EnableAsync;
@SpringBootApplication
@EnableAsync      //开启基于注解异步任务支持
public class Chapter09Application {
    public static void main(String[] args) {
        SpringApplication.run(Chapter09Application.class, args);
    }
}
```

重新启动项目，在浏览器访问 http://localhost:8080/async，很短时间内完成了主流程的执行，并向页面响应返回的字符串，如图 9-5 所示，控制台输出内容如图 9-6 所示。

需要说明的是，无返回值异步方法在被主流程方法调用时，主流程方法不会阻塞，而是继续向下执行主流程方法内容，直接向页面响应结果，而调用的异步方法会作为一个子线程

图 9-5 无返回值异步任务调用浏览器显示

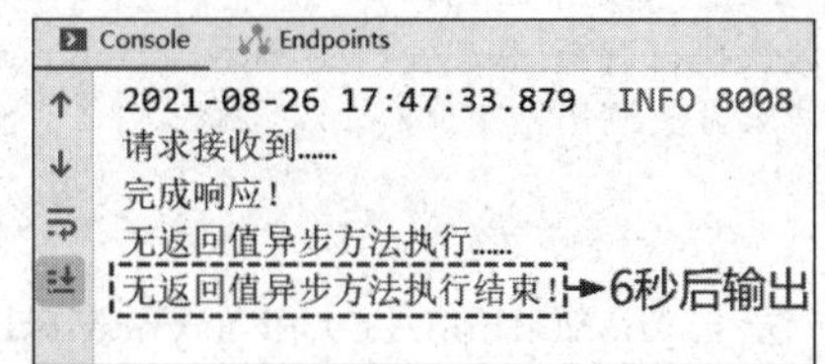

图 9-6 无返回值异步任务调用控制台输出

单独执行，直到异步方法执行完成。

2. 有返回值异步任务调用

在实际开发中，项目可能会涉及有返回值的异步任务调用。有返回值必须用 Future 泛型封装，如 new AsyncResult<Integer>(count)，当想获取封装的值时用 Future 对象调用其相应方法。下面通过示例来介绍有返回值的异步任务调用。

(1) 在 1.(2)创建的 AsyncService 异步任务处理类中，添加两个有返回值的异步任务业务处理方法，代码如下。

```
//有返回值的异步方法
@Async
public Future<Integer> asyncMethodA() throws Exception{
    System.out.println("有返回值异步方法 A 执行……");
    Thread.sleep(6000);
    System.out.println("有返回值异步方法 A 执行结束!");
    return new AsyncResult<Integer>(6);
}
@Async
public Future<Integer> asyncMethodB() throws Exception{
    System.out.println("有返回值异步方法 B 执行……");
    Thread.sleep(9000);
    System.out.println("有返回值异步方法 B 执行结束!");
    return new AsyncResult<Integer>(9);
}
```

上述两个方法都是用@Async 注解标记为异步方法，两个方法都会有一定的处理时间，并且返回一个 Future<Integer>结果。

(2) 在 1.(3)创建的异步任务业务处理类 AsyncController 中，编写业务数据计算的请求处理方法 asyncReturnValue()，代码如下。

```
@GetMapping("/asyncReturnValue")
public String asyncReturnValue() throws Exception{
    Long startTime = System.currentTimeMillis();
    System.out.println("请求接收到……");
    Future<Integer> fA = asyncService.asyncMethodA();
    Future<Integer> fB = asyncService.asyncMethodB();
    int sum = fA.get() + fB.get();
    System.out.println("完成响应!最终合计结果为: " + sum);
    Long endTime = System.currentTimeMillis();
    System.out.println("此次执行共耗时: " + (endTime - startTime));
    return "有返回值异步请求方法调用结束!";
}
```

在该方法中调用异步方法 asyncMethodA()和 asyncMethodB()输出主流程的耗时时长。

(3) 启动项目,在浏览器访问 http://localhost:8080/asyncReturnValue,测试异步任务请求,会发现浏览器上响应输出“有返回值异步请求方法调用结束!”信息需要一段时间,如图 9-7 所示,此时查看控制台内容如图 9-8 所示。

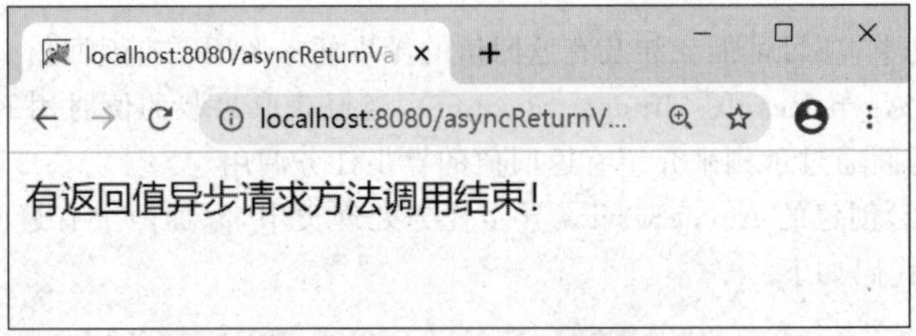

图 9-7 有返回值异步任务调用浏览器显示

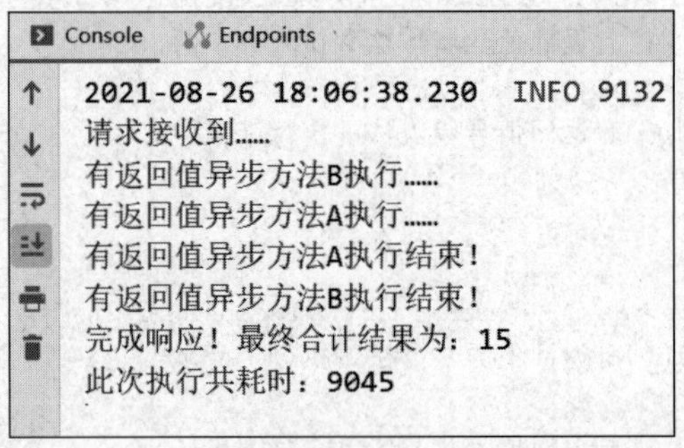

图 9-8 有返回值异步任务调用控制台输出

从图 9-8 的调试结果看,执行 asyncReturnValue() 方法并调用异步方法处理业务数据计算时,需要耗费一定的时间(9045 毫秒)完成主流程的执行,并向页面响应结果,在主流程输出结果之前一直等待 asyncMethodA()和 asyncMethodB()两个异步方法的异步调用处理和结果计算。

需要说明的是,上述的异步方法有返回值,当返回值较多时,主流程在执行异步方法时会有短暂阻塞,需要等待并获取异步方法的返回结果,而调用的两个异步方法会作为两个子线程并行执行，直到异步方法执行完成并返回结果，这样主流程会在最后一个异步方法返回结果后跳出阻塞状态。

9.2 定时任务

在日常的项目开发中，经常会涉及一些需要做到定时执行的代码，例如，自动将超过24小时未付款的订单改为取消状态，自动将超过14天客户未签收的订单改为已签收状态，服务器数据定时在晚上零点备份等。通常可以使用 Spring 框架提供的 scheduling Tasks 实现定时任务的处理。

1. 定时任务介绍

Spring 框架的定时任务调度功能支持配置和注解两种方式，Spring Boot 不仅继承了 Spring 框架定时任务调度功能，而且可以更好地支持注解方式的定时任务，在实现定时任务时需要了解几种和定时任务相关的注解，具体如下。

(1) @Enablescheduling 注解用于开启基于注解方式的定时任务支持，该注解是 Spring 框架提供的，主要用在项目启动类上。

(2) @Scheduled 注解用于配置定时任务的执行规则，该注解主要用在定时业务方法上。

@Scheduled 注解提供有多个属性，主要属性如下。

- cron 属性：类似 cron 表达式，可以定制定时任务触发的秒、分钟、小时、月中的日、月、周中的日。
- zone 属性：主要与 cron 属性配合使用，解析 cron 属性值的时区。
- fixedDelay 和 fixedDelayString 属性：两种属性作用类似，表示上一次任务执行完毕时间点之后多长时间再执行。
- fixedRate 和 fixedRateString 属性：两种属性作用类似，表示上一次开始执行时间点之后多长时间再执行。
- initialDelay 和 initialDelayString 属性：两种属性作用类似，主要是与 fixedDelay 或者 fixedRate 属性配合使用，指定定时任务第一次延迟多长时间后再执行，然后再按照各自相隔时间重复执行任务。

2. 定时任务示例

通过在 Spring Boot 框架中实现简单的定时任务，步骤如下。

(1) 在 chapter09 项目的 com.yzpc.service 的包下新建一个定时任务管理的业务处理类 ScheduleService，并在其中添加对应的定时任务处理方法，代码如下。

```
package com.yzpc.service;
import org.springframework.scheduling.annotation.Scheduled;
import org.springframework.stereotype.Service;
import java.util.Date;
```

```
@Service
public class ScheduleService {
    private Integer number1 = 1;
    private Integer number2 = 1;
    private Integer number3 = 1;
    @Scheduled(fixedRate = 60000)
    public void taskMethod1() {
        System.out.println("fixedRate 第" + (number1++) + "次执行,当前时间为: " + new
Date());
    }
    @Scheduled(fixedDelay = 60000)
    public void taskMethod2() throws InterruptedException {
        System.out.println("fixedDelay 第" + (number2++) + "次执行,当前时间为: " + new
Date());
        Thread.sleep(10000);
    }
    @Scheduled(cron = "0 * * * * *")
    public void taskMethod3(){
        System.out.println("cron 第" + (number3++) + "次执行,当前时间为: " + new Date());
    }
}
```

使用@Scheduled注解声明了3个定时任务方法,执行规则基本相同,都是每隔1分钟重复执行一次定时任务。在使用fixedDelay属性的方法taskMethod2()中,模拟该定时任务处理耗时为10秒。Spring Boot使用定时任务相关注解时,必须引入Spring框架依赖,因该项目已经引入了Web依赖,可以直接使用相关注解。

(2) 为使Spring Boot基于注解方式的定时任务生效,还需在项目启动类上使用@EnableScheduling注解开启基于注解的定时任务支持,添加的注解代码如下。

```
@EnableScheduling  //开启基于注解的定时任务支持
```

(3) 启动项目,项目启动过程中仔细查看控制台输出,如图9-9所示。

```
Console  Endpoints
2021-08-27 13:10:13.453  INFO 8268 --- [           main] w.s.c.ServletWebServerApplicationContext :
2021-08-27 13:10:13.944  INFO 8268 --- [           main] o.s.b.w.embedded.tomcat.TomcatWebServer  :
fixedRate第1次执行，当前时间为: Fri Aug 27 13:10:13 CST 2021
fixedDelay第1次执行，当前时间为: Fri Aug 27 13:10:13 CST 2021
2021-08-27 13:10:13.968  INFO 8268 --- [           main] com.yzpc.Chapter09Application            :
cron第1次执行，当前时间为: Fri Aug 27 13:11:00 CST 2021
fixedRate第2次执行，当前时间为: Fri Aug 27 13:11:13 CST 2021
fixedDelay第2次执行，当前时间为: Fri Aug 27 13:11:23 CST 2021
cron第2次执行，当前时间为: Fri Aug 27 13:12:00 CST 2021
fixedRate第3次执行，当前时间为: Fri Aug 27 13:12:13 CST 2021
fixedDelay第3次执行，当前时间为: Fri Aug 27 13:12:33 CST 2021
```

图9-9 定时任务调用

从图9-9的输出结果看出,项目启动后,配置@Scheduled属性的fixedRate和fixedDelay属性的定时方法会立即执行一次,配置@cron属性的定时方法会在整数分钟时间点首次执行；配置 fixedRate 和 cron 属性的方法会每隔 1 分钟重复执行一次定时任务，配置 fixedDelay 属性的方法是在上一次方法执行完成后再相隔 1 分钟重复执行一次定时任务。

9.3 邮件任务

对于开发者来说，经常会遇到需要发送 Email 邮件的情况，如注册、找回密码、发送验证码、向客户发送邮件、定期向系统维护人员发送报告等。在早期的 Java 开发中，通常使用 JavaMail 相关 API 实现邮件发送，但是配置比较烦琐，Spring 中提供了 JavaMailSender 用来简化配置，Spring Boot 框架对 Spring 提出的邮件发送服务也进行了整合支持。下面就针对 Spring Boot 框架整合支持的邮件任务进行介绍。

1. 发送邮件的配置准备

邮件的发送和接收需要遵循相关协议，发送邮件通过 SMTP 协议将邮件发送至邮件服务器，MIME 协议是对 SMTP 协议的一种补充，如发送图片、附件等，收件人通过 POP 协议从邮件服务器收取邮件。各大邮件运营商都有其对应的安全系统，必须获得其对应的客户端授权码才能使用。获得授权码后，在项目中配置 SMTP 服务协议及主机账户，在项目中就可以使用各大邮件运营商进行邮件发送。

下面以 163 邮箱为例介绍发送邮件的配置。首先登录 163 邮箱，进入邮件页面，单击“设置”按钮下的 POP3/SMTP/IMAP 选项，找到 POP3/SMTP 服务，单击后面的“开启”按钮，如图 9-10 所示。

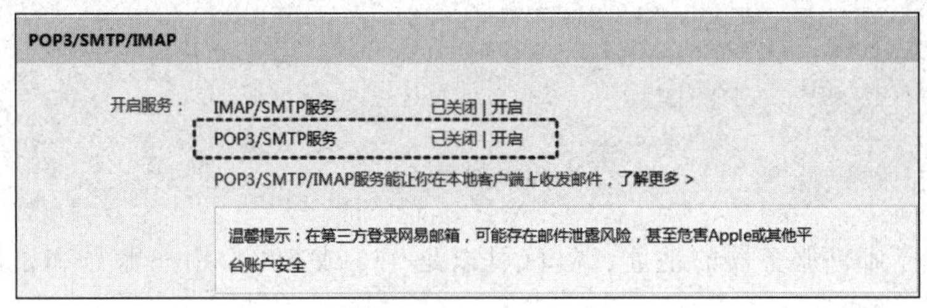

图 9-10 开启 POP3/SMTP 服务

在图 9-10 中，单击“开启”按钮后，会进入验证过程，根据引导步骤发送短信，验证成功后即可得到自己 163 邮箱的客户端授权码，保存授权码以便后期使用，如图 9-11 所示。

其他邮箱的授权码获取操作大同小异。

2. 发送普通邮件

普通邮件是指纯文本邮件，在定制普通邮件时，只需要指定收件人邮箱账号、邮件标题和邮件内容即可。下面使用 Spring Boot 框架实现普通邮件的发送。

(1) 在 chapter09 项目的 pom. xml 依赖文件中，添加 Spring Boot 整合支持的邮件服务依赖启动器 spring-boot-starter-mail，代码如下。

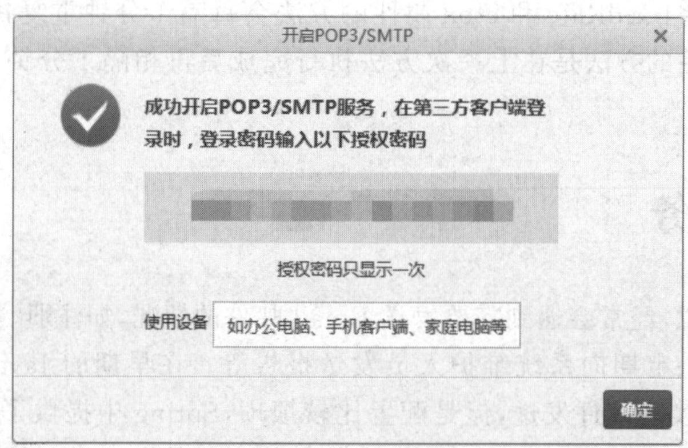

图 9-11 客户端授权码

```
<dependency>
    <groupId>org.springframework.boot</groupId>
    <artifactId>spring-boot-starter-mail</artifactId>
</dependency>
```

(2) 在项目的 application.properties 配置文件中，添加发件人的邮箱服务器的基本信息配置，代码如下。

```
# 发件人邮箱服务器的配置
spring.mail.host=smtp.163.com
spring.mail.port=465
# 配置邮箱的账号和密码(密码为加密后的授权码)
spring.mail.username=shikham66@163.com
spring.mail.password=XRP*QSL*TKO*YBI*        # 邮箱的客户端授权码
spring.mail.default-encoding=UTF-8
# 启用SSL认证
spring.mail.properties.mail.smtp.ssl.enable=true
```

这里配置了邮件服务器的地址、端口(可以是 465 或 587)、用户账号和密码(客户端授权码)以及默认编码等。

(3) 在项目的 com.yzpc.service 的包下新建一个 MailService 的邮件业务处理类，并在该类中编写发送普通邮件的业务方法，代码如下。

```
package com.yzpc.service;
import org.springframework.beans.factory.annotation.Autowired;
import org.springframework.beans.factory.annotation.Value;
import org.springframework.mail.MailException;
import org.springframework.mail.SimpleMailMessage;
import org.springframework.mail.javamail.JavaMailSender;
import org.springframework.stereotype.Service;
@Service
public class MailService {
    //提供发送邮件的简单抽象,使用JavaMailSender接口,这里直接注入使用
```

```
    @Autowired
    private JavaMailSender mailSender;
    @Value("${spring.mail.username}")
    private String from;
    public void sendSimpleMail(String to,String subject,String content){
        //创建SimpleMailMessage对象
        SimpleMailMessage message = new SimpleMailMessage();
        message.setFrom(from);                          //邮件发送
        message.setTo(to);                              //邮件接收
        message.setSubject(subject);                    //邮件主题
        message.setText(content);                       //邮件内容
        try {
            //通过JavaMailSender的send方法把邮件发送出去
            mailSender.send(message);
            System.out.println("普通邮件发送成功!");
        } catch (MailException e) {
            System.out.println("普通邮件发送失败!" + e.getMessage());
            e.printStackTrace();
        }
    }
}
```

Spring Boot 提供发送邮件的抽象，使用 JavaMailSender 接口，这里直接注入使用。普通邮件通过 SimpleMailMessage 对象封装打包数据，通过 JavaMailSender 类将数据发送出去。

(4) 在项目的测试类中添加一个测试方法，调用 MailService 类中发送普通邮件的方法，代码如下。

```
package com.yzpc;
import com.yzpc.service.MailService;
import org.junit.jupiter.api.Test;
import org.springframework.beans.factory.annotation.Autowired;
import org.springframework.boot.test.context.SpringBootTest;
@SpringBootTest
public class Chapter09ApplicationTests {
    @Autowired
    private MailService mailService;
    @Test
    public void sendSimpleMail(){
        mailService.sendSimpleMail("15737337@qq.com","主题: 普通邮件","这是一封普通邮件
发送的测试……");
    }
}
```

注意：测试类中的类和方法都要设置成 public 才可以运行，执行 sendSimpleMail()测试方法，即可在控制台看到“普通邮件发送成功!”的提示，如图 9-12 所示。打开接收邮件的邮箱，可以看到已接收到测试邮件，如图 9-13 所示。

```
Chapter09ApplicationTests.sendSimpleMail
✔ Tests passed: 1 of 1 test – 768 ms

2021-08-28 17:58:39.889  INFO 11148 --- [
2021-08-28 17:58:39.892  INFO 11148 --- [
2021-08-28 17:58:48.351  INFO 11148 --- [
普通邮件发送成功!

Process finished with exit code 0
```

图 9-12　控制台提示发送成功

主题：普通邮件
发件人：shikham66 <shikham66@163.com>
时　间：2021年8月28日（星期六）下午5：58
收件人：shi <15737337@qq.com>

这是一封普通邮件发送的测试……

图 9-13　查看邮箱普通邮件发送成功

3. 发送带附件邮件

（1）很多时候，在发送邮件时，需要附带附件一起发送，通过 addAttachment()方法即可添加附件，在 MailService 类中添加 sendAttachmentMail()方法，代码如下。

```
public void sendAttachmentMail(String to,String subject,String content,
String filePath){
    //定义复杂邮件信息 MimeMessage
    MimeMessage message = mailSender.createMimeMessage();
    try {
        //使用 MimeMessageHelper 帮助类,并设置 multipart 多部件使用为true
        MimeMessageHelper helper = new MimeMessageHelper(message,true);
        helper.setFrom(from);
        helper.setTo(to);
        helper.setSubject(subject);
        helper.setText(content,true);
        //设置邮件附件,添加附件,发送邮件
        FileSystemResource file = new FileSystemResource(new File(filePath));
        String fileName = file.getFilename();
        helper.addAttachment(fileName,file);
        mailSender.send(message);
        System.out.println("带附件邮件发送成功!");
    } catch (Exception e) {
        System.out.println("带附件邮件发送失败!" + e.getMessage());
        e.printStackTrace();
    }
}
```

这里使用的 MimeMessageHelper 简化了邮件配置，它的构造方法的第二个参数 true 表示构造一个 multipart message 类型的邮件，包含多个正文、附件以及内嵌资源，邮件的表现形式更加丰富，通过 addAttachment()方法添加附件。

（2）在项目测试类中添加测试方法，调用 MailService 类中发送带附件邮件的方法，代码如下。

```
@Test
public void sendAttachmentMail(){
```

```
    String content = "< html >< body >< h3 > < font color = \"blue\">" + "这是带附件的邮件,请查
看附件" + "</font ></h3 ></body ></html >";
    String filePath = "D:\\开学注意事项.docx";
    mailService.sendAttachmentMail("15737337@qq.com","主题: 带附件的邮件测试",content,
filePath);
}
```

(3) 执行 sendAttachmentMail()测试方法,可以在控制台看到“带附件邮件发送成功!”的提示,如图 9-14 所示。打开接收邮件的邮箱,可以看到已接收到带有附件的测试邮件,如图 9-15 所示。

```
✔ Tests passed: 1 of 1 test – 2 sec 607 ms
带附件邮件发送成功!

Process finished with exit code 0
```

图 9-14 控制台提示发送成功

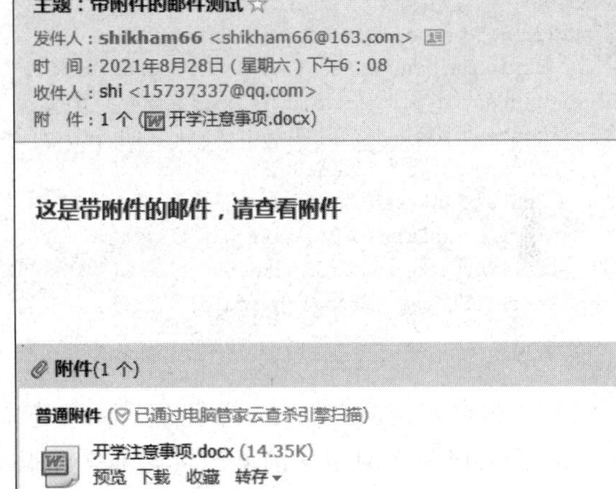

图 9-15 带附件邮件发送成功

4. 发送带图片邮件

(1) 在发送邮件的时候,有时需要在正文中插入图片,使用 FileSystemResource 可以实现这一功能,在 MailService 类中添加 sendInlineResourceMail()方法,代码如下。

```
public void sendInlineResourceMail(String to,String subject,
String content,String rscPath,String rscId){
    MimeMessage message = mailSender.createMimeMessage();
    try {
        MimeMessageHelper helper = new MimeMessageHelper(message,true);
        helper.setFrom(from);
        helper.setTo(to);
        helper.setSubject(subject);
        helper.setText(content,true);
        // 设置邮件静态资源
        FileSystemResource resource = new FileSystemResource(new File(rscPath));
        helper.addInline(rscId,resource);        // 重复使用可添加多张图片
        mailSender.send(message);
        System.out.println("带图片邮件发送成功!");
```

```
        } catch (Exception e) {
            System.out.println("带图片邮件发送失败!" + e.getMessage());
            e.printStackTrace();
        }
    }
```

设置邮件内嵌静态资源的方法为 addInLine(String rscId, Resource resource),rscId 为资源唯一标识,resource 为静态资源文件。

(2) 在项目测试类中添加测试方法,调用 MailService 类中发送带图片邮件的方法,代码如下。

```
@Test
public void sendInlineResourceMail(){
    StringBuilder content = new StringBuilder();
    content.append("<html><head></head>");
    content.append("<body><h3>新学期开始</h3>");
    String rscId = "image001";
    content.append("<img src = 'cid: " + rscId + "'/></body>");
    String rscPath = "D:\\new.jpg";
    mailService.sendInlineResourceMail("15737337@qq.com","主题: 带图片的邮件测试",
content.toString(),rscPath,rscId);
}
```

编写内嵌静态资源文件时,cid 为嵌入式静态资源文件关键字的固定写法,如果改变,将无法识别; rscId 属于自定义的静态资源唯一标识,一个邮件内容中可能会包括多个静态资源,该属性用于区别唯一性。

(3) 执行 sendInlineResourceMail()测试方法,控制台输出"带图片邮件发送成功!"的提示,如图 9-16 所示。打开接收邮件的邮箱,可以看到已接收到带有图片的测试邮件,如图 9-17 所示。

```
✔ Tests passed: 1 of 1 test – 2 sec 433 ms
带图片邮件发送成功!

Process finished with exit code 0
```

图 9-16 控制台提示发送成功

图 9-17 带图片邮件发送成功

5. 发送模板邮件

在前面几个案例中，每次发送邮件都必须手动定制邮件内容，在一些特定任务发送中则比较麻烦，例如用户注册、重置密码等，给每个用户发送的内容基本一样，只是一些动态的用户名、验证码、激活码等有所不同。所以，很多时候可以使用模板引擎将各类邮件设置成模板，这样只需在发送时替换变化部分的参数。在 Spring Boot 中使用模板引擎实现模板化的邮件发送比较容易，下面以 Thymeleaf 为例介绍实现模板化的邮件发送。

(1) 在 chapter09 项目的 pom. xml 依赖文件中，添加 Spring Boot 整合支持的 Thymeleaf 模板引擎依赖启动器 spring-boot-starter-thymeleaf，代码如下。

```
<dependency>
    <groupId>org.springframework.boot</groupId>
    <artifactId>spring-boot-starter-thymeleaf</artifactId>
</dependency>
```

(2) 在项目的模板页面文件夹 templates 中新建发送用户注册验证码的模板页面，代码如下。

```
<!DOCTYPE html>
<html lang="zh" xmlns:th="http://www.thymeleaf.org">
<head>
    <meta charset="UTF-8">
    <title>邮件模板</title>
</head>
<body>
    <div><span th:text="${username}">某某</span> 先生/女士,您好:</div>
    <P style="text-indent: 2em">您的新用户验证码为
    <span th:text="${code}" style="color: blue">654321</span>,请您妥善保管.</P>
</body>
</html>
```

(3) 在 MailService 业务处理类中添加一个 sendTemplateMail()方法，用来发送 HTML 模板邮件，代码如下。

```
public void sendTemplateMail(String to,String subject,String content){
    MimeMessage message = mailSender.createMimeMessage();
    try {
        MimeMessageHelper helper = new MimeMessageHelper(message,true);
        helper.setFrom(from);
        helper.setTo(to);
        helper.setSubject(subject);
        helper.setText(content,true);
        mailSender.send(message);
        System.out.println("模板邮件发送成功!");
    } catch (Exception e) {
        System.out.println("模板邮件发送失败!" + e.getMessage());
        e.printStackTrace();
    }
}
```

sendTemplateMail()方法主要用于处理 HTML 内容(包括 Thymeleaf 邮件模板)的邮件发送,使用了 MimeMessageHelper 类对邮件信息进行封装处理。

(4) 在项目测试类中添加测试方法 sendTemplateMail(),在该方法中调用已编写的 emailTemplate 模板邮件,代码如下。

```
// 使用 TemplateEngine 来对模板进行渲染
@Autowired
private TemplateEngine templateEngine;
@Test
public void sendTemplateMail(){
    // 向 Thymeleaf 模板传值,并解析成字符串
    Context context = new Context();
    context.setVariable("username", "王力");
    context.setVariable("code", "987654");
    // 使用 TemplateEngine 设置要处理的模板页面
    String mailContent = templateEngine.process("emailTemplate", context);
    // 发送模板邮件
    mailService.sendTemplateMail("15737337@qq.com","主题: 模板邮件测试", mailContent);
}
```

先使用@Autowired 注解引入 Thymeleaf 提供的模板引擎解析器 TemplateEngine,然后定制模板邮件发送所需的参数。

(5) 启动 sendTemplateMail()测试方法,控制台输出"模板邮件发送成功!"的提示,如图 9-18 所示。打开接收邮件的邮箱,可以看到已接收到测试邮件,如图 9-19 所示。

```
✔ Tests passed: 1 of 1 test – 2 sec 607 ms
模板邮件发送成功!

Process finished with exit code 0
```

图 9-18 控制台提示发送成功

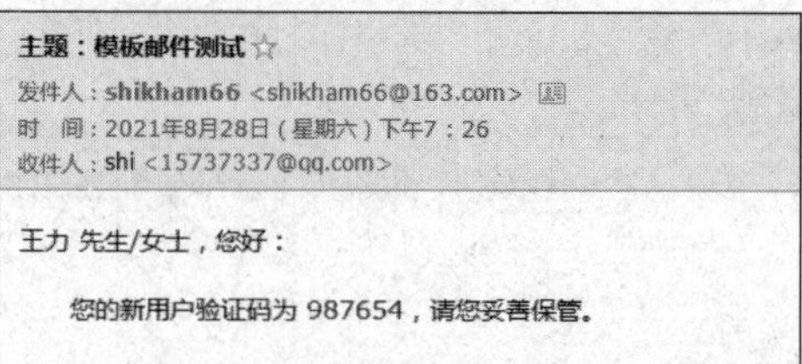

图 9-19 模板邮件发送成功

在图 9-19 中,指定的收件邮箱正确接收了定制的模板邮件,并且模板中涉及的 username 和 code 变量都被动态赋值,说明模板邮件业务实现成功。

在前面几个示例中,都只演示了一个收件人的情况,如果想一次指定多个收件人,将收件人地址转为字符串类型的数组即可,如 String[] ps = new String[]{"15737337@qq.com", "shikham@sina.com"};。需要注意的是,在邮件发送中可能会出现各种问题,例如,邮件发送过于频繁或者多次大批量发送,邮件可能会被邮件服务器拦截并识别为垃圾邮件,甚至被拉入黑名单。

9.4 小结

本章主要针对实际开发中可能涉及的项目辅助性质的功能任务进行了介绍,并结合 Spring Boot 框架进行了整合使用,这些常用任务包括异步任务、定时任务和邮件任务。

第10章 Spring Boot安全管理

在 Web 应用开发中,安全十分重要,一般项目都会有严格的认证和授权机制。例如,对于一些重要的操作,有些请求需要用户验证身份后才可以执行,还有一些请求需要用户具有特定的权限才可以执行。这样做的意义在于不仅可以保护项目安全,还可以控制项目访问效果。在 Java 开发领域,常见的安全框架为 Spring Security。

10.1 Spring Security 概述

Spring Security 是一个专注于为 Java 应用程序提供身份验证和访问控制的框架。它是 Spring 组织设计的安全管理框架,充分利用了 Spring 框架的依赖注入和 AOP 功能,为 Spring 应用系统提供安全访问控制解决方案。Spring Security 与所有 Spring 项目一样,可以很容易地被扩展以满足定制需求。

Spring Security 的前身是 Acegi Security,在被收纳为 Spring 子项目后正式更名为 Spring Security。在 Java 应用安全领域, Spring Security 是首先被推崇的安全解决方案。为了方便 Spring Boot 项目管理,Spring Boot 对 Spring Security 安全框架进行了整合支持,并提供了自动化配置方案,可在项目中零配置使用 Spring Security,实现了 Spring Security 安全框架中包含的多数安全管理功能。

在 Spring Security 安全框架中,有两个重要概念:认证(Authentication)和授权(Authorization)。认证即确认用户访问当前系统的身份;授权即确定用户在当前应用系统下所拥有的功能权限。认证的过程是识别用户身份是否合法;而授权是给予已经通过认证的用户功能权限的过程,授权发生在认证之后。

Spring Security 是一个强大的、高度自定义的认证和访问控制框架,其核心是一组过滤器链相互配合完成认证与授权,项目启动后将会自动配置。Spring Security 最核心的是 Basic Authentication Filter,用来认证用户的身份。在 Spring Security 中,一种过滤器处理一种认证方式,如图 10-1 所示。

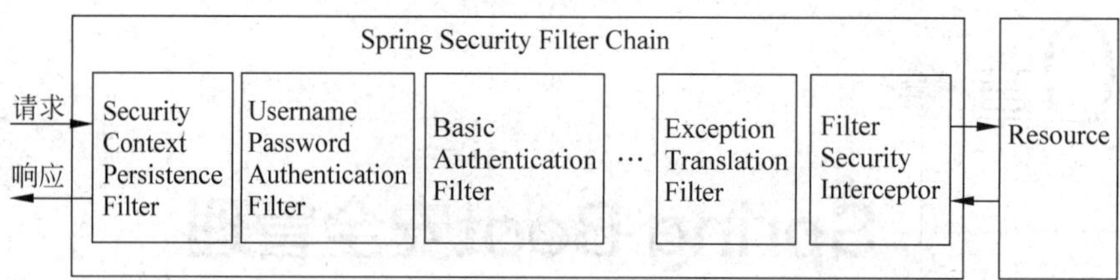

图 10-1 Spring Security 的基本原理

10.2 Spring Security 快速入门

10.2.1 入门案例

下面通过一个简单案例来快速了解 Spring Security,具体步骤如下。

(1) 新建一个 Spring Boot 工程 chapter10,Group 和 Package name 为 com.yzpc,在 Dependencies 依赖中选择 Web 节点下的 Spring Web 依赖,Security 节点下的 Spring Security 依赖,单击 Finish 按钮,如图 10-2 所示。

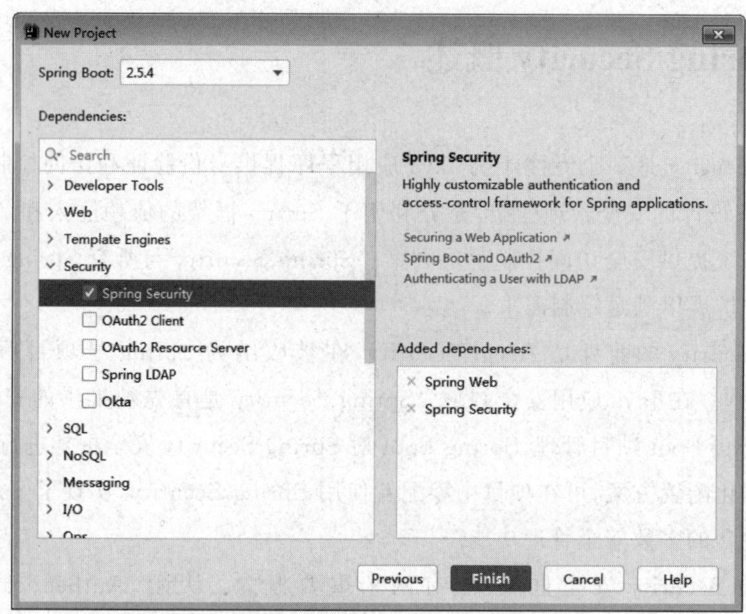

图 10-2 创建基于 Spring Security 的项目

pom.xml 中自动添加的依赖代码如下。

```
<dependency>
    <groupId>org.springframework.boot</groupId>
    <artifactId>spring-boot-starter-security</artifactId>
</dependency>
<!-- 省略Web 中Spring Web 依赖代码 -->
```

一旦项目引入 spring-boot-starter-security 启动器,相应的安全功能会立即生效。

(2) 在项目的 src/main/java/ 路径下的 com.yzpc 包中,新建一个 controller 包,并在该包中新建 HelloController 类,代码如下。

```
package com.yzpc.controller;
import org.springframework.web.bind.annotation.GetMapping;
import org.springframework.web.bind.annotation.RestController;
@RestController
public class HelloController {
    @GettMapping("/hello")
    public String hello(){
        return "Hello! Welcome to Spring Security! ";
    }
}
```

(3) 启动项目,在浏览器访问 http://localhost:8080/hello,此时会自动跳转到登录页面。这是 Spring Security 提供的登录页面,如图 10-3 所示。在登录页面随意输入一个错误的用户名和密码,会出现错误提示,如图 10-4 所示。

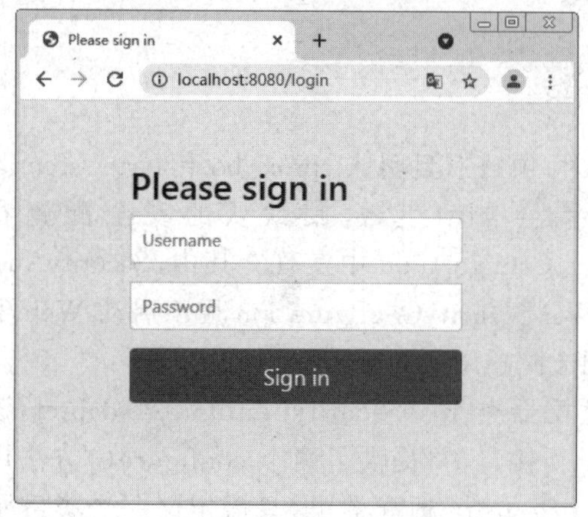

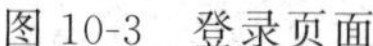

图 10-3　登录页面

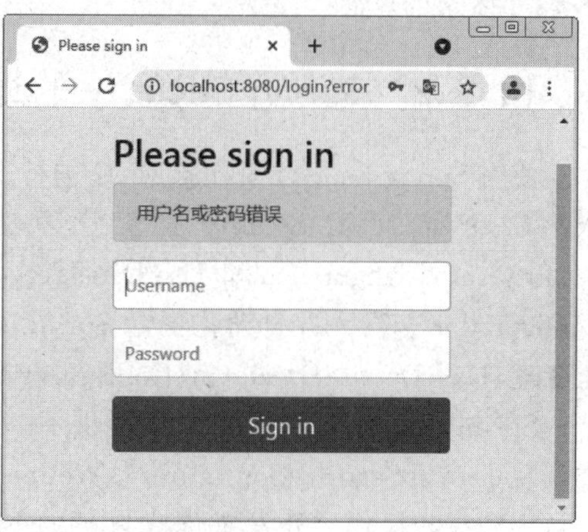

图 10-4　登录错误

图 10-4 说明 Spring Security 的安全认证已经发挥作用,该登录页面是 Spring Security 自带的默认登录页面。项目实现了 Spring Security 的自动化配置,并且具备了一些默认的安全管理功能。Spring Security 默认的用户名为 user,启动的时候会生成默认密码,在启动日志中可以看到该密码(每次启动时,默认密码随机生成),如图 10-5 所示。

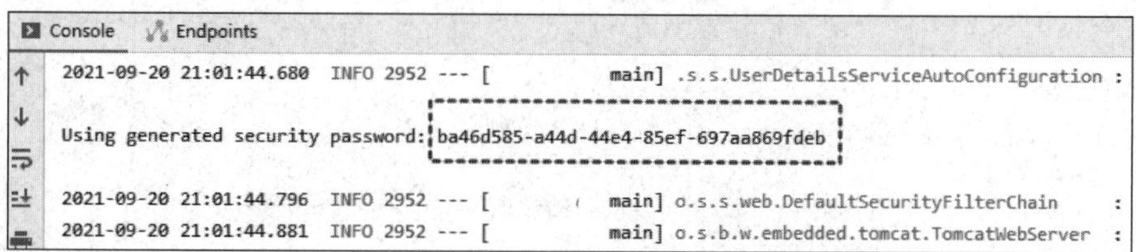

图 10-5　控制台输出默认密码

在登录页输入正确的用户名和随机生成的密码，项目登录成功，效果如图 10-6 所示。

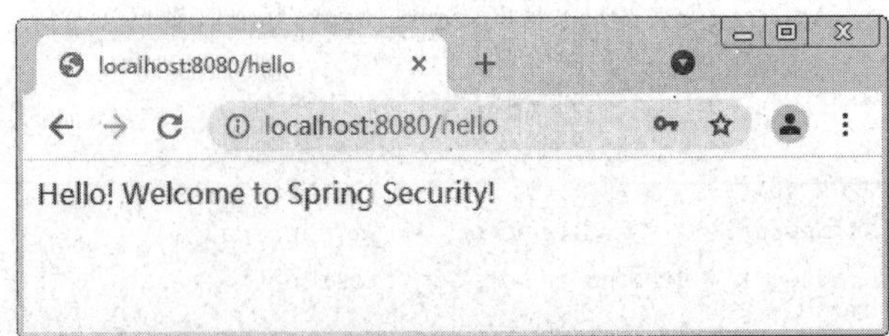

图 10-6 登录成功

(4) 也可以自己设置用户名和密码，在 application.properties 中添加如下配置。

```
# 添加用户名和密码
spring.security.user.name = admin
spring.security.user.password = 123
```

重新启动项目，控制台不再输出随机生成的密码，访问 http://localhost: 8080/hello，自动跳转到 Spring Security 默认的登录页面，输入用户名 admin 和密码 123 后成功跳转到图 10-6 所示的界面。

10.2.2 Spring Security 的适配器

使用 Spring Boot 与 Spring MVC 进行开发时，项目一旦引入 spring-boot-starter-security 依赖启动器，MVC Security 安全管理功能就会自动生效，其默认的安全配置在 SecurityAutoConfiguration 和 UserDetailServiceAutoConfiguration 中实现。其中，SecurityAutoConfiguration 会导入并自动化配置 SpringBootWebSecurityConfiguration 用于启动 Web 安全管理，UserDetailServiceAutoConfiguration 用于配置用户身份信息。

Spring Security 为 Web 应用提供了一个适配器类 WebSecurityConfigurerAdapter，该类实现了 WebSecurityConfigurer<WebSecurity>接口，并提供了两个 configure()方法用于认证和授权操作。开发者创建自己的 Spring Security 适配器类，只需定义一个类继承 WebSecurityConfigurerAdapter 的类，并在该类中使用@Configuration 注解，就可以通过重写两个 configure()方法来配置所需要的安全配置。自定义适配器类的示例代码如下。

```
@Configuration
public class MySecurityConfig extends WebSecurityConfigurerAdapter {
    // 用户认证，定制用户认证管理器来实现用户认证
    @Override
    protected void configure(AuthenticationManagerBuilder auth) throws Exception {
        super.configure(auth);
    }
    // 请求授权，定制基于 HTTP 请求的用户访问控制
    @Override
    protected void configure(HttpSecurity http) throws Exception {
        super.configure(http);
    }
}
```

10.2.3 角色访问控制

通常情况下，应用需要实现特定资源只能由特定角色访问的功能，这里假设有 ADMIN 和 USER 两个角色，ADMIN 可以访问所有资源，USER 只能访问特定资源。下面通过示例介绍角色访问控制功能。

(1) 在 chapter10 项目的 src/main/java/路径下的 com. yzpc. controller 包中，分别新建 UserController 类和 AdminController 类。

UserController 类，增加/user/ ** 接口代表用户信息方面的资源，代码如下。

```
package com.yzpc.controller;
import org.springframework.web.bind.annotation.RequestMapping;
import org.springframework.web.bind.annotation.RestController;
@RestController
@RequestMapping("/user")
public class UserController {
    // /user/ ** 接口代表用户信息方面的资源(USER 可以访问)
    @RequestMapping("/hello")
    public String hello(){
        return "Hello, user! ";
    }
}
```

AdminController 类，增加/admin/ ** 接口代表管理员方面的资源，代码如下。

```
@RestController
@RequestMapping("/admin")
public class AdminController {
    // /admin/ ** 接口代表管理员方面的资源(ADMIN 才能访问)
    @RequestMapping("/hello")
    public String hello(){
        return " Hello, admin! ";
    }
}
```

(2) 在项目的 src/main/java/ 路径下的 com. yzpc 包中，新建一个 config 包，并在该包中新建 MySecurityConfig 配置类，重写 configure(AuthenticationManagerBuilder auth)方法，在该方法中使用 AuthenticationManagerBuilder 的 inMemoryAuthentication()方法可以添加在内存中的用户，实现内存身份认证，并给用户指定角色权限，代码如下。

```
@Configuration
public class MySecurityConfig extends WebSecurityConfigurerAdapter {
    @Bean
    PasswordEncoder passwordEncoder(){
        return NoOpPasswordEncoder.getInstance();
    }
```

```
    @Override      //用户认证
    protected void configure(AuthenticationManagerBuilder auth) throws Exception {
        //使用内存用户信息,作为测试使用
        auth.inMemoryAuthentication()
            .withUser("admin").password("admin").roles("ADMIN","USER")
            .and()
            .withUser("tom").password("tom").roles("USER");
    }
}
```

上述代码中,定义用户认证信息时,设置了两个用户,包括用户名、密码和角色。管理员用户 admin,密码 admin,具备 ADMIN 角色和 USER 角色;普通用户 tom,密码 tom,具备 USER 角色。本例使用 NoOpPasswordEncoder,即不对密码进行加密,Spring Security 提供了多种密码编码器,包括 BcryptPasswordEncoder、Pbkdf2PasswordEncoder、ScryptPasswordEncoder 等。

(3) 重新运行项目,在浏览器访问 http://localhost:8080/user/hello,输入普通用户的用户名和密码,如图 10-7 所示;访问 http://localhost:8080/admin/hello,输入管理员的用户名和密码,如图 10-8 所示。

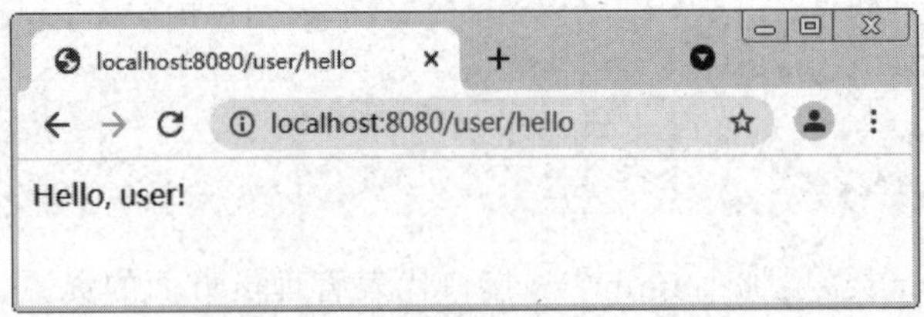

图 10-7　普通用户登录

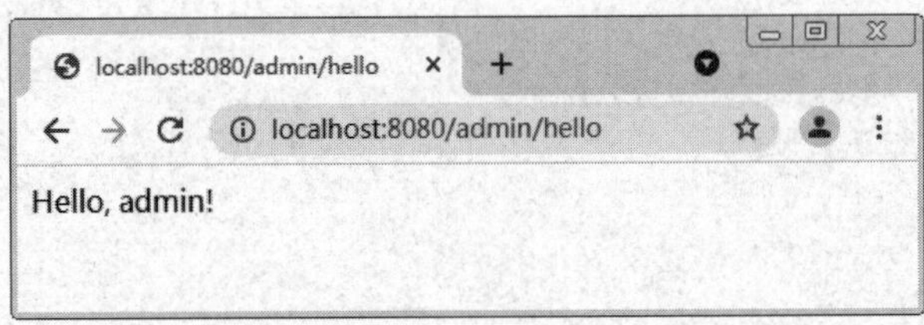

图 10-8　管理员登录

在访问 http://localhost:8080/admin/hello 时,输入普通用户的用户名和密码也可以访问,这就不符合角色访问控制的要求,需要在配置类中,添加请求授权。

(4) 在 MySecurityConfig 配置类中,重写 WebSecurityConfigurerAdapter 中的请求授权方法,代码如下。

```
// 请求授权
@Override
protected void configure(HttpSecurity http) throws Exception {
    http.authorizeRequests()
            .antMatchers("/user/**").hasRole("USER")          // 普通用户访问的URL
            .antMatchers("/admin/**").hasRole("ADMIN")        // 管理员访问的URL
```

```
        .anyRequest().authenticated()          //其他路径都必须认证
        .and()
        .formLogin()
        .loginProcessingUrl("/login")
        .permitAll()                       //访问/login 接口不要进行身份认证,防止重定向死循环
        .and()
        .csrf().disable();                     //关闭 csrf
}
```

重新运行项目,使用 tom 登录,用户具有访问/user/ ** 接口的权限;使用 admin 登录,可以访问所有接口。例如,访问 http://localhost:8080/admin/hello,若使用 tom 用户的用户名和密码登录,则会出现 403 的错误,原因是定义的角色名称不匹配,如图 10-9 所示。

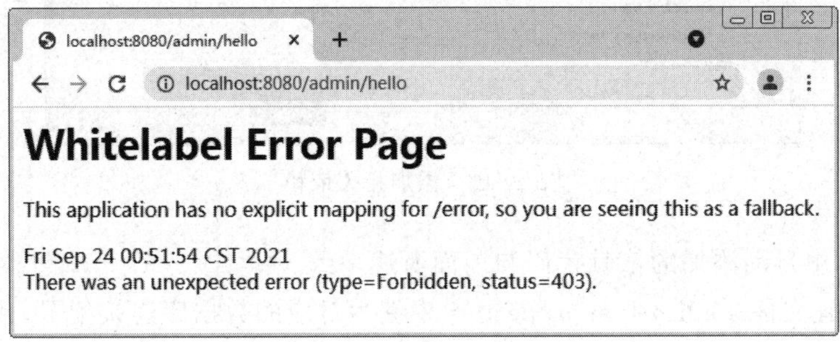

图 10-9　访问的角色名称不匹配

无论是在配置文件中设置用户还是内存式用户,存在如下问题:用户不在数据库中,无法动态增删改用户;密码以明文的方式写在项目中。对此,可以通过使用数据库对用户进行存储以实现动态增、删、改,再搭配合适的加密技术以实现密码的密文存储来解决。

10.3　用户认证

在自定义的适配器类中,重写 configure(AuthenticationManagerBuilder auth)方法,可以自定义用户认证。Spring Security 提供了多种自定义认证方式,除了内存身份认证外,还有 JDBC 身份认证、UserDetailsService 身份认证等多种方式。

10.3.1　JDBC 身份认证

在 10.2 小节中,用户登录系统的用户名和密码定义在内存中,在实际开发中,用户的基本信息以及角色都通过查询数据库进行认证和授权。JDBC 身份认证就是通过 JDBC 连接数据库对已有用户的身份进行认证,下面通过示例介绍 JDBC 身份认证的实现方式。

(1) 选中 chapter10 的项目,右击选择 New 菜单下的 Module 选项,Name 为 jdbcAuthentication,Group 和 Package name 为 com.yzpc,在 Dependencies 依赖中选择 Web 节点下的 Spring Web、Template Engines 节点下的 Thymeleaf、Security 节点下的 Spring Security 和 SQL 节点下的 JDBC API、MyBatis Framework、MySQL Driver,单击 Finish 按钮,如图 10-10 所示。

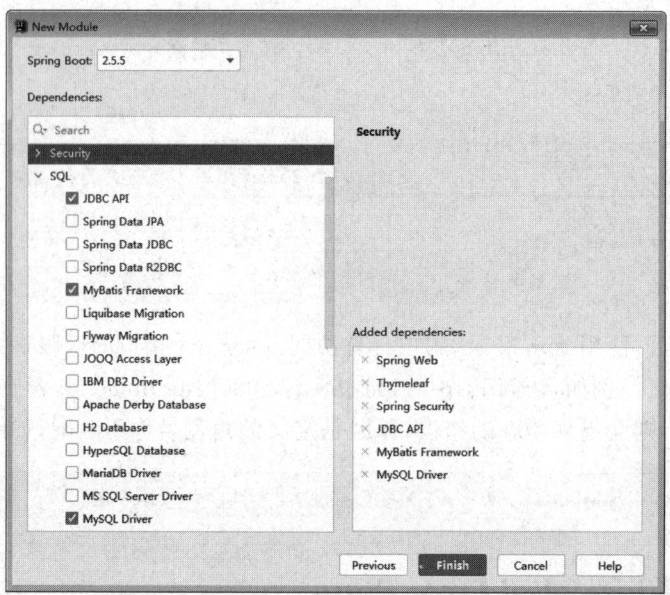

图 10-10 添加相关依赖

pom.xml 中自动添加的依赖代码与前面所述一致。

在全局配置文件 application.properties 中编写对应的数据库连接配置，代码如下。

```
# MySQL 8.0 数据库连接配置
spring.datasource.driver-class-name = com.mysql.cj.jdbc.Driver
spring.datasource.url = jdbc:mysql://localhost:3306/chapter10?&serverTimezone = UTC
spring.datasource.username = root
spring.datasource.password = 123456
```

(2) 在 jdbcAuthentication 中的 src/main/resources/templates 目录中，新建资源文件，结构如图 10-11 所示。

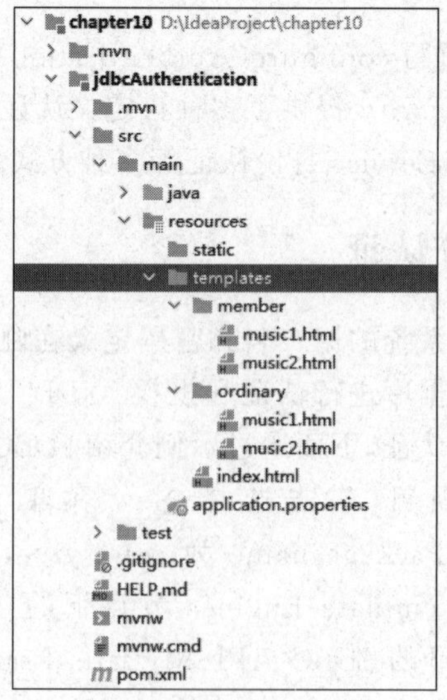

图 10-11 资源文件结构图

在图 10-11 中，index.html 文件是首页面，member 和 ordinary 文件夹中分别对应会员用户和普通用户可访问的页面。

index.html 首页通过标签分类展示了普通音乐和会员音乐，并且这些音乐通过超链接链接到具体的音乐详情路径，代码如下。

```
<!DOCTYPE html>
<html lang="en" xmlns:th="http://www.thymeleaf.org">
<head>
    <meta charset="UTF-8">
    <title>音乐网</title>
</head>
<body>
    <h2 align="center">欢迎来到音乐网站首页</h2>
    <h3>普通音乐</h3>
    <ul>
        <li><a th:href="@{/ordinary/music1}">南泥湾</a></li>
        <li><a th:href="@{/ordinary/music2}">朋友</a></li>
    </ul>
    <h3>会员音乐</h3>
    <ul>
        <li><a th:href="@{/member/music1}">可可托海的牧羊人</a></li>
        <li><a th:href="@{/member/music2}">时光背面的我</a></li>
    </ul>
</body>
</html>
```

在 templates 文件夹下，ordinary 和 member 文件夹中放置的 HTML 文件是对应音乐的介绍信息，这里以 member 文件夹下 music1.html 文件为例，代码如下。

```
<!DOCTYPE html>
<html lang="en" xmlns:th="http://www.thymeleaf.org">
<head>
    <meta http-equiv="Content-Type" content="text/html; charset=UTF-8">
    <title>音乐详情</title>
</head>
<body>
    <a th:href="@{/}">返回</a> <!-- th:href="@{/}"用于返回项目首页连接 -->
    <h2>可可托海的牧羊人</h2>
    <p style="width: 350px">
        词：王琪<br>曲：王琪<br>
        ..............<br>那夜的雨也没能留住你<br>
        山谷的风它陪着我哭泣<br>你的驼铃声仿佛还在我耳边响起<br>
        …<!-- 省略此处的歌词 -->
    </p>
</body>
</html>
```

(3) 在 jdbcAuthentication 中的 src/main/java/ 路径下的 com.yzpc 包中，新建一个

controller 包，并在该包下创建请求处理的控制类，代码如下。

```
package com.yzpc.controller;
import org.springframework.stereotype.Controller;
import org.springframework.web.bind.annotation.GetMapping;
import org.springframework.web.bind.annotation.PathVariable;
@Controller
public class MusicController {
    @GetMapping("/{type}/{path}")
    public String introduce(@PathVariable("type") String type, @PathVariable("path")
String path){
        return type + "/" + path;
    }
}
```

在上述代码中，只有一个向歌曲详情页面请求跳转的方法 introduce()，没有涉及用户登录提交以及退出操作的控制方法。

(4) 在 MySQL 数据中，新建一个名为 chapter10 的数据库，在该数据库中创建 user 用户表、role 角色表、user_role 用户角色关联表，插入几条测试数据，其 SQL 语句如下。

```
# 创建数据库
CREATE DATABASE chapter10;
# 选择使用的数据库
USE `chapter10`;
# 创建 role 表并插入数据
DROP TABLE IF EXISTS `role`;
CREATE TABLE `role` (
  `id` int NOT NULL AUTO_INCREMENT,
  `name` varchar(30) DEFAULT NULL,
  `nameZh` varchar(30) DEFAULT NULL ,
  PRIMARY KEY (`id`)
) ENGINE = InnoDB AUTO_INCREMENT = 3 DEFAULT CHARSET = utf8;
insert into `role`(`id`,`name`,`nameZh`) values (1,'ROLE_member','会员角色');
insert into `role`(`id`,`name`,`nameZh`) values (2,'ROLE_ordinary','普通角色');
# 创建 user 表并插入数据
DROP TABLE IF EXISTS `user`;
CREATE TABLE `user` (
  `id` int NOT NULL AUTO_INCREMENT,
  `username` varchar(30) DEFAULT NULL COMMENT '用户名',
  `password` varchar(200) DEFAULT NULL COMMENT '密码',
  `valid` tinyint(1) NOT NULL DEFAULT '1' COMMENT '校验用户身份',
  PRIMARY KEY (`id`)
) ENGINE = InnoDB AUTO_INCREMENT = 3 DEFAULT CHARSET = utf8;
insert into `user`(`id`,`username`,`password`,`valid`) values (1,'admin', '$2a$10$ByPK1poz.
JRGS1O/F9CiqexXW2VEgeaScjuQPLPqksLRGv8xtd5uy',1);
insert into `user`(`id`,`username`,`password`,`valid`) values (2,'tom', '$2a$10$ByPK1poz.
JRGS1O/F9CiqexXW2VEgeaScjuQPLPqksLRGv8xtd5uy',1);
# 创建 user_role 表并插入数据
DROP TABLE IF EXISTS `user_role`;
```

```
CREATE TABLE `user_role` (
  `id` int NOT NULL AUTO_INCREMENT,
  `uid` int DEFAULT NULL,
  `rid` int DEFAULT NULL,
  PRIMARY KEY (`id`)
) ENGINE = InnoDB AUTO_INCREMENT = 3 DEFAULT CHARSET = utf8;
insert  into `user_role`(`id`,`uid`,`rid`) values (1,1,1);
insert  into `user_role`(`id`,`uid`,`rid`) values (2,2,2);
```

使用 JDBC 身份认证方式创建数据表及初始化数据时，应注意以下几点。

- 创建用户表 user 时，用户名 username 必须唯一，因为 Security 在进行用户查询时先通过 username 定位是否存在唯一用户。
- 初始化用户表 user 数据时，插入的用户密码 password 必须对应编码器编码后的密码，本例中密文密码对应的原始密码为 123456。
- 初始化角色表 role 数据时，角色 role 值必须带有 ROLE_前缀，而默认的用户角色值则是对应角色值去掉 ROLE_前缀。

(5) 在 com.yzpc 包中，新建一个 config 包，并在该包中新建 WebSecurityConfig 配置类，重写 configure(AuthenticationManagerBuilder auth)方法，在该方法中使用 JDBC 身份认证的方式进行认证，代码如下。

```
package com.yzpc.config;
import org.springframework.beans.factory.annotation.Autowired;
import org.springframework.security.config.annotation.authentication.builders.
AuthenticationManagerBuilder;
import org.springframework.security.config.annotation.web.configuration.*;
import org.springframework.security.crypto.bcrypt.BCryptPasswordEncoder;
import javax.sql.DataSource;
@EnableWebSecurity    //开启 MVC Security 安全支持
public class WebSecurityConfig extends WebSecurityConfigurerAdapter {
    //JDBC 身份认证
    @Autowired
    private DataSource dataSource;
    @Override
    protected void configure(AuthenticationManagerBuilder auth) throws Exception {
        // 密码需要设置编码器
        BCryptPasswordEncoder encoder = new BCryptPasswordEncoder();
        // 使用JDBC 进行身份认证
        String userSql = "select username,password,valid from user where username = ?";
        String roleSql = "select u.username,r.name from user u,role r, user_role ur where ur.
uid = u.id and ur.rid = r.id and u.username = ?";
        auth.jdbcAuthentication().passwordEncoder(encoder)
                .dataSource(dataSource)
                .usersByUsernameQuery(userSql)
                .authoritiesByUsernameQuery(roleSql);
    }
}
```

上述代码中,@Autowired 注解装配了 DataSource 数据源。在 JDBC 身份认证时,首先需要对密码进行编码设置(必须与数据库中用户密码加密方式一致);而后加载 JDBC 进行认证连接的数据源 DataSource;最后,执行 SQL 语句,通过用户名 username 查询用户信息和用户角色。

(6) 启动 jdbcAuthentication 项目进行测试,在浏览器中访问 http://localhost:8080/ 查看首页面,会自动跳转到用户登录页面 http://localhost:8080/login,如图 10-12 所示。

图 10-12　登录页面

在图 10-12 中,输入错误的用户信息,则出现 Bad credentials 的提示,如图 10-13 所示;输入正确的用户信息,则进入项目的首页面,如图 10-14 所示。

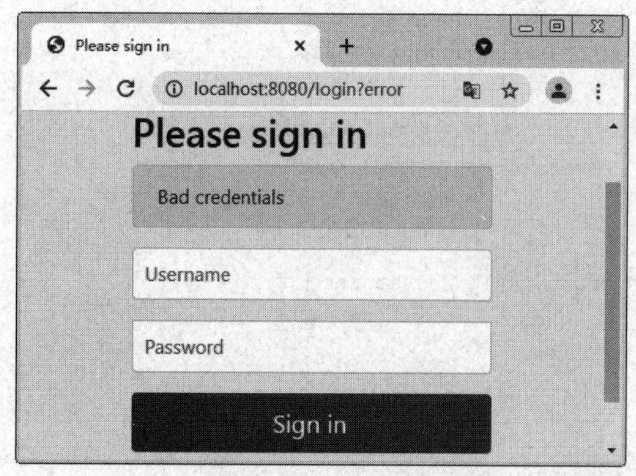

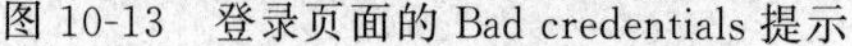
图 10-13　登录页面的 Bad credentials 提示

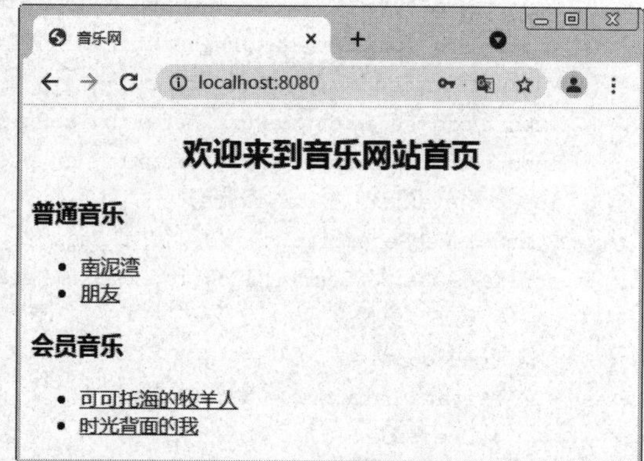

图 10-14　首页访问效果

10.3.2　UserDetailsService 身份认证

Spring Security 中进行身份验证的是 AuthenticationManager 接口,ProviderManager 是它的一个默认实现,但它并不用来处理身份认证,而是委托给配置好的 AuthenticationProvider,每个 AuthenticationProvider 会轮流检查身份认证。检查后或者返回 Authentication 对象或者抛出异常。

验证身份就是加载响应的 UserDetails,检查是否和用户输入的账号、密码、权限等信息相匹配。此步骤由实现 AuthenticationProvider 的 DaoAuthenticationProvider(它利用

UserDetailsService 验证用户名、密码和授权）处理，包含 GrantedAuthority 的 UserDetails 对象在构建 Authentication 对象时填入数据。

对于项目来说，频繁使用 JDBC 进行数据查询不仅麻烦，还会降低响应速度。如果项目中某些业务已实现用户信息查询的服务，则无须使用 JDBC 进行身份验证。下面通过使用 UserDetailsService 进行自定义用户身份认证。

（1）在 10.3.1 小节 jdbcAuthentication 项目的 com.yzpc 包中，新建一个 domain 包，根据前面创建的数据表的结构，创建与角色表、用户表对应的 Role 类和 User 类。

角色类 Role，代码如下。

```
package com.yzpc.domain;
import lombok.Data;
@Data   //@Data 注解,使用Lombok 插件实现,可以省略 getter/setters 等方法
public class Role {
    private Integer id;
    private String name;
    private String nameZh;
}
```

提示：@Data 注解可以为类提供读写功能，不用写 getter、setter 方法，并提供 equals()、hashCode()、toString()等方法。使用@Data 注解，需要安装 Lombok 插件，单击 File 菜单下的 Settings 选项，在 Settings 对话框的左侧，选择 plugins 选项，搜索 Lombok 进行安装，如图 10-15 所示。

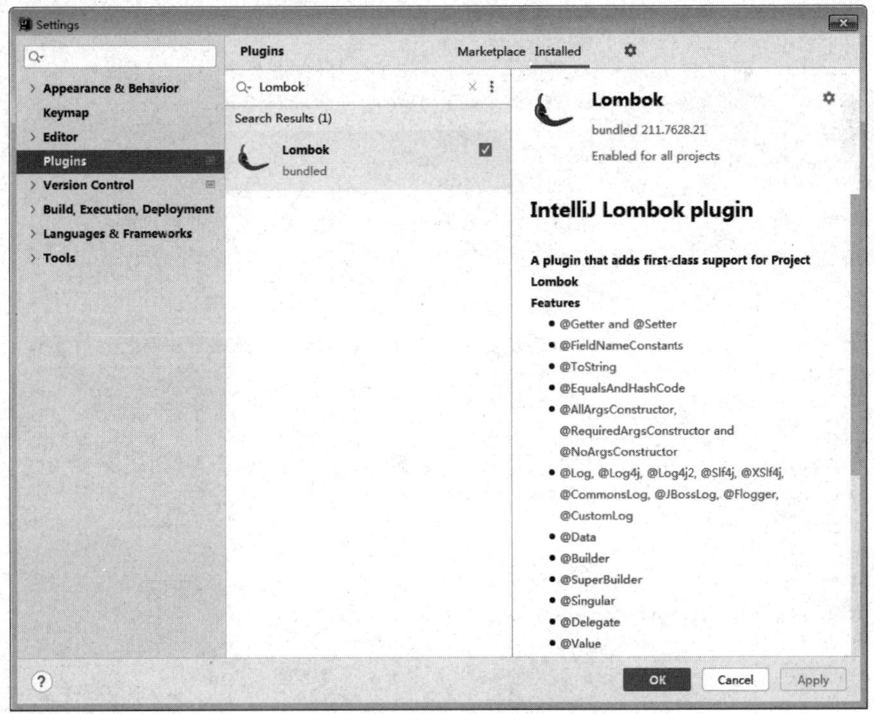

图 10-15　安装 Lombok 插件

使用 Lombok 的@Data 注解，需在 pom.xml 文件中添加相应的依赖代码如下。

```
<dependency>
    <groupId>org.projectlombok</groupId>
    <artifactId>lombok</artifactId>
</dependency>
```

用户类 User 实现 UserDetails 接口，代码如下。

```
package com.yzpc.domain;
import lombok.Data;
import org.springframework.security.core.GrantedAuthority;
import org.springframework.security.core.authority.SimpleGrantedAuthority;
import org.springframework.security.core.userdetails.UserDetails;
import java.util.ArrayList;
import java.util.Collection;
import java.util.List;
@Data
public class User implements UserDetails {
    private Integer id;
    private String username;
    private String password;
    private int valid;
    private List<Role> roles;
    //得到用户权限,交给 security 的权限,放在 UserDetailsService 进行处理
    @Override
    public Collection<? extends GrantedAuthority> getAuthorities() {
        Collection<GrantedAuthority> authorities = new ArrayList<>();
        for (Role role: roles) {
            //数据库 roles 表字段中是以 ROLE_开头的,所以此处不必再加 ROLE_
            authorities.add(new SimpleGrantedAuthority(role.getName()));
        }
        return authorities;
    }
    @Override              //指示用户的账户是否已过期,无法验证过期的账户
    public boolean isAccountNonExpired() {
        return true;
    }
    @Override              //指示用户是锁定还是解锁,无法对锁定的用户进行身份验证
    public boolean isAccountNonLocked() {
        return true;
    }
    @Override              //提示用户的凭证(密码)是否已过期,在有效期返回 true
    public boolean isCredentialsNonExpired() {
        return true;
    }
    @Override              //指示用户是启用还是禁用,启用则返回 true
    public boolean isEnabled() {
        return true;
    }
}
```

(2) 使用 MyBatis 实现数据库访问，在 com.yzpc 包中，新建一个 mapper 包，并在该包

中新建 UserMapper 接口，添加通过用户名查询用户的方法 getUserByUsername(String username)、通过用户 id 查询用户角色的方法 getUserRolesByUid(int id)，代码如下。

```
package com.yzpc.mapper;
import com.yzpc.domain.Role;
import com.yzpc.domain.User;
import org.apache.ibatis.annotations.Mapper;
import java.util.List;
@Mapper
public interface UserMapper {
    User getUserByUsername(String username);
    List<Role> getUserRolesByUid(int id);
}
```

在 src/main/resources 路径下，新建一个 mapper 的文件夹，并在其中新建 UserMapper.xml 的映射文件，代码如下。

```
<?xml version="1.0" encoding="UTF-8" ?>
<!DOCTYPE mapper
        PUBLIC "-//mybatis.org//DTD Mapper 3.0//EN"
        "http://mybatis.org/dtd/mybatis-3-mapper.dtd">
<!-- 命名空间必须和UserMapper 全类名相同 -->
<mapper namespace="com.yzpc.mapper.UserMapper">
    <!-- 此处与接口方法名对应,指定参数类型与返回结果类型 -->
    <select id="getUserByUsername" resultType="com.yzpc.domain.User">
        select * from user where username = #{username}
    </select>
    <select id="getUserRolesByUid" resultType="com.yzpc.domain.Role">
        select * from role r,user_role ur where r.id = ur.rid and ur.uid = #{id}
    </select>
</mapper>
```

上述配置文件，也可以直接在接口的方法上添加相应注解来实现。

在 application.properties 配置文件中，添加 MyBatis 扫描 mapper 路径的配置，代码如下。

```
mybatis.mapper-locations=classpath:mapper/*.xml
mybatis.type-aliases-package=com.yzpc.mapper
```

(3) UserDetailsService 是 Security 提供的用于封装认证用户信息的接口，在 com.yzpc 包中，新建 service 包，并在该包中新建 UserService 类实现 UserDetailsService 接口并重写 loadUserByUsername()方法，该方法将在用户登录时自动调用，参数是登录时的用户名，通过该用户名去数据库查找。若用户不存在，就抛出“用户不存在”的异常，如果查找到用户，就会将用户及角色信息返回，代码如下。

```
package com.yzpc.service;
import com.yzpc.mapper.UserMapper;
```

```
import com.yzpc.domain.User;
import org.springframework.beans.factory.annotation.Autowired;
import org.springframework.security.core.userdetails.UserDetails;
import org.springframework.security.core.userdetails.UserDetailsService;
import org.springframework.security.core.userdetails.UsernameNotFoundException;
import org.springframework.stereotype.Service;
@Service
public class UserService implements UserDetailsService {
    @Autowired
    private UserMapper userMapper;
    @Override
    public UserDetails loadUserByUsername(String username) throws UsernameNotFoundException {
        //通过用户名从数据库获取用户信息
        User user = userMapper.getUserByUsername(username);
        if (user == null){
            throw new UsernameNotFoundException("用户不存在");
        }
        user.setRoles(userMapper.getUserRolesByUid(user.getId()));
        return user;
    }
}
```

（4）配置 Spring Security，在 WebSecurityConfig 配置类中，注释或删除已实现的 10.3.1(5)中“1. JDBC 身份认证”部分的代码，添加 UserDetailsService 身份认证，代码如下。

```
// 2. UserDetailsService 身份认证
@Autowired
private UserService userService;
@Override
protected void configure(AuthenticationManagerBuilder auth) throws Exception {
    // 密码需要设置编码器
    BCryptPasswordEncoder encoder = new BCryptPasswordEncoder();
    // 使用 UserDetailsService 进行身份认证
    auth  //从数据库读取用户进行身份认证
          .userDetailsService(userService)
          .passwordEncoder(encoder);
}
```

上述代码中，通过@Autowired 注解引入了 UserDetailsService 接口实现类 UserService，在重写 configure(AuthenticationManagerBuilder auth)方法中使用 UserDetailsService 身份认证的方式自定义了认证用户信息，可直接调用 userDetailsService(T userService)对 UserService 实现类进行认证，并且需要对密码进行编码处理。

（5）重新启动项目，通过浏览器访问 http://localhost：8080/，访问项目首页时，同样自动跳转到用户登录页面，效果与图 10-12 一致，此时输入错误或正确的用户信息，效果与前面示例中的效果一致。

Spring Boot 整合 Spring Security 中的自定义用户认证的方式，内存身份认证最简单，主要用于测试和体验。JDBC 认证和 UserDetailsService 身份认证在实际开发中使用较多，主要根据实际开发中已有业务的支持来确定使用何种认证方式。

10.4 用户授权

授权即对系统资源进行权限设置，判断用户是否有权限访问。一个系统中的不同用户一般具有不同的操作权限。

10.4.1 用户访问控制

在10.2小节的快速入门案例中，介绍Spring Security的适配器时，通过重写WebSecurityConfigurerAdapter类的configure(HttpSecurity http)方法实现了入门案例的角色访问控制，HttpSecurity参数类型对象http的authorizeRequests()方法提供了Http请求的限制以及权限、CSRF跨站请求问题等方法，主要方法如下。

- authorizeRequests()：允许基于使用HttpServletRequest限制访问。
- formLogin()：指定支持基于表单的身份验证。
- httpBasic()：配置基于HTTP请求的Basic认证登录。
- logout()：允许配置退出登录。
- sessionManagement()：允许配置Session会话管理。
- rememberMe()：允许配置"记住我"的验证。
- csrf()：允许配置跨站请求。

先对authorizeRequests()方法的返回值做进一步查看，其中涉及用户访问控制的主要方法如下。

- antMatchs(java.lang.String…antPatterns)：开启Ant风格的路径匹配。
- mvcMatchs(java.lang.String…patterns)：开启MVC风格的路径匹配。
- regexMatchs(java.lang.String regexPatterns)：开启正则表达式的路径匹配。
- and()：功能连接符。
- anyRequest()：表示所有请求。
- rememberMe()：开启"记住我"功能。
- access(String attribute)：匹配给定的SpEL表达式计算结果是否为true。
- hasAnyRole(String…roles)：在用户拥有任意一个指定的角色权限时返回true。
- hasAuthority(String authority)：在用户拥有指定的权限时返回true。
- hasAnyAuthority(String…authorities)：在用户拥有任意一个指定的权限时返回true。
- authenticated()：当前用户不是匿名用户时返回true。
- fullyAuthenticated()：当前用户既不是匿名用户也不是rememberMe时返回true。
- hasIpAddress(String ipaddress)：通过参数匹配请求发送的IP，匹配时返回true。
- permitAll()：无条件对请求进行放行。

下面在前面认证案例的基础上，配置用户访问控制，步骤如下。

(1) 打开jdbcAuthentication项目中的WebSecurityConfig自定义配置类，重写configure (HttpSecurity http)方法进行用户访问控制，代码如下。

```
// 请求授权
@Override
protected void configure(HttpSecurity http) throws Exception {
    // 1. 自定义用户访问控制
    http.authorizeRequests()
        .antMatchers("/").permitAll()          // 路径为"/"的请求直接放行
        .antMatchers("/ordinary/**").hasRole("ordinary")
        .antMatchers("/member/**").hasRole("member")
        .anyRequest().authenticated()                  // 其他路径都必须认证
        .and().formLogin();
}
```

configure()方法设置了用户访问权限,其中,定义为/路径的请求直接放行;定义为/ordinary/**路径及其下子路径的请求,需要具备ordinary(即ROLE_ordinary)角色才允许访问;定义为/member/**路径及其下子路径的请求,需要具备member(即ROLE_member)角色才允许访问;其他请求则要求用户必须先进行登录认证。

(2)重启项目,在浏览器访问http://localhost:8080/,进入首页面,如图10-16所示,可以看出自定义的用户访问控制中,对/的请求直接放行,说明自定义用户访问控制生效。在首页单击普通音乐或者会员音乐查询详情,则跳转到登录页面,如图10-17所示,

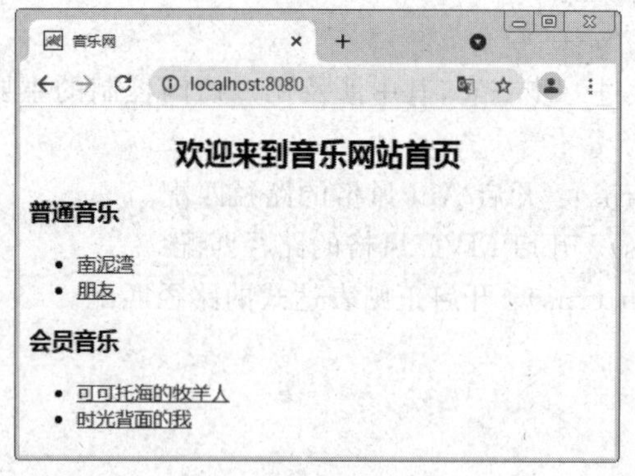

图10-16 首页访问效果

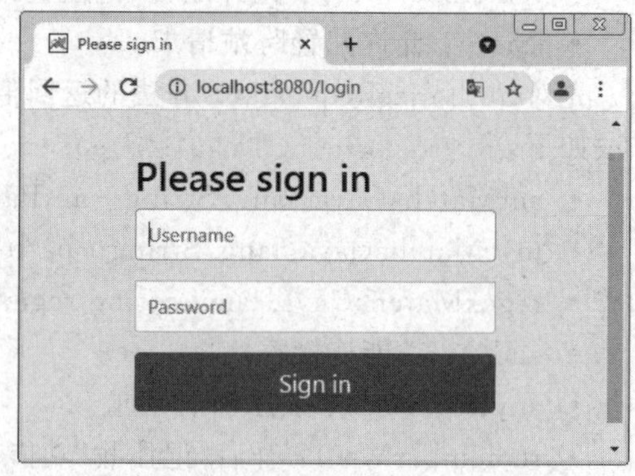

图10-17 未登录访问音乐详情效果

在首页访问音乐详情页,实质是/ordinary/music1的URL跳转,会直接被自定义的访问控制拦截并转到默认用户登录页。在图10-17的登录界面输入正确的用户名和密码,如访问普通音乐,用户名tom,密码123456,则会跳转到之前将要访问的音乐详情页,说明当前登录用户tom,有查看普通音乐的权限,如图10-18所示。

单击图10-18详情页左上角的"返回"链接,会再次回到首页面。此时,之前登录的普通用户tom还处于登录状态,再次单击会员音乐下的音乐名称查看详情,结果如图10-19所示。

登录过的tom普通用户,查看会员音乐详情时,页面出现了403 Forbidden(禁止访问)的错误信息,而控制台没有报错。说明示例中配置的用户访问控制对不同的请求拦截生效了。当前示例还没有用户注销,所以登录一个用户后要切换到其他用户的话需要重新启动浏览器,再次使用新账号登录。

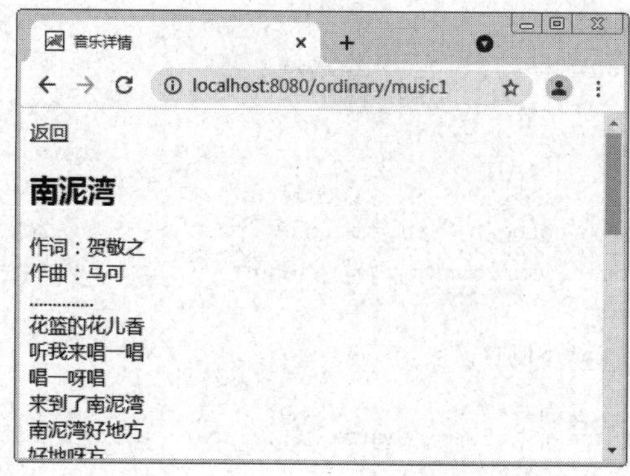

图 10-18 登录后访问音乐详情效果

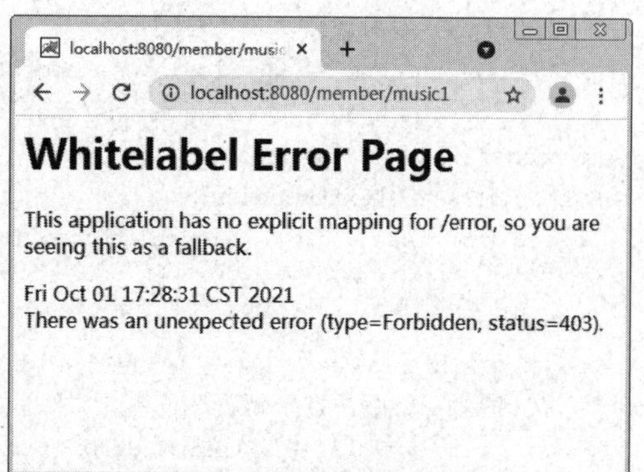

图 10-19 普通用户访问会员音乐效果

10.4.2 用户登录

通常情况下，Spring Security 提供的默认登录页面不一定符合开发需求，有时需要自定义用户登录页面。下面围绕 formLogin()方法来介绍自定义用户登录的具体实现。formLogin() 用户登录方法中涉及用户登录的主要方法介绍如下。

- loginPage(String loginPage)：登录页面跳转路径，默认为 get 请求的/login。
- successForwardUrl(String forwardUrl)：登录成功后的重定向地址。
- successHandler(AuthenticationSuccessHandler successHandler)：登录成功后的处理。
- defaultSuccessUrl(String defaultSuccessUrl)：直接登录后默认跳转地址。
- failureForwardUrl(String forwardUrl)：登录失败后的重定向地址。
- failureUrl(String authenticationFailureUrl)：登录失败后的跳转地址，默认/login?error。
- failureHanlder(AuthenticationFailureHandler handler)：登录失败后的错误处理。
- usernameParameter(String usernameParameter)：用户的用户名参数，默认 username。
- passwordParameter(String passwordParameter)：用户的密码参数，默认 password。
- loginProcessingUrl(String loginProcessingUrl)：表单提交路径，默认 post 请求的/login。
- permitAll()：无条件对请求进行放行。

下面在上一个自定义用户访问控制案例的基础上，实现自定义用户登录，步骤如下。

(1) 在项目 jdbcAuthentication 的 resources/templates 目录下新建一个 login 文件夹，在该文件夹中创建一个 login.html 的用户登录页面，代码如下。

```
<!DOCTYPE html>
<html xmlns:th="http://www.thymeleaf.org">
<head>
    <meta http-equiv="Content-Type" content="text/html; charset=UTF-8">
```

```
    <title>用户登录界面</title>
    <link th:href="@{/login/css/bootstrap.min.css}" rel="stylesheet">
    <link th:href="@{/login/css/signin.css}" rel="stylesheet">
</head>
<body class="text-center">
    <form class="form-signin" th:action="@{/toLogin}" th:method="post">
        <img class="mb-4" th:src="@{/login/img/login.jpg}" width="220px" height=
"50px">
        <h1 class="h3 mb-3 font-weight-normal">请登录</h1>
        <!-- 用户登录错误信息提示框 -->
        <div th:if="${param.error}" style="color:blue; height:40px; text-align:left;
font-size:1.1em">
            <img th:src="@{/login/img/loginError.jpg}" width="15px">用户名或密码错
误,请重新登录!
        </div>
        <input type="text" name="name" class="form-control" placeholder="用户名"
required="" autofocus="">
        <input type="password" name="pwd" class="form-control" placeholder="密码"
required="">
        <button class="btn btn-lg btn-primary btn-block" type="submit">登录</button>
        <p class="mt-5 mb-3 text-muted">Copyright? 2020-2021</p>
    </form>
</body>
</html>
```

上述代码中,定义了一个用户登录页面,数据以 POST 方式通过/toLogin 路径进行提交,用户名参数、密码参数和提交路径可自定义。其中有一个专门用来存储登录错误后返回错误信息的<div>块,使用 th:if="${param.error}" 来判断请求中是否带有一个 error 参数,从而判断是否登录成功,该参数是 Spring Security 默认的,用户也可自行定义。

此外,在项目中还引入了两个 css 样式文件和两个 img 图片文件,存放于项目 resources 下的 static 目录下,将其放到新建的 login 文件夹中,结构如图 10-20 所示。

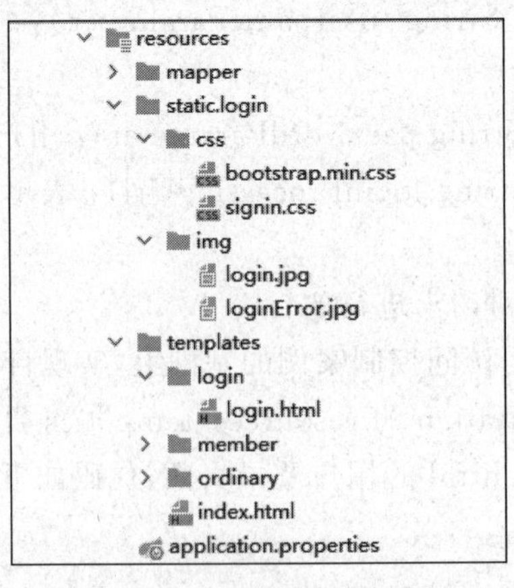

图 10-20 引入静态资源文件

(2) 在 10.3.1(3)MusicController 类中添加一个跳转到登录页面的方法,代码如下。

```
@GetMapping("/toLogin")
public String toLogin(){
    return "login/login";
}
```

在该 toLogin()方法中，配置了请求路径为/toLogin 的 Get 请求，并向静态资源目录下的 login 文件夹中的 login.html 页面跳转。Spring Security 默认采用 Get 方式的/login 请求用于向登录页面跳转，使用 Post 方式的/login 请求用于对登录后的数据处理。

(3) 打开项目中的 WebSecurityConfig 自定义配置类，重写 configure (HttpSecurity http)方法，实现自定义用户登录控制，代码如下。

```
@Override     // 请求授权
protected void configure(HttpSecurity http) throws Exception {
    // 1. 自定义用户访问控制
    http.authorizeRequests()
            .antMatchers("/").permitAll()
            // 对static 文件夹下静态资源进行统一放行
            .antMatchers("/login/ ** ").permitAll()
            .antMatchers("/ordinary/ ** ").hasRole("ordinary")
            .antMatchers("/member/ ** ").hasRole("member")
            .anyRequest().authenticated();
    // 2. 自定义用户登录
    http.formLogin()
            .loginPage("/toLogin").permitAll()
            .usernameParameter("name").passwordParameter("pwd")
            .defaultSuccessUrl("/")
            .failureUrl("/toLogin?error");
}
```

在上述代码中，定义一个 http 形式的 formLogin()方法，用于实现用户登录控制，具体介绍如下。

- loginPage("/toLogin")方法指定了自定义登录页跳转的请求路径，并使用 permitAll()方法对进行登录跳转的请求进行放行。
- usernameParameter("name")和 passwordParameter("pwd")方法用来接收登录时提交的用户名和密码。这里的 name 和 pwd 参数必须与 login.html 登录页中用户名、密码中的 name 属性值保持一致。
- defaultSuccessUrl("/")方法指定了用户登录成功后默认跳转到项目首页。
- failureUrl("/toLogin? error")方法用来控制用户登录认证失败后的跳转路径，其中/toLogin 为向登录页面跳转的映射，error 是一个错误标识，作用是登录失败后在登录页面进行接收判断。
- antMatchers("/login/ ** ").permitAll()方法作用是对项目 static 文件下 login 文件夹及其子文件夹中的静态资源文件进行统一放行处理。若未放行，则页面无法加载页面关联的静态资源文件。

(4) 重新启动项目,在浏览器访问 http://localhost:8080/,会直接进入首页面,单击访问音乐详情时,会被 Spring Security 拦截并跳转到自定义的登录页面 login.html,如图 10-21 所示。在登录页面输入错误的账号信息后,效果如图 10-22 所示。

从图 10-22 可以看出,在登录页面输入错误信息,会返回到当前登录页面,此时的请求路径上已经携带了 error 错误标识,并且登录页面也有错误提示,说明自定义登录失败设置成功。使用正确的账户进系统登录,查看详情页面,访问效果与之前成功案例一样。

图 10-21 自定义登录用户页面

图 10-22 自定义登录用户登录失败

10.4.3 用户退出

在使用过程中,如果有多个账号,需要进行账号切换。一般需要用户先注销当前的登录用户,然后再登录另一个账号。这时,需要使用到退出登录的操作,在 Spring Security 中默认调用接口/logout 进行用户退出操作,一般退出成功后会设置为自动跳转到登录页面。HttpSecurity 类的 logout()方法用来处理用户退出,默认处理路径为/logout 的 Post 类型请求,同时也会清除 Session 和 Remember Me(记住我)等默认用户配置。

logout()方法中涉及用户退出的主要方法介绍如下。

- logoutUrl(String logoutUrl):用户退出处理控制 URL,默认为 post 请求的/logout。
- logoutSuccessUrl(String logoutSuccessUrl):用户退出成功后的重定向地址。
- logoutSuccessHandler(LogoutSuccessHandler handler):用户退出成功后的处理器设置。
- deleteCookies(String cookieNamesToClear):用户退出后删除指定 Cookie。
- invalidateHttpSession(boolean invalidateHttpSession):用户退出后是否立即清除 Session(默认为 true)。
- clearAuthentication(boolean clearAuthentication):退出后是否立即清除 Authentication 用户认证信息(默认为 true)。

下面在前面案例的基础上实现自定义用户退出功能。

(1) 在页面上实现用户退出功能,必须先在页面上定义用户退出按钮或者链接。这里在项目首页 index.html 上增加一个用户退出的按钮链接，添加的代码如下。

```
<body>
    <div align="right">
        <form th:action="@{/mylogout}" method="post">
            <input th:type="submit" th:value="退出">
        </form>
    </div>
    <h2 align="center">欢迎来到音乐网站首页</h2>
    <!-- ……此处省略已实现代码 -->
</body>
```

在首页面新增了一个< form >标签进行退出控制，且定义的退出表单 action 为/mylogout（默认为/logout），方法为 post。需要注意的是，Spring Boot 项目中引入 Spring Security 框架后会自动开启 CSRF 防护功能（后面小节介绍），用户退出时必须使用 POST 请求；如果关闭了 CSRF 防护功能，那么可以使用任意方式的 HTTP 请求进行注销退出。

（2）页面定义好用户退出的链接后，不需要在 Controller 控制层中定义用户退出方法，可直接在 Security 中定制 logout()方法实现退出功能。打开 WebSecurityConfig 类，重写 configure(HttpSecurity http)方法，实现用户退出控制，代码如下。

```
@Override    //请求授权
protected void configure(HttpSecurity http) throws Exception {
    //省略前面已实现的自定义访问控制和自定义用户登录控制代码
    //3.自定义用户退出
    http.logout()
            .logoutUrl("/mylogout")
            .logoutSuccessUrl("/");
}
```

在上述代码中，在 configure(HttpSecurity http)方法中使用 logout()及其相关方法实现用户退出功能。其中，logoutUrl("/mylogout")方法指定了用户退出的请求路径，这个路径与首页面 index.html 退出表单中 action 的值保持一致，如果退出表单使用了/logout 请求，则此方法可以省略；logoutSuccessUrl("/")方法指定了用户退出成功后重定向到/地址（即首页面）。在用户退出后，用户会话信息则会默认清除。

（3）重新启动项目，在浏览器访问 http://localhost:8080/，进入首页，在页面的右上方已经出现了新添加的用户退出按钮，如图 10-23 所示。

为了演示自定义的用户退出功能，先访问音乐详情，会跳转到自定义用户登录页面，输入正确的用户名和密码后，进入详情页面，如图 10-24 所示。

单击图 10-24 详情页中的“返回”链接回到项目首页（此时用户仍处于登录状态），单击图 10-23 首页面中的“退出”按钮进行用户退出。用户退出后，根据自定义设置的重定向到首页，如果此时再次访问详情页则又会被拦截到用户登录页面，说明自定义用户退出功能已经实现。

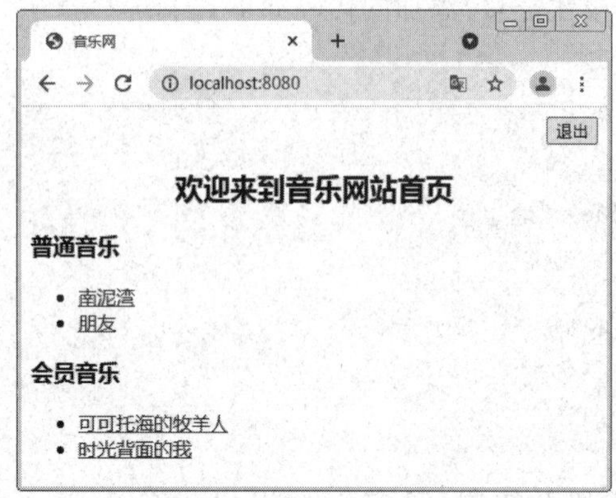

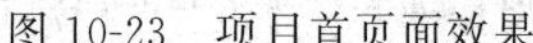
图 10-23　项目首页面效果

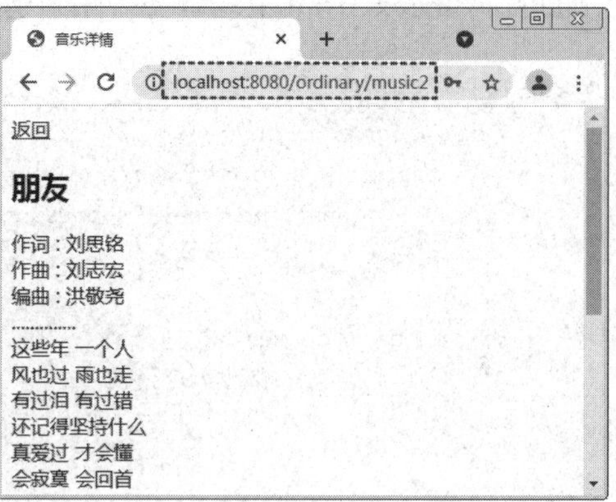

图 10-24　访问音乐详情效果

10.4.4　获取登录用户信息

在前面的案例中，使用整合 Spring Security 进行用户授权后，并没有显示登录后的用户信息，可以通过 HttpSession 和 SecurityContextHolder 两种方式来获取登录后的用户信息。

(1) 使用 HttpSession 获取用户信息。在 10.3.1(3)创建的 MusicController 控制类中新增用于获取当前会话用户信息的 getUserInfomation()方法，代码如下。

```
@GetMapping("/getUserBySession")
@ResponseBody
public void getUserInfomation(HttpSession session){
    // 从当前 HttpSession 获取绑定到此会话的所有对象的名称
    Enumeration<String> names = session.getAttributeNames();
    while(names.hasMoreElements()){
        // 获取 HttpSession 中会话名称
        String element = names.nextElement();
        // 获取 HttpSession 中的应用上下文
            SecurityContextImpl attribute = (SecurityContextImpl) session.getAttribute
(element);
        System.out.println("element: " + element);
        System.out.println("attribute: " + attribute);
        // 获取用户相关信息
        Authentication authentication = attribute.getAuthentication();
        UserDetails principal = (UserDetails)authentication.getPrincipal();
        System.out.println(principal);
        System.out.println("username: " + principal.getUsername());
    }
}
```

上述代码中，在 getUserInformation(HttpSession session)方法中通过获取当前 HttpSession 的相关方法遍历并获取了会话中的用户信息。通过 getAttribute(element)获取会话对象时，默认返回一个 Object 对象，其本质是一个 SecurityContextImpl 类，为了方便查看对象数据，所以强制转换为 SecurityContextImpl；在获取用户认证信息时，使用了 Authentication 的 getPrincipal()方法，默认返回一个 Object 对象，其本质是封装用户信息的 UserDetails 封装类，其中包括有用户名、密码、权限、是否过期等信息。

以 Debug 模式重启项目，在浏览器访问 http://localhost:8080/首页，查看一首歌的音乐详情进行用户登录。登录成功后，在保证当前浏览器未关闭的情况下，在同一浏览器中访问 http://localhost:8080/getUserBySession 来获取用户详情，控制台获取到的用户详情如图 10-25 所示。

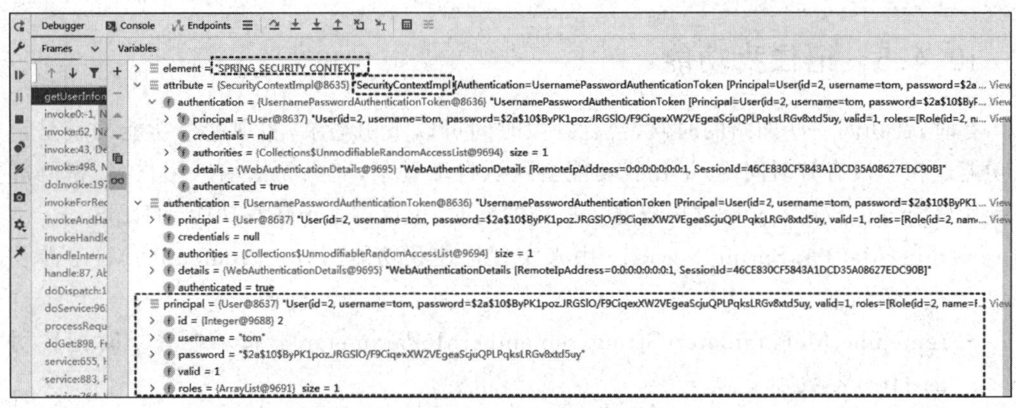

图 10-25 Debug 模式下控制台获取到的用户详情效果

(2) 使用 SecurityContextHolder 获取用户信息。在 Spring Security 中，针对拦截的登录用户专门提供了一个 SecurityContextHolder 类，用来获取 Spring Security 的上下文 SecurityContext，进而获取封装的用户信息。下面通过 SecurityContextHolder 类来获取登录的用户信息。在 10.3.1(3)MusicController 控制类中新增一个获取当前会话用户信息的 getUser()方法，代码如下。

```
@GetMapping("/getUserByContext")
@ResponseBody
public void getUser(HttpSession session){
    //获取应用上下文
    SecurityContext context = SecurityContextHolder.getContext();
    System.out.println("UserDetails: " + context);
    //获取用户相关信息
    Authentication authentication = context.getAuthentication();
    UserDetails principal = (UserDetails) authentication.getPrincipal();
    System.out.println(principal);
    System.out.println("username: " + principal.getUsername());
}
```

重新启动项目，在浏览器访问 http://localhost: 8080/，进入首页面，访问某首歌的音乐详情，会跳转到自定义的用户登录页面，用户登录成功后，在保证当前浏览器未关闭的情况下访问 http://localhost: 8080/getUserByContext 来获取用户信息，控制台输出信息如图 10-26 所示。

```
Console  Endpoints
2021-10-05 17:28:39.595  INFO 6592 --- [nio-8080-exec-1] com.zaxxer.hikari.HikariDataSource       : HikariPool-1 - Start completed.
UserDetails: SecurityContextImpl [Authentication=UsernamePasswordAuthenticationToken [Principal=User(id=2, username=tom, password=$2a$10$ByPK1poz.JRGSlO/F9CiqexXW2VEgea
User(id=2, username=tom, password=$2a$10$ByPK1poz.JRGSlO/F9CiqexXW2VEgeaScjuQPLPqksLRGv8xtdSuy, valid=1, roles=[Role(id=2, name=ROLE_ordinary, nameZh=普通角色)])
username: tom
```

图 10-26　控制台获得的用户信息

在以上两种方法中，HttpSession 的方式比较传统，而且必须引入 HttpSession 对象；而 Security 提供的 SecurityContextHolder 则相对简单，也是在 Security 项目中推荐的使用方式。

10.4.5　记住我功能

有时，在访问一些网站的时候，会有一段时间可以不用登录，实现这个功能的就是记住我功能。在一段有效时间内，记住我功能会默认自动登录，免去再次输入用户名、密码等登录操作。Spring Security 提供用户登录控制的同时，也提供了记住我功能，HttpSecurity 类的 rememberMe()是 Spring Security 用来实现记住我功能的方法，rememberMe()涉及的主要方法介绍如下。

- rememberMeParameter(String rememberMeParameter)：指示在登录时记住用户的 Http 参数。
- key(String key)：记住我认证生成的 Token 令牌标识。
- tokenValiditySeconds(int tokenValiditySeconds)：记住我 Token 令牌有效期，单位为秒(s)。
- tokenRepository(PersistentTokenRepository tokenRepository)：指定要使用的 PersistentTokenRepository，用来配置持久化 Token 令牌。
- alwaysRemember(boolean alwaysRemember)：是否应该始终创建记住我 Cookie，默认为 false。
- clearAuthentication(boolean clearAuthentication)：是否设置 Cookie 为安全的，如果设置为 true，则必须通过 https 进行连接请求。

需要说明的是，Spring Security 针对记住我功能提供了两种实现方式：一种是简单地使用加密来保证基于 Cookie 中 Token 安全；另一种是通过数据库或者其他持久化机制来保存生成的 Token。下面分别对这两种记住我功能的实现进行介绍。

1. 基于简单加密 Token 的方式

Spring Security 基于简单加密 Token 方式实现 rememberMe()功能的基本原理如下。

(1) 当用户登录的时候，除了用户名、密码，还可以选择"记住我"复选框。

(2) 如果用户选择了"记住我"复选框，当用户登录成功之后，Spring Security 服务端将会生成一个 Cookie 并发送给客户端浏览器，这个 Cookie 的名字默认是 remember-me；值是一个 Token 令牌。

(3) 当用户在有效期内再次访问应用时，经过 RememberMeAuthenticationFilter，读取 Cookie 中的 token 进行验证。若验证通过，则不需要再次登录就可以访问应用。

Spring Security 服务端生成的 Cookie，值由下面方式组合后加密而成。

```
base64(username + ": " + expirationTime + ": " +
    md5Hex(username + ": " + expirationTime + ": " + password + ": " + key))
```

上述加密生成方式中，username 为登录用户名；password 为登录密码；+expirationTime 为记住我中的 Token 的失效日期，单位为毫秒；key 为方式修改 Token 的标识。

在项目的登录页 login.html 中的< form >表单的密码框下方，新增一个记住我功能的复选框，代码如下。

```
<!-- 省略前面已实现的代码 -->
< input type = "password" name = "pwd" class = "form - control" placeholder = "密码"
required = "">
< div class = "checkbox mb - 3">
    < label >< input type = "checkbox" name = "rememberme"> 记住我 </label >
</div >
```

上述修改的登录页中，“记住我”复选框的 name 属性值设为 rememberme，而 Spring Security 提供的记住我功能的 name 属性值默认为 remember-me。

打开 WebSecurityConfig 类，重写 configure（HttpSecurity http）方法，实现记住我功能，代码如下。

```
@Override    // 请求授权
protected void configure(HttpSecurity http) throws Exception {
    // 省略前面已实现的自定义访问控制、用户登录、用户退出的代码
    // 4.定制 remember - me 记住我功能
    http.rememberMe()
            .rememberMeParameter("rememberme")
            .tokenValiditySeconds(200);
}
```

在上述代码中，在之前实现的 configure(HttpSecurity http)方法中使用 rememberMe()及相关方法实现了记住我功能。其中，rememberMeParameter("rememberme")方法指定了“记住我”复选框的 name 属性值，若使用默认的 remember-me，则该方法可省略；tokenValiditySeconds(200)方法设置了记住我功能中 Token 的有效期为 200 秒。

重新启动项目，在浏览器访问 http://localhost：8080/，进入首页面，访问某首歌的音乐详情，会跳转到自定义的用户登录页面，用户登录成功后，在保证当前浏览器未关闭的情况下访问 http://localhost：8080/toLogin 进行登录，在此登录页面上输入正确的用户名和密码，同时选择“记住我”复选框，会默认跳转到首页面，如图 10-27 所示。

为了演示记住我功能的实现效果，重新打开浏览器访问项目首页，单击刚才登录用户相应权限的音乐，效果如图 10-28 所示。

可见，选择“记住我”复选框后，在设置的Token有效期内进行访问不需要重新登录认证。如果 Token 失效后，再次访问，则需要重新登录认证。

在基于简单加密 Token 的方式中，使用 Cookie 存储虽然很方便，但是 Cookie 保存在客户端，存在安全隐患。Token 令牌是一个 MD5 Hash 字符串，包含 username、expirationTime、passwod 和一个预定义的 key，并将他们经过 MD5 加密。如果 Cookie 被劫持，则别人就可

以用这个字符串在有效期内访问用户应用。但是不存在密码被破解为明文的可能性，MD5 Hash 是不可逆的。

图 10-27　访问 toLogin 登录

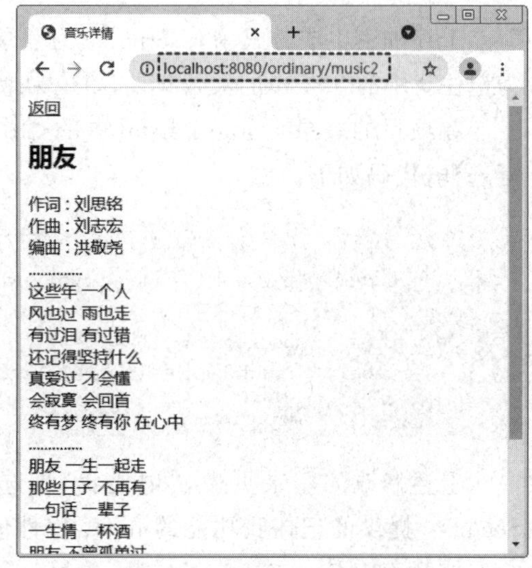

图 10-28　访问音乐详情页面

2. 基于持久化 Token 的方式

Spring Security 还提供了另一种相对更安全的实现机制。在客户端的 Cookie 中，仅保存一个无意义的加密串(与用户名、密码等敏感数据无关)，在数据库中保存该加密串-用户信息的对应关系。自动登录时，用 Cookie 中的加密串，到数据库中进行验证，如果验证通过，则可以自动登录。

基于持久化 Token 的方式实现记住我功能的大致过程如下：当用户选择"记住我"复选框并登录成功后，Spring Security 会生成一个 Token 标识，然后将该 Token 标识持久化到数据库，并且生成一个与该 Token 相对应的 Cookie 返回给浏览器。当用户过段时间再次访问系统时，如果该 Cookie 没有过期，Spring Security 便会根据 Cookie 包含的信息从数据库中获取相应的 Token 信息，从而帮用户自动完成登录操作。基于持久化 Token 方式的实现原理如图 10-29 所示。

用户发送请求到 UsernamePasswordAuthenticationFilter，当用户认证成功以后，会调用 RemeberMeService 服务。该服务中的 TokenRepository 会生成一个 Token，并将 Token 写入到浏览器的 Cookie 中，同时，TokenRepository 把生成的 Token 写入到数据库中(包括用户名)。在 Token 有效期内，用户访问系统不再需要登录，直接访问某一个受保护的服务，这个请求在经过过滤器链时会经由 RemberMeAuenticationFilter(读取 Cookie 中的 Token)给 RemberMeService，RemberMeService 会根据 Token 到数据库中进行查找。如果有记录，就会读取 Username 并调用 UserDetailsService，获取用户信息，然后把用户信息放入到 SecurityContext 中。

持久化 Token 的方式比简单加密 Token 的方式更安全。使用简单加密的 Token 方式，一旦用户的 Cookie 被盗用，在 Token 有效期内，盗用者可无限制地自动登录应用进行

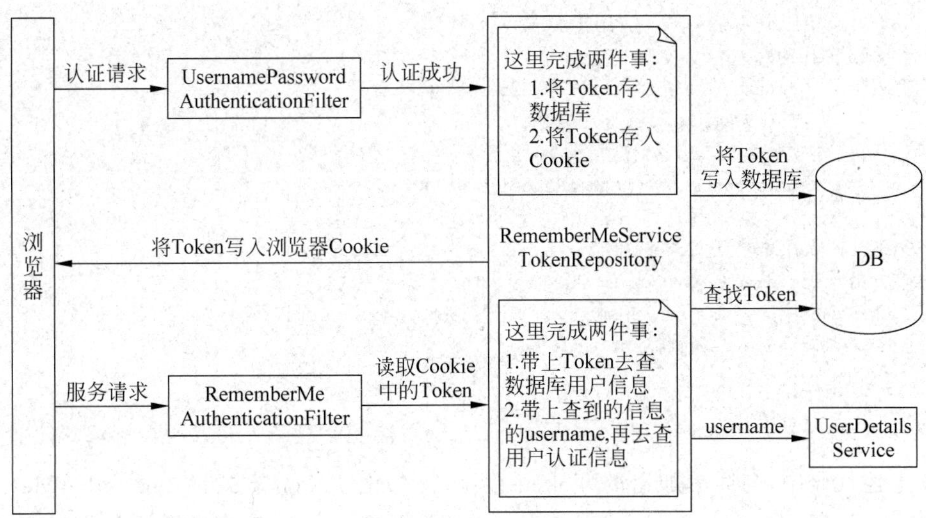

图 10-29 基于持久化的记住我基本原理

操作，直至本人发现并修改密码；而使用持久化 Token 的方式相对安全，用户每登录一次应用都会生成新的 Token 和 Cookie，但用户进行第 2 次登录前也可能存在账户被盗用的问题，只有用户在第 2 次登录并更新 Token 和 Cookie 时，才会避免这种问题。总体来讲，对于安全性要求很高的应用，不推荐使用记住我功能。

下面结合前面介绍的 rememberMe()相关方法来实现持久化 Token 方式的记住我功能。为了对持久化 Token 进行存储，在数据库中创建一个存储 Cookie 信息的持久登录用户表 persistent_logins，相关语句如下。

```
create table persistent_logins (
    username varchar(64) not null,
    series varchar(64) primary key,
    token varchar(64) not null,
    last_used timestamp not null);
```

在 persistent_logins 数据表中，username 用于存储用户名，series 用于存储随机生成的序列号，token 用于存储每次访问更新的 Token，last_used 是最近登录日期。在默认情况下基于持久化 Token 方式会使用上述官方提供的用户表 persistent_logins 进行持久化 Token 的管理，不需要自定义存储 Cookie 信息的用户表。

完成存储 Cookie 信息的用户表创建以及页面记住我功能复选框设置后，打开 WebSecurityConfig 类，重写 configure (HttpSecurity http)方法进行记住我功能配置，代码如下。

```
@Autowired
private DataSource dataSource;
@Override                    // 请求授权
protected void configure(HttpSecurity http) throws Exception {
    // 省略前面已实现的自定义访问控制、用户登录、用户退出的代码
```

```
        // 4.定制 remember - me 记住我功能
        http.rememberMe()
                .rememberMeParameter("rememberme")
                .tokenValiditySeconds(200)
                // 对 Cookie 信息进行持久化管理
                .tokenRepository(tokenRepository());
    }
    @Bean       // 持久化 Token 存储
    public JdbcTokenRepositoryImpl tokenRepository(){
        JdbcTokenRepositoryImpl jtr = new JdbcTokenRepositoryImpl();
        jtr.setDataSource(dataSource);
        return jtr;
    }
```

在上述代码中，与简单加密的 Token 相比，持久化 Token 方式的 rememberMe()示例中加入了 tokenRepository(tokenRepository())方法对 Cookie 信息进行持久化管理。其中的 tokenRepository()参数会返回一个设置 dataSource 数据源的 JdbcTokenRepositoryImpl 实现类对象，该对象包含操作 Token 的各种方法。

重新启动项目，在浏览器访问 http://localhost：8080/toLogin ，进入登录页，输入正确的用户名和密码，同时选择"记住我"复选框，登录后跳转到首页面 index.html。查看数据库中的 persistent_logins 数据表，如图 10-30 所示。

1 信息　2 表数据　3 信息

	username	series	token	last_used
☐	admin	uUAvZVmo6RU5twHr+hP++g==	v3Ov/1M2MdhK8XkU9W659g==	2021-10-07 00:56:20
*	(NULL)	(NULL)	(NULL)	(NULL)

图 10-30　首次查看登录用户表

关闭当前浏览器，重新打开浏览器访问项目首页面，并直接查看某首歌的音乐详情(打开与之前登录用户权限对应的音乐)，此时无须重新登录即可直接访问。再次查看 persistent_logins 数据表，如图 10-31 所示。

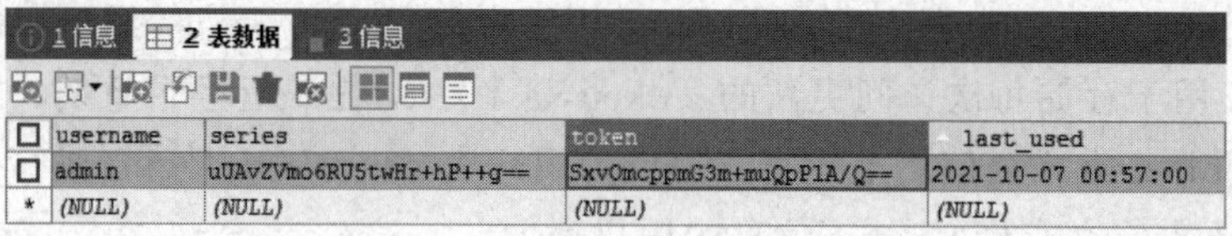

1 信息　2 表数据　3 信息

	username	series	token	last_used
☐	admin	uUAvZVmo6RU5twHr+hP++g==	SxvOmcppmG3m+muQpP1A/Q==	2021-10-07 00:57:00
*	(NULL)	(NULL)	(NULL)	(NULL)

图 10-31　再次查看登录用户表

对比图 10-30 和图 10-31 中的 Token 字段，发现在 Token 有效期内再次登录时，数据库中的 Token 会更新而其他数据不变。返回首页面，单击右上角的"退出"按钮，在 Token 有效期内进行用户手动退出。此时，再次查看 persistent_logins 数据表，如图 10-32 所示。

图 10-32 显示，登录用户手动实现用户退出后，数据库中 persistent_logins 表的持久化用户信息也会随之删除。如果用户是在 Token 有效期后自动退出的，那么 persistent_logins 表的持久化用户信息不会随之删除，当用户再次进行访问登录时，则在表中新增一条持久化用户信息。

图 10-32 退出登录后查看登录用户表

10.5 小结

本章主要介绍了 Spring Boot 的 MVC Security 安全管理。首先介绍了 Spring Security 安全框架以及 Spring Boot 支持的安全管理，通过快速入门案例体验了 Spring Security 默认的安全管理；其次介绍了 Spring Security 自定义用户认证以及授权管理。希望通过本章的学习，读者能够初步掌握 Spring Boot 的安全管理机制，并灵活运用在实际开发中，提升项目的安全性。

第11章 Vue前端框架

前端开发通过相关技术，创建 Web 页面或 App 等前端界面，将界面呈现给用户，与用户进行交互。前端开发的核心技术包括 HTML、CSS 和 JavaScript 等。为了加快 Web 开发速度，节约时间，出现了许多前端框架。本章将介绍一种优秀的前端框架 Vue，它能够更高效、便捷地实现与页面的交互。

11.1 Vue 简介

与知名前端框架 Angular 一样，Vue. js 在设计上也使用 MVVM（Model-View-ViewModel）模式。MVVM 模式本质上是 MVC 的改进版，View 绑定到 ViewModel，然后执行一些命令，再向它请求一个动作，View 和 ViewModel 之间通过双向绑定建立联系。ViewModel 跟 Model 通讯，告诉它更新以响应 UI（UserInterface，用户界面），这样便使得为应用构建 UI 非常容易。

11.2 Vue 脚手架

vue-cli 是 Vue 的脚手架工具，它大大降低了webpack 的使用难度，支持热更新，有 webpack-dev-server 的支持，相当于启动了一个请求服务器，搭建了一个测试环境。

使用 vue-cli 前，需安装最新版本的 Node. js 和 NPM。由于篇幅所限，此处不再介绍，读者可以参照本节配套视频学习安装过程。

下面介绍如何使用 vue-cli 构建一个项目。首先需要创建自己的工作空间，并在命令端口切换至刚刚创建好的工作空间，这里以 D 盘根目录为工作空间。

（1）安装 vue-cli，可以直接在 cmd 命令端口中输入如下命令。

```
D:\> npm install -g @vue/cli
```

这个命令是全局安装 vue-cli，只需运行一次即可完成 vue-cli 的安装。安装完成后，可以在 cmd 命令端口中输入如下命令，查看 vue-cli 的版本。

```
D:\> vue - V
@vue/cli 4.5.13
```

(2) 在命令端口中输入如下命令,创建一个基于 webpack 模板的新项目。

```
D:\> vue create vue3 - demo
```

vue3-demo 是项目名称,输入命令并按下回车键后,会让用户选择一个预设,如图 11-1 所示。

(3) 在图 11-1 中,选择 Default (Vue 3) ([Vue 3] babel, eslint),按下回车键,开始安装 CLI 插件,安装成功后,控制台显示如图 11-2 所示。

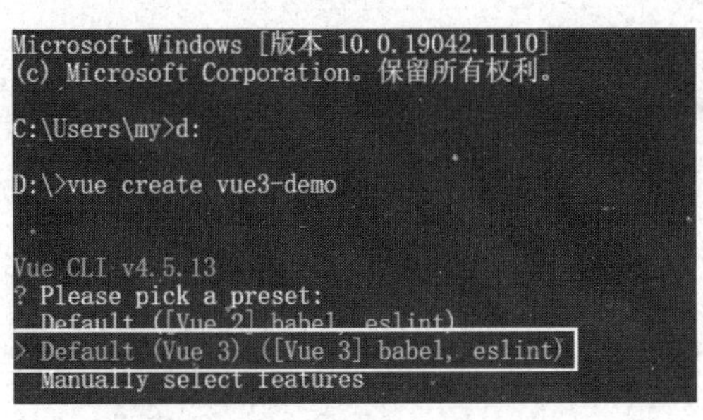

```
Microsoft Windows [版本 10.0.19042.1110]
(c) Microsoft Corporation。保留所有权利。

C:\Users\my>d:

D:\>vue create vue3-demo

Vue CLI v4.5.13
? Please pick a preset:
  Default ([Vue 2] babel, eslint)
> Default (Vue 3) ([Vue 3] babel, eslint)
  Manually select features
```

图 11-1　选择一个预设

```
Vue CLI v4.5.13
□  Creating project in D:\vue3-demo.
□□  Installing CLI plugins. This might take a while...


added 1278 packages in 32s
◆  Invoking generators...
◆  Installing additional dependencies...


added 78 packages in 4s
□  Running completion hooks...

◆  Generating README.md...

◆  Successfully created project vue3-demo.
◆  Get started with the following commands:

 $ cd vue3-demo
 $ npm run serve


D:\>
```

图 11-2　安装 CLI 插件成功

(4) 在命令端口中,切换到项目路径,命令如下。

```
D:\> cd vue3 - demo
```

(5) 输入命令,启动该项目,命令如下。

```
D:\vue3 - demo > npm run serve
```

项目启动成功后,控制台显示如图 11-3 所示。

```
 DONE  Compiled successfully in 2589ms

  App running at:
  - Local:   http://localhost:8080/
  - Network: http://192.168.1.102:8080/

  Note that the development build is not optimized.
  To create a production build, run npm run build.
```

图 11-3　启动项目成功

(6) 打开浏览器,输入地址 http://localhost:8080,页面输出如图 11-4 所示。

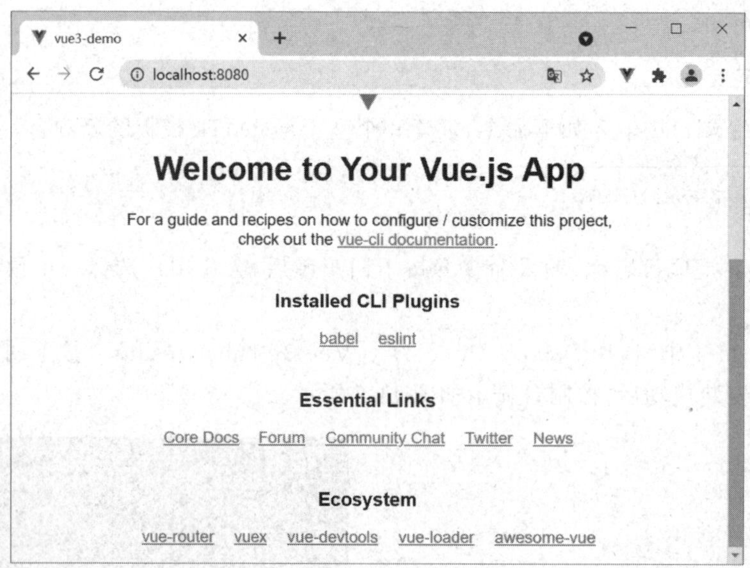

图 11-4 浏览 vue-cli 创建的项目

11.3 目录结构

使用 Visual Studio Code 打开项目 vue3-demo，可以查看项目的目录结构，如图 11-5 所示。

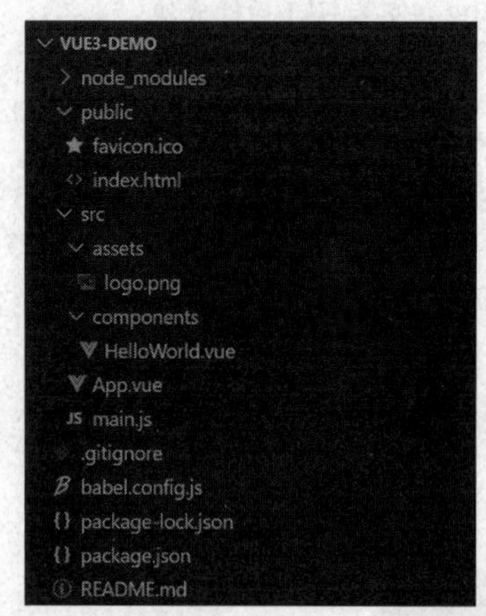

图 11-5 项目 vue3-demo 目录结构

vue3-demo 的目录格式介绍如下。

node_modules 目录用于存放项目的依赖项。public 目录用于存放公共静态资源，该目录下的 index. html 是项目的主页，即入口 html 页面。src 是项目的核心目录，assets 子目录用于存放静态资源文件，如字体、图标、图片等；components 子目录用于存放公共组件，使用 vue-cli 构建项目时，默认会在该目录下创建组件 HelloWorld. vue。App. vue 是根组件，main. js 是项目执行的入口 js。. gitignore 用来指定 git 提交时需要忽略的文件格式，babel. config. js 是 babel 配置文件，package. json 用来保存项目依赖项的版本信息，package-lock. json 用来保存 node_modules 目录下所有依赖项的具体来源、版本号和其他信息，README. md 是项目的说明文档。

main. js 是项目执行的入口 js 文件，初始时，该文件中只有如下三行代码。

```
import { createApp } from 'vue'
import App from './App.vue'
createApp(App).mount('#app')
```

在 main. js 文件中，通过 import... from...（ES6 语法中的新特性）先引入一个名为 createApp 的工厂函数，再引入 App. vue 根组件。接下来通过 createApp(App)创建一个应用实例对象，createApp 方法的参数是根组件对象 App，即创建了一个根实例对象，并调用 mount 方法将这个实例对象挂载到某个 html 的 DOM 节点上。一般通过 id 选择器的形式，指定挂载的 DOM 元素，这里的＃app 表示挂载到 public 目录下 index. html 这个文件中 id 为 app 的 div 节点上。

打开 index. html 文件，可以看到 id 为 app 的 div 节点。

```
<!DOCTYPE html>
<html lang="">
  <head>
    <meta charset="utf-8">
    <meta http-equiv="X-UA-Compatible" content="IE=edge">
    <meta name="viewport" content="width=device-width,initial-scale=1.0">
    <link rel="icon" href="<%= BASE_URL %>favicon.ico">
    <title><%= htmlWebpackPlugin.options.title %></title>
  </head>
  <body>
    <noscript>
      <strong>We're sorry but <%= htmlWebpackPlugin.options.title %> doesn't work
properly without JavaScript enabled. Please enable it to continue.
</strong>
    </noscript>
    <div id="app"></div>
    <!-- built files will be auto injected -->
  </body>
</html>
```

启动项目并通过浏览器访问，首先会访问该 index. html 文件。因为有一个 id 为 app 的挂载点，根实例对象就会挂载到这个挂载点上。此时，页面会显示根组件 App. vue 中的内容，如图 11-4 所示。

初始时，根组件 App. vue 内容如下。

```
<template>
  <img alt="Vue logo" src="./assets/logo.png">
  <HelloWorld msg="Welcome to Your Vue.js App"/>
</template>
<script>
import HelloWorld from './components/HelloWorld.vue'
export default {
  name: 'App',
  components: {
    HelloWorld
  }
}
</script>
<style>
```

```
#app {
  font-family: Avenir, Helvetica, Arial, sans-serif;
  -webkit-font-smoothing: antialiased;
  -moz-osx-font-smoothing: grayscale;
  text-align: center;
  color: #2c3e50;
  margin-top: 60px;
}
</style>
```

一个 Vue 组件主要包括模板、行为和样式三个部分。以根组件 App. vue 为例，模板是组件中 template 中的内容（将替代原来挂载点的内容），行为是 script 中的内容，样式是 style 中的内容。

在 script 中，使用 export default 将根组件 App. vue 导出为一个对象，并通过 name 属性将导出名称指定为 App。在其他组件中，就可以使用 import 引入这个导出的对象了。此外，通过 components 属性局部注册组件 HelloWorld，该组件需要事先通过 import 进行引入，代码如下。

```
import HelloWorld from './components/HelloWorld.vue'
```

同时，在根组件 App. vue 的 template 中，需要使用 HelloWorld 组件，代码如下。

```
<HelloWorld msg="Welcome to Your Vue.js App"/>
```

因此，图 11-4 中显示的内容除了根组件 App. vue 自己的 img 元素外，还包含组件 HelloWorld. vue 的内容。初始的 HelloWorld. vue 组件内容比较繁杂，可将其进行以下简化。

```
<template>
  <div class="hello">
    <h1>{{ msg }}</h1>
  </div>
</template>
<script>
export default {
  name: 'HelloWorld',
  props: {
    msg: String
  }
}
</script>
<!-- Add "scoped" attribute to limit CSS to this component only -->
<style scoped>
</style>
```

有关 Vue 组件的更多内容，将在后面进行介绍。重新启动项目并通过浏览器访问，页面输出如图 11-6 所示。

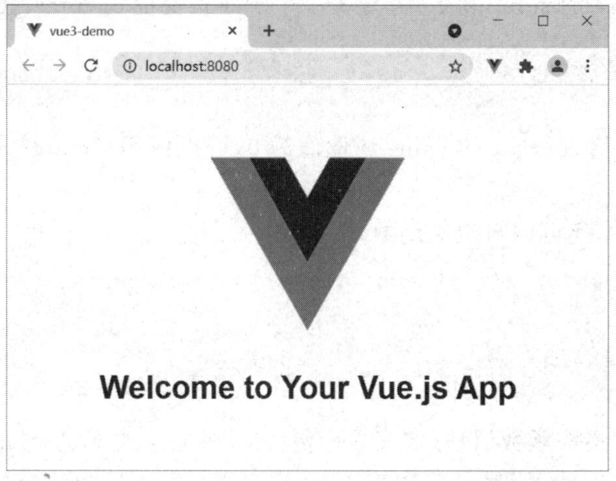

图 11-6 组件 HelloWorld. vue 简化后页面

11.4 初识 setup 和 ref

setup 是 Vue3 组合式 API 中的一个函数，在 setup 函数中，可以定义变量和方法等。但是，需要将这些变量和方法 return 出去，否则无法在模板中使用。

在组件 HelloWorld. vue 中，添加 setup 函数，代码如下。

```
export default {
  name: 'HelloWorld',
  props: {
    msg: String
  },
  setup() {
    // 定义变量
    …
    // 定义自定义函数
    …
    // 将变量和自定义函数 return 出去，供组件中 template 模板使用
    return{
        …
    }
  },
};
```

ref 也是 Vue3 组合式 API 中的一个函数，用来定义响应式变量。使用 ref 函数时，需要在组件的 script 中引入，代码如下。

```
import { ref } from "vue";
```

ref 函数接收参数并返回一个对象，并将这个参数包装在返回对象的 value 属性中。在组件 HelloWorld. vue 的 setup 函数中，定义一个响应式变量 counter，代码如下。

```
var counter = ref(0);
```

这里，响应式变量 counter 的 value 属性值为 0，可以使用 counter. value 访问或修改响应式变量的值。

在 setup 函数中，还可以自定义函数，代码如下。

```
const add = () => {
  counter.value++;
};
```

这里，使用 ES6 箭头函数语法，定义了函数 add，对响应式变量 counter 进行++操作。箭头函数省略了 function 关键字，采用箭头=>来定义函数。函数的参数放在箭头前面的括号中，函数体跟在箭头后面的花括号中。

在 setup 函数中，将响应式变量 counter 和自定义函数 add 通过 return 返回出去，这样才能在 template 模板中使用，代码如下。

```
return{
  counter,
  add
}
```

在组件 HelloWorld. vue 的 template 模板中，使用响应式变量 counter 和自定义函数 add，代码如下。

```
<template>
  …
  <h1>{{ counter }}</h1>
  <button @click = "add"> + 1 </button>
</template>
```

在 template 模板中，使用了 Mustache 语法（双大括号）的文本插值，{{ counter }}将会被替代为组件实例中响应式变量 counter 的值。无论何时，绑定的组件实例上响应式变量 counter 的值发生了改变，插值处的内容都会更新。还添加了< button >标签，并通过@click 给< button >标签绑定单击事件。单击这个按钮时，会调用自定义函数 add，将响应式变量 counter 的 value 值加 1。此时，组件 HelloWorld. vue 的代码如下。

```
<template>
  <div class = "hello">
    <h1>{{ msg }}</h1>
    <hr>
    <h1>{{ counter }}</h1>
    <button @click = "add"> + 1 </button>
```

```
  </div>
</template>
<script>
import { ref } from "vue";
export default {
  name: "HelloWorld",
  props: {
    msg: String,
  },
  setup() {
    // 定义响应式变量 counter
    var counter = ref(0);
    // 定义一个自定义函数 add
    const add = () => {
      counter.value++;
    };
    return {
      counter,
      add,
    };
  },
};
</script>
<!-- Add "scoped" attribute to limit CSS to this component only -->
<style scoped>
</style>
```

启动项目并通过浏览器访问，多次单击+1 按钮，计数不断变化，如图 11-7 所示。

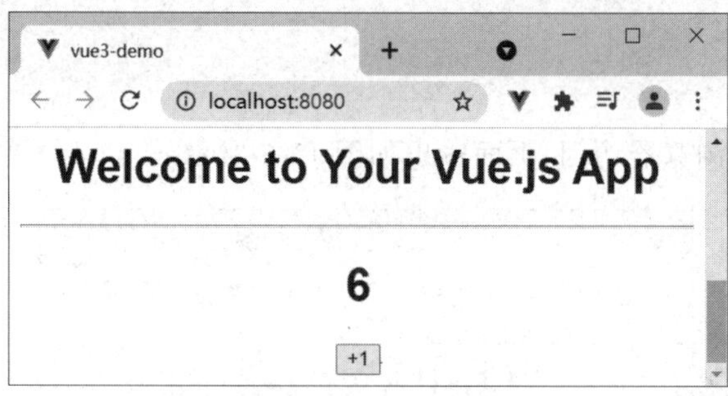

图 11-7　页面效果

11.5　模板语法

Vue.js 使用了基于 HTML 的模板语法，可以采用简洁的模板语法来声明式地将数据渲染进 DOM。通过结合响应系统，在应用状态改变时，能智能地计算重新渲染组件的最小代价并应用到 DOM 操作上。

11.5.1 插值

(1) 文本插值。数据绑定最常见的形式是使用{{...}}(双大括号)的文本插值,在组件HelloWorld.vue的template中,文本插值代码如下。

```
<h1>{{ counter }}</h1>
```

(2) 原始HTML。双大括号会将数据解释为普通文本,而非HTML代码。为了输出真正的HTML,需要使用v-html指令。

在组件HelloWorld.vue的setup函数中,先定义一个响应式变量msg1,代码如下。

```
var msg1 =  ref("<font color = 'red' size = '14'>Hello Vue!</font>");
```

再将变量msg1通过return返回出去,代码如下。

```
return{
  …,
  msg1
}
```

接着,在template模板中使用v-html指令,代码如下。

```
<template>
  …
  <hr>
  <div v-html = "msg1"></div>
  <div>{{msg1}}</div>
</template>
```

启动项目并通过浏览器访问,页面输出如图11-8所示。

图11-8 v-html指令示例

使用v-html="msg",输出内容Hello Vue!,颜色为红色,字体为14。使用{{ msg }},输出内容为<font color='red' size='14'> Hello Vue! </font>。

(3) 属性。如果要动态更新HTML元素上的属性,比如id、class和style等,可以使用v-bind指令。在组件HelloWorld.vue的setup函数中,先定义一个响应式变量styleObject,代码如下。

```
var styleObject = ref({
  color: "red",
  fontSize: "25px",
});
```

再将变量 styleObject 通过 return 返回出去，代码如下。

```
return {
  …,
  styleObject
};
```

接着，在 template 模板中使用 v-bind：style，给<div>标签动态绑定样式，将其直接绑定到样式对象 styleObject，代码如下。

```
<template>
  …
  <hr>
  <div v-bind:style="styleObject">Hello v-bind!</div>
</template>
```

启动项目并通过浏览器访问，页面输出如图 11-9 所示。

图 11-9　v-bind：style 动态绑定样式示例

从图 11-9 可以看到，<div>标签得到了样式对象 styleObject 中 color 和 fontSize 属性指定的样式。v-bind：style 的对象语法非常直观，它是一个 JavaScript 对象，类似 CSS。

（4）表达式。Vue.js 提供了完全的 JavaScript 表达式支持，在组件 HelloWorld.vue 的 setup 函数中，先定义三个响应式变量，代码如下。

```
var flag = ref(true);
var message = ref("Hello Vue!");
var id = ref(10);
```

再将这三个变量通过 return 返回出去，代码如下。

```
return {
  …,
  flag,
  message,
  id
};
```

接着，在 template 模板中，使用表达式，代码如下。

```
<template>
  …
  <hr />
  {{ 5 + 10 }}<br />
  {{ flag ? "YES" : "NO" }}<br />
  {{ message.split("").reverse().join("") }}
  <div v-bind:id="'list-' + id">Hello Vue!</div>
</template>
```

在上述代码中，使用了四个表达式。第一个是加法表达式，第二个是条件运算符，第三个表达式将把字符串翻转重组成字符串，第四个表达式将字符串 list-和变量 id 值拼接成一个新的字符串 list-10，作为这个< div >标签的 id 属性值。

启动项目并通过浏览器访问，页面输出如图 11-10 所示。

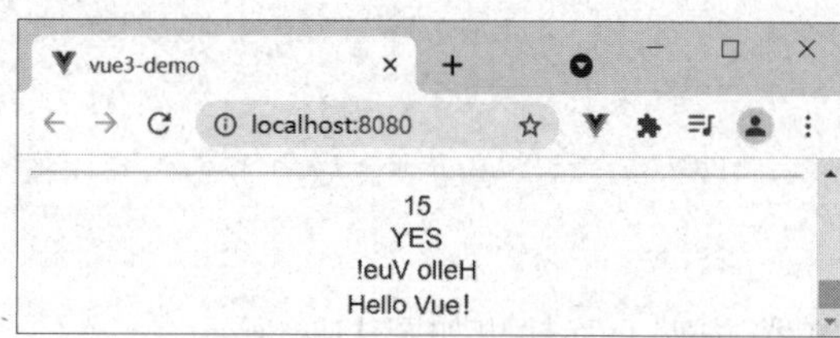

图 11-10 表达式使用示例

从图 11-10 可以看出，这些表达式会在所属 Vue 实例的数据作用域下，作为 JavaScript 被解析。

11.5.2 指令

指令(Directives)是带有 v-前缀的特殊特性，用于在表达式的值改变时，将某些行为应用到 DOM 上。

在组件 HelloWorld.vue 的 setup 函数中，先定义一个响应式变量 show，代码如下。

```
var show = ref(true);
```

再将变量 show 通过 return 返回出去，代码如下。

```
return {
  …,
  show
};
```

接着，在 template 模板中，使用 v-if 指令，代码如下。

```
<template>
  …
  <hr>
```

```
    <p v-if="show">这是一段文本</p>
</template>
```

启动项目并通过浏览器访问,页面输出如图 11-11 所示。

图 11-11 v-if 指令示例

当变量 show 的值为 true 时,<p>元素会被插入,页面输出内容为"这是一段文本"。当变量 show 的值为 false 时,<p>元素会被移除,页面不显示这段文本。

一些指令还可以接收一个参数,在指令名称之后以冒号表示。例如 v-bind: href,这里 href 是参数,告知 v-bind 指令将该元素的 href 特性与表达式 url 的值绑定。

在组件 HelloWorld.vue 的 setup 函数中,先定义一个响应式变量 url,代码如下。

```
var url = ref("http://www.baidu.com");
```

再将变量 url 通过 return 返回出去,代码如下。

```
return {
  ...
  url
};
```

接着,在 template 模板中,使用 v-bind: href,代码如下。

```
<hr><pre><a v-bind: href="url">百度一下</a></pre>
```

启动项目并通过浏览器访问,页面中显示"百度一下"超链接,如图 11-12 所示,单击该连接,跳转到百度官网。

图 11-12 指令参数示例

11.5.3 用户输入

在 input 输入框中,可以使用 v-model 指令来实现双向数据绑定。在组件 HelloWorld.vue 的 template 模板中,使用 v-model,代码如下。

```
<hr><p>{{ message }}</p>
<input v-model="message">
```

启动项目并通过浏览器访问，页面输出如图 11-13 所示。在输入框中修改内容，和它绑定的值也会发生变化，页面效果如图 11-14 所示。

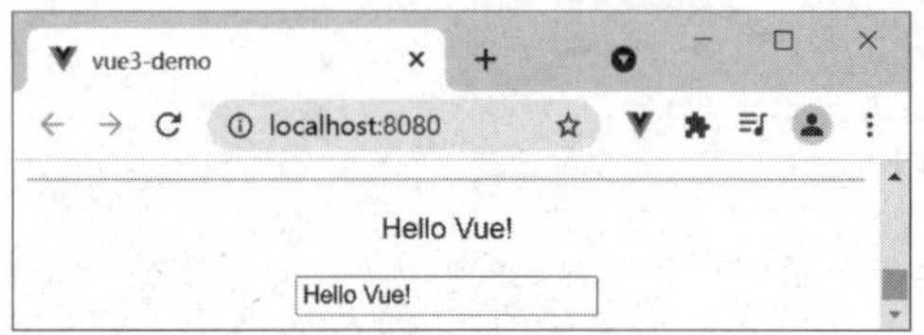

图 11-13 使用 v-model 效果一

图 11-14 使用 v-model 效果二

v-model 指令可以自动让原生表单组件的值自动和用户输入的值绑定，在这个示例中输入框的值和数据 message 是绑定的，输入框的值变化，和它绑定的值也会发生变化。

11.5.4 缩写

Vue.js 为 v-bind 和 v-on 这两个最常用的指令提供了特定简写，v-bind 指令缩写前后的对比如下。

```
<!-- 完整语法 -->
<a v-bind:href="url"></a>
<!-- 缩写 -->
<a :href="url"></a>
```

v一on 指令缩写前后的对比如下。

```
<!-- 完整语法 -->
<a v-on:click="doSomething"></a>
<!-- 缩写 -->
<a @click="doSomething"></a>
```

11.6 生命周期钩子

每个 Vue 实例在被创建时都要经过一系列的初始化过程，例如，需要设置数据监视、编译模板、将实例挂载到 DOM 并在数据变化时更新 DOM 等。同时在这个过程中也会运行一些叫作生命周期钩子的函数，用户可以在不同阶段添加自己的处理业务逻辑的代码。

在组件的 setup 函数内，可以调用的生命周期钩子包括 onBeforeMount、onMounted、onBeforeUpdate、onUpdated、onBeforeUnmount 和 onUnmounted 等。

onBeforeMount 在实例挂载开始之前被调用，onMounted 在实例被挂载后调用，onBeforeUpdate 在 DOM 更新前调用，onUpdated 在 DOM 更新后调用，onBeforeUnmount

在卸载组件实例之前调用，onUnmounted 在卸载组件实例后调用。

要使用生命周期钩子，需要先引入。在组件 HelloWorld. vue 的 script 中，使用 import 引入 6 个生命周期钩子，代码如下。

```
import { onBeforeMount, onMounted, onBeforeUpdate, onUpdated, onBeforeUnmount, onUnmounted, }
from "vue";
```

在组件 HelloWorld. vue 的 setup 函数中，添加这些生命周期钩子，并输入提示信息，代码如下。

```
console.log("setup");
onBeforeMount(() => {
  console.log("onBeforeMount");
});
onMounted(() => {
  console.log("onMounted");
});
onBeforeUpdate(() => {
  console.log("onBeforeUpdate");
});
onUpdated(() => {
  console.log("onUpdated");
});
onBeforeUnmount(() => {
  console.log("onBeforeUnmount");
});
onUnmounted(() => {
  console.log("onUnmounted");
});
```

启动项目并通过浏览器访问，再按下 F12 键，打开开发者工具，在 Console 中可以看到 setup、onBeforeMount 和 onMounted 这三个函数被依次调用，如图 11-15 所示。单击页面中的+1 按钮，可以看到 onBeforeUpdate 和 onUpdated 这两个函数被调用，如图 11-16 所示。

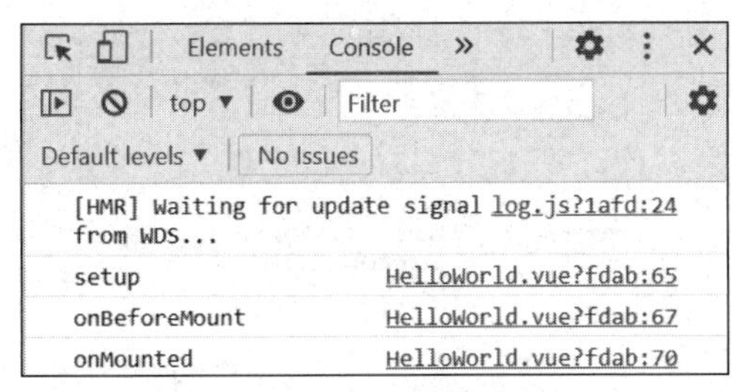

图 11-15 查看生命周期钩子一

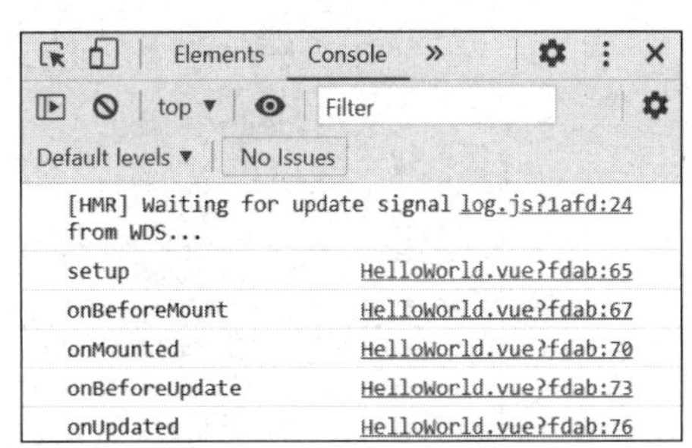

图 11-16 查看生命周期钩子二

为了演示 onBeforeUnmount 和 onUnmounted 这两个生命周期钩子，先在根组件 App. vue 的 script 中引入 ref 函数，代码如下。

```
import { ref } from "vue";
```

再添加 setup 函数,并定义一个响应式变量 isShow,并通过 return 返回出去,代码如下。

```
export default {
  name: "App",
  components: { HelloWorld, },
  setup() {
    var isShow = ref(true);
    return {
      isShow
    };
  },
};
```

接着,在根组件 App. vue 的 template 中,给HelloWorld元素添加一个条件渲染指令 v-if,当变量 isShow 的值为 true 时,渲染 HelloWorld 元素,否则不渲染。

最后添加一个用于隐藏和显示 HelloWorld 元素的 button 元素,并通过@click 给 button 元素绑定单击事件,将变量 isShow 值取反。

```
<template>
  <img alt = "Vue logo" src = "./assets/logo.png" />
  <HelloWorld msg = "Welcome to Your Vue.js App" v-if = "isShow" />
  <hr /><div @click = "isShow = !isShow">隐藏/显示 HelloWorld 组件</div>
</template>
```

启动项目并通过浏览器访问,再按下 F12 键,打开开发者工具,在 Console 中可以看到 setup、onBeforeMount 和 onMounted 这三个函数被依次调用,如图 11-17 所示。此时,单击图 11-17 中的"隐藏/显示 HelloWorld 组件"按钮,在 Console 中可以看到 onBeforeUnmount 和 onUnmounted 这两个生命周期钩子被调用了,如图 11-18 所示。

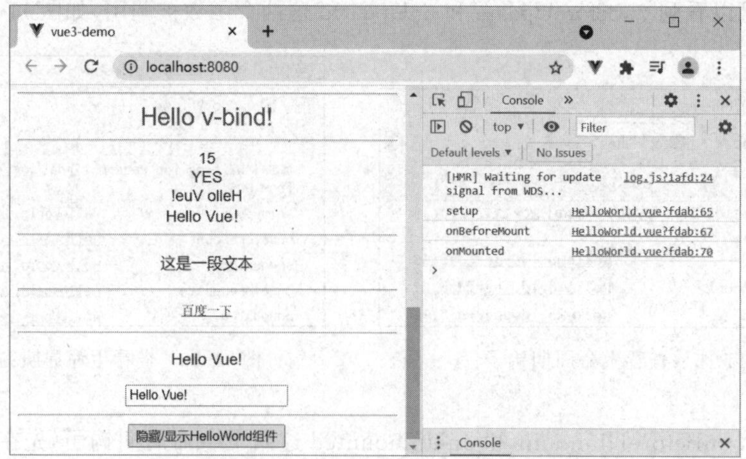

图 11-17 隐藏组件 HelloWorld 前

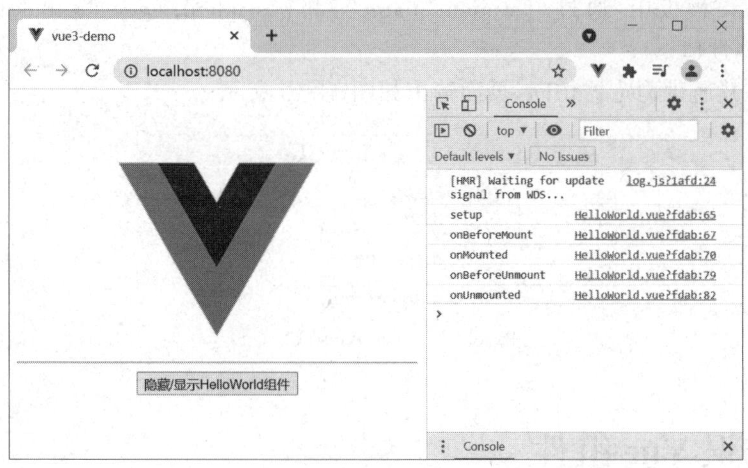

图 11-18　隐藏组件 HelloWorld 后

11.7　reactive 函数

reactive 函数用来定义一个对象类型的响应式变量，不能用来定义基本类型的响应式变量。使用 reactive 函数前，需要先引入。在组件 HelloWorld. vue 的 script 中，引入 reactive 函数，代码如下。

```
import { ref, reactive } from "vue";
```

在组件 HelloWorld. vue 的 setup 函数中，定义一个对象类型的响应式变量 student，代码如下。

```
var student = reactive({
  name: "张三",
  age: 21,
});
```

在 setup 函数中，将响应式变量 student 通过 return 返回出去，代码如下。

```
return{
  …,
  student
}
```

在组件 HelloWorld. vue 的 template 模板中，使用响应式变量 student，代码如下。

```
<template>
  …
  <hr /><p>姓名：{{student.name}}，年龄：{{student.age}}</p>
</template>
```

在 template 模板中，通过{{student. name}}和{{student. age}}输出 student 对象的 name 和 age 属性值。

启动项目并通过浏览器访问，页面输出如图 11-19 所示。

图 11-19 reactive 函数示例

11.8 初识 Vue 组件

在使用 vue-cli 构建的项目 vue3-demo 中，默认创建了根组件 App. vue 和组件 HelloWorld. vue。

组件(Component)是 Vue. js 最强大的功能之一，组件可以扩展 HTML 元素，封装可重用的代码，通过可复用的小组件可以构建大型应用。组件需要注册后才能使用，注册有全局注册和局部注册两种方式。

在组件 HelloWorld. vue 的 script 中，使用 export default 将组件导出为一个对象，并通过 name 属性将导出名称指定为 HelloWorld，代码如下。

```
export default {
  name: "HelloWorld",
  props: {
    msg: String,
  },
  setup() {
    …
    return {
      …
    };
  },
};
```

在根组件 App. vue 的 script 中，就可以使用 import 引入这个导出的 HelloWorld 对象，代码如下。

```
import HelloWorld from "./components/ HelloWorld.vue";
```

在根组件 App. vue 中，使用组件 HelloWorld. vue，还需要通过 components 属性局部注册组件 HelloWorld，代码如下。

```
export default {
  name: "App",
```

```
  components: {
    HelloWorld
  },
  setup() {
    …
  },
};
```

这样，就可以在根组件 App. vue 的 template 模板中，通过< HelloWorld >标签使用组件 HelloWorld. vue。

为了完整演示一个组件的创建和使用过程，在项目 vue3-demo 的 components 目录下，新建一个组件 Child. vue，代码如下。

```
<template>
  <h1>Child 组件</h1>
</template>
<script>
export default {
  name: "Child"
}
</script>
<style>
</style>
```

在根组件 App. vue 的 script 中，先通过 import 引入组件 Child. vue，代码如下。

```
import Child from "./components/Child.vue";
```

再通过 export default 中的 components 属性，采用局部注册的方式，注册组件 Child. vue，代码如下。

```
export default {
  name: "App",
  components: {
    HelloWorld,
    Child
  },
  setup() {
    …
  },
};
```

在根组件 App. vue 的 template 模板中，添加< Child />标签，就可以使用组件 Child. vue 了，代码如下。

```
<template>
  <img alt="Vue logo" src="./assets/logo.png" />
```

```
    <HelloWorld msg = "Welcome to Your Vue.js App" v-if = "isShow" />
    <Child />
    …
</template>
```

启动项目并通过浏览器访问，页面输出如图 11-20 所示。

图 11-20 Vue 组件示例

11.9 深入 setup

在 11.4 节，简单介绍了 setup 函数，该函数还可以接收 props 和 context 这两个参数，下面介绍这两个参数的使用。

1. props 参数

props 参数包含父组件传递给子组件的数据，在子组件中，则需要通过 props 属性指定要接收的数据。

为了演示根组件 App.vue 传递数据给子组件 Child.vue，在根组件 App.vue 的 template 中，修改<Child>标签，添加 title 和 content 两个属性。title 属性值为 hello child，content 属性值为"这是父组件 App.vue 传递的数据"，即通过 title 和 content 向子组件 Child.vue 传递两个静态的值，代码如下。

```
<template>
  …
  <Child title = "hello child" content = "这是父组件 App.vue 传递的数据" />
  …
</template>
```

在子组件 Child.vue 的 export default 中，先添加 props 属性，指定要从父组件接收的属性名为 title 和 content，代码如下。

```
export default {
  name: "Child",
  props: ['title','content'],
  …
}
```

再修改子组件 Child.vue 的 setup 函数，添加一个 props 参数，该参数包含了从父组件传递来的 title 和 content 属性值，经过 toRefs 函数处理后 return 出去，代码如下。

```
export default {
  name: "Child",
  props: ['title','content'],
  setup(props) {
    return {
      ...toRefs(props)
    };
  },
};
```

setup 函数的第一个参数 props 是一个响应式的对象。如果直接解构 props 对象，会失去响应式。使用 toRefs 函数可以将响应式对象转换为普通对象，且这个对象上的每个属性都是 ref 类型的响应式数据。

“...”是 ES6 的扩展语法，将对象中的属性取出来。这里，使用“...”从 props 中取出所有属性，通过 return 返回出去，在组件的模板中就可以直接使用对象的属性了。

使用 toRefs 函数前，需要引入。在子组件 Child.vue 的 script 中，引入 toRefs 函数，代码如下。

```
import { toRefs } from 'vue'
```

最后，在子组件 Child.vue 的 template 模板中，通过文本插值，将从父组件 App.vue 接收到的 title 和 content 属性值显示出来，代码如下。

```
<template>
  <h1>Child 组件</h1>
  <h2>{{title}}: {{content}}</h2>
</template>
```

启动项目并通过浏览器访问，页面输出如图 11-21 所示。

图 11-21　props 传递数据示例

2. context 参数

setup 函数的第二个参数是 context，表示上下文。context 是一个普通的 JavaScript 对象，包含 attrs、slots 和 emit 三个属性。

(1) attrs 属性。在父组件向子组件传递数据时，子组件需要通过 props 属性来声明父组件传递的属性名，这样才能从 props 获取相应的属性值。如果 props 属性中没有声明，则属性值会出现在 context 对象的 attrs 里。props 属性中声明的属性名，attrs 里就不会出现相应的属性值。props 支持 string 以外的类型，而 attrs 只有 string 类型。

在子组件 Child. vue 的 export default 中，修改 props 属性，将其数组中声明的属性名称 content 移除，只保留 title，代码如下。

```
export default {
  name: "Child",
  props: ['title'],
  setup(props) {
    return {
      ...toRefs(props)
    };
  },
};
```

启动项目并通过浏览器访问，页面输出如图 11-22 所示。

图 11-22 使用 attrs 属性前效果

由于 props 属性中没有指定 content 属性名，因此子组件 Child. vue 无法从 props 中获取属性 content 的值。此时，从 context 对象的 attrs 里可以获得属性 content 的值。修改 setup 函数，给 setup 函数添加第二个参数 context，使用 toRefs 函数对 context 对象的 attrs 进行处理，代码如下。

```
setup(props, context) {
    return {
     ...toRefs(props),
     ...toRefs(context.attrs),
    };
},
```

启动项目并通过浏览器访问，页面显示出 content 属性值，如图 11-23 所示。

(2) slots 属性。context 对象的 slots 属性用于接收父组件中定义的插槽内容，为了演示 slots 的作用，在根组件 App. vue 的 template 模板中，先将自结束标签< Child />修改为包含开始和结束的 Child 标签，并在 Child 标签体部添加一个 span 元素，代码如下。

图 11-23 使用 attrs 属性后效果

```
<Child title="hello child" content="这是父组件 App.vue 传递的数据">
    <span>我是嵌在子组件内的 span 标签</span>
</Child>
```

直接在父组件 App. vue 的< Child >标签体部定义的内容，在子组件 Child. vue 中不会被渲染，可以使用插槽解决这个问题。在子组件 Child. vue 的 template 模板中，添加一个< slot >元素占位（默认插槽）用于显示父组件< Child >标签下的内容，代码如下。

```
<template>
  <h1>Child 组件</h1>
  <h2>{{title}}: {{content}}</h2>
  <!-- 默认插槽 -->
  <slot></slot>
</template>
```

在子组件 Child. vue 中的默认插槽< slot ></slot >中，还可以提供一个默认内容，如果父组件的< Child >标签中没有为这个插槽提供内容，就会显示默认的内容，否则默认的内容会被替换掉。启动项目并通过浏览器访问，页面输出如图 11-24 所示。

图 11-24 默认插槽示例

如果有多个插槽时，可以使用具名插槽。在子组件 Child. vue 的 template 模板中，定义两个具名插槽，通过< slot >标签的 name 属性，将这两个插槽的名称分别指定为 header 和 footer，代码如下。

```
<template>
  ...<br>
  <!-- 具名插槽 header -->
  <slot name="header"></slot><br>
```

```
    <!-- 具名插槽 footer -->
    <slot name="footer"></slot>
</template>
```

此外，还需要在根组件 App.vue 的 template 中修改< Child >标签，在< Child >标签体部添加两个 template 节点，代码如下。

```
<Child title="hello child" content="这是父组件 App.vue 传递的数据">
    <span>我是嵌在子组件默认插槽内的 span 标签</span>
    <template v-slot:header>
        <span>
          我是嵌在子组件具名插槽 header 内的 span 标签
        </span>
    </template>
    <template v-slot:footer>
        <span>
          我是嵌在子组件具名插槽 footer 内的 span 标签
        </span>
    </template>
</Child>
```

在这两个新添加的 template 节点上，使用了 v-slot，冒号后边的值分别与子组件 Child.vue 内的两个< slot >标签的 name 属性对应，即具名插槽的名称。

启动项目并通过浏览器访问，页面输出如图 11-25 所示。

图 11-25　具名插槽示例

如果想在输出插槽内容时获取到插槽作用域的值，可以使用作用域插槽。也就是说，在子组件的< slot >标签上绑定属性值，通过作用域插槽，在父组件的插槽模板中可以得到子组件绑定的这些属性值，从而实现子组件向父组件的数据传递。

为了演示作用域插槽，先修改子组件 Child.vue 的 template 模板中名称为 header 的具名插槽，添加一个 userName 属性，将其值指定为张三，代码如下。

```
<slot name="header" userName="张三"></slot>
```

然后，在根组件 App.vue 的 template 中，修改< Child >标签体部相应的 template 节点，代码如下。

```
<Child title = "hello child" content = "这是父组件 App.vue 传递的数据">
    <span>我是嵌在子组件默认插槽内的 span 标签</span>
    <template v-slot: header = "header_slot_scope">
      <span>
        我是嵌在子组件具名插槽 header 内的 span 标签
        {{ header_slot_scope.userName }}
      </span>
    </template>
    …
</Child>
```

在子组件 Child. vue 的插槽模板< template v-slot：header＝"header_slot_scope">中，等号后面指定插槽作用域的名称，这里指定为 header_slot_scope，它相当于一个对象，这个对象里的数据就是子组件插槽绑定传递来的数据，可以通过文本插值{{header_slot_scope.userName}}显示出来。

启动项目并通过浏览器访问，页面输出如图 11-26 所示。

图 11-26　作用域插槽示例

(3) emit。在子组件中，可以使用 context. emit 向父组件发送事件。为了演示如何使用 emit，先在子组件 Child. vue 的 template 模板中，添加一个< button >标签，并给其绑定 click 事件，单击该按钮时，会触发 postMsg 函数，代码如下。

```
<button @click = "postMsg">
    我是子组件,向父组件发送事件
</button>
```

在子组件 Child. vue 的 setup 函数中，添加一个 postMsg 函数。在 postMsg 函数中，通过 context. emit 向父组件 App. vue 发送自定义事件。emit 方法的第一个参数是自定义事件名称，这里设置为 postMsgClick；第二个参数为需要传递的参数，这里设置为一个 js 对象，该对象中包含一个属性名 userName，其值为 admin，代码如下。

```
function postMsg(){
  context.emit('postMsgClick',{userName: 'admin'})
}
```

在子组件 Child. vue 的 setup 函数的 return 中，将函数 postMsgClick 返回出去，代码如下。

```
return {
  …,
  postMsg
};
```

接着，在父组件 App. vue 的 template 模板中，修改< Child >标签，给其绑定自定义事件 postMsgClick，该事件由子组件发送而来。当 postMsgClick 事件触发时，会执行 showMsg 函数，< Child >标签代码如下。

```
< Child @postMsgClick = "showMsg" title = "hello child" content = "这是父组件 App. vue 传递的数据">
```

在父组件 App. vue 的 setup 函数中，添加一个 showMsg 函数。在 showMsg 函数中，通过 js 的 alert 函数，弹出一个警告框，显示一段提醒信息和从子组件传递来的参数 userName 的值，showMsg 函数代码如下。

```
const showMsg = (e) =>{
  alert(`你触发了 postMsgClick 事件,收到子组件传递的参数 userName 是: ${e.userName}`); }
```

在父组件 App. vue 的 setup 函数的 return 中，将函数 showMsg 返回出去，代码如下。

```
return {
  isShow,
  showMsg
};
```

启动项目并通过浏览器访问，单击“我是子组件，向父组件发送事件”按钮，弹出警告框，如图 11-27 所示。

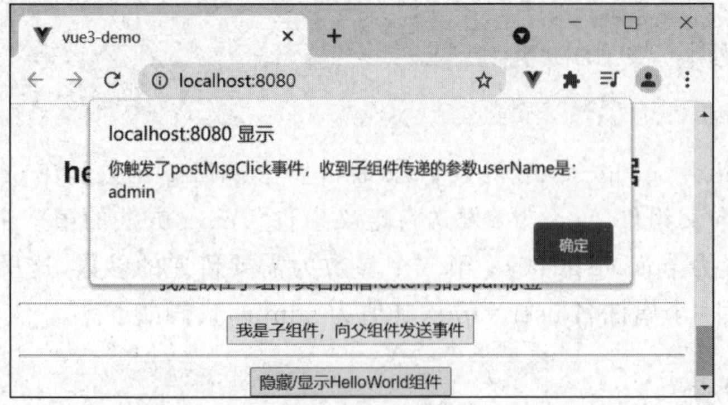

图 11-27 使用 emit 发送事件示例

11.10 计算属性

模板内的表达式非常便利，常用于简单运算。但在模板中放入大量的逻辑会让模板难以维护。为了解决这个问题，可以使用计算属性。使用计算属性，需要先引入。在子组件Child. vue的script中，修改import，先引入computed函数，再引入ref函数，用于声明响应式变量，代码如下。

```
import { toRefs, ref, computed } from "vue";
```

然后，在子组件Child. vue的template中，添加三个< input >标签，用于输入姓、名和全名。在input输入框中，使用v-model指令来实现双向数据绑定，代码如下。

```
< hr />姓: < input v - model = "firstName" />< br />< br />
名: < input v - model = "lastName" />< br />< br />
全名: < input v - model = "fullName" />< br />< br />
```

接着，在子组件Child. vue的setup函数中，使用ref声明响应式变量firstName和lastName，并给它们指定初始值，代码如下。

```
let firstName = ref("李");
let lastName = ref("四");
```

定义一个计算属性fullName，在computed函数中，提供了get和set方法，代码如下。

```
let fullName = computed({
      get() {
        return firstName.value + " " + lastName.value;
      },
      set(value) {
        let names = value.split(" ");
        firstName.value = names[0];
        lastName.value = names[1];
      },
});
```

计算属性默认只有get方法，当读取fullName这个值时，get方法会被调用，将姓和名通过空格拼接起来。当设置fullName值时，set方法会被执行，根据空格将全名字符串分割成字符串数组，数组的第一个元素赋值给响应式变量firstName，第二个元素赋值给响应式变量lastName。

最后，再将firstName、lastName和fullName这三个变量return出去，代码如下。

```
return { ..., firstName, lastName, fullName, };
```

启动项目并通过浏览器访问，页面输出如图11-28所示。

图 11-28 计算属性示例效果(一)

在图 11-28 中,将“姓:”文本框内容修改为“张”,此时全名会发生变化,如图 11-29 所示。

将“全名:”文本框内容修改为“诸葛 四”,则“姓:”文本框内容也会随之发生变化,如图 11-30 所示。

图 11-29 计算属性示例效果(二)

图 11-30 计算属性示例效果(三)

11.11 条件渲染

与 JavaScript 的条件语句 if、else、else if 类似,Vue. js 提供的条件渲染指令 v-if、v-else、v-else-if 可以根据表达式的值将 DOM 中元素或组件渲染或销毁。

为了演示条件渲染,先在子组件 Child. vue 的 template 模板中,添加若干< p >标签,分别使用 v-if、v-else-if 和 v-else 添加渲染指令,代码如下。

```
<p v-if="score >= 90">优秀</p>
<p v-else-if="score >= 80">良好</p>
<p v-else-if="score >= 70">中等</p>
<p v-else-if="score >= 60">及格</p>
<p v-else>不及格</p>
```

v-else-if 要紧跟在 v-if 之后,v-else 要紧跟在 v-if 或 v-else-if 之后。哪个< p >标签上的条件成立,则渲染该标签,而其他不满足条件的< p >标签就会被销毁。

然后,在子组件 Child. vue 的 setup 函数中,声明一个响应式变量 score,并指定初始值为 85,代码如下。

```
let score = ref(85);
```

最后，将 score 变量 return 出去，代码如下。

```
return { ..., score };
```

启动项目并通过浏览器访问，页面输出如图 11-31 所示。

图 11-31　条件渲染示例效果

11.12　列表渲染

如果想要循环显示一个数组或一个对象属性时，可以使用 v-for 指令。为了演示 v-for 指令，先在子组件 Child.vue 的 setup 函数中，使用 reactive 函数声明一个响应式变量 students，并指定初始值，代码如下。

```
let students = reactive([
  { id: 1, name: "zhangsan" },
  { id: 2, name: "lisi" },
  { id: 3, name: "wangwu" },
  { id: 4, name: "zhaoliu" },
]);
```

为了使用 reactive 函数，需要在子组件 Child.vue 的 script 中通过 import 引入，代码如下。

```
import { toRefs, ref, computed, reactive } from "vue";
```

然后，将 students 变量 return 出去，代码如下。

```
return { ..., students };
```

接着，在子组件 Child.vue 的 template 模板中，添加一个< ul >标签，在< ul >标签体部再添加一个< li >标签，在< li >标签上添加 v-for 指令，代码如下。

```
<ul>
    <li v-for = "student in students" :key = "student.id">{{ student.name }}</li>
</ul>
```

v-for 指令会循环渲染< li >标签，在 v-for 指令的表达式中，students 是源数据数组，student 是数组元素迭代的别名，通过文本插值显示数组中各个元素的 name 属性值。v-for 指令还需要动态绑定一个 key 属性，比较好的 key 值是每项都有唯一的 id，这里将 key 值设置为 student. id。

启动项目并通过浏览器访问，页面输出如图 11-32 所示。

图 11-32 数组渲染示例

除了数组外，对象的属性也可以使用 v-for 指令进行渲染。为了演示 v-for 指令渲染对象的属性，先在子组件 Child. vue 的 setup 函数中，使用 reactive 函数声明一个响应式变量 stu，并指定初始值，代码如下。

```
let stu = reactive({
  name: "zhangsan", age: 21, gender: "男", });
```

然后，将 students 变量 return 出去，代码如下。

```
return { ..., stu };
```

接着，在子组件 Child. vue 的 template 模板中，添加一个< ul >标签，在< ul >标签体部再添加一个< li >标签，在< li >标签上添加 v-for 指令，代码如下。

```
<ul>
    <li v-for="(value, key, index) in stu" :key="index">
      {{ index }}-{{ key }}: {{ value }}
    </li>
</ul>
```

在 v-for 指令的表达式中，stu 是一个对象。遍历对象属性时，参数 key 是键名，表示对象中的属性名，即 name、age、gender。value 是属性值，即 zhangsan、21、男。index 是索引，即 0、1、2。此外，v-for 指令还需要动态绑定一个 key 属性，这里将 key 值设置为 index。启动项目并通过浏览器访问，页面输出如图 11-33 所示。

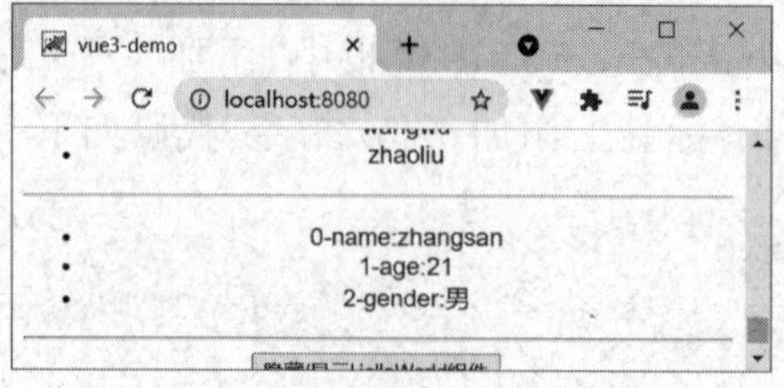

图 11-33 对象的属性渲染示例

11.13 watch 监视

watch 函数可以用来监视响应式数据的变化，并且可以得到 newValue（新值）和 oldValue（旧值）。

（1）监视 ref 定义的数据

使用 watch 函数，需要先导入。在子组件 Child.vue 的 script 中通过 import 引入 watch 函数，代码如下。

```
import { toRefs, ref, computed, reactive, watch } from "vue";
```

在子组件 Child.vue 的 setup 中，使用 ref 函数声明一个响应式变量 counter，并指定初始值，代码如下。

```
let counter = ref(0);
```

然后，将 counter 变量 return 出去，代码如下。

```
return { ..., counter };
```

在子组件 Child.vue 的 setup 函数中，定义一个函数 changeCounter，用于更新 counter 变量的值，代码如下。

```
const changeCounter = () => {
    counter.value += 1;
};
```

将 changeCounter 函数 return 出去，代码如下。

```
return { ..., counter, changeCounter };
```

在子组件 Child.vue 的 setup 函数中，定义一个 watch 函数，用来监视 counter 值的变化，代码如下。

```
watch(counter, (newValue, oldValue) => {
    console.log("counter 值改变了", newValue, oldValue);
});
```

在子组件 Child.vue 的 template 模板中，添加一个<button>标签“改变 counter”，并绑定 click 事件。单击“改变 counter”按钮时，会调用自定义函数 changeCounter，代码如下。

```
<hr /><button @click="changeCounter">改变 counter</button>
```

启动项目并通过浏览器访问，页面输出如图 11-34 所示。

图 11-34　watch 监视 ref 定义的数据示例

在图 11-34 中，按下 F12 键，打开开发者工具，再多次单击“改变 counter”按钮。此时，在 Console 中可以看到输出信息，如图 11-35 所示。

```
counter值改变了, 1 0          Child.vue?b541:78
counter值改变了, 2 1          Child.vue?b541:78
counter值改变了, 3 2          Child.vue?b541:78
```

图 11-35　开发者工具 Console 输出信息

由图 11-35 可知，每次单击“改变 counter”按钮后，都会调用自定义函数 changeCounter 修改 counter 变量的值，watch 函数监视到 counter 变量值发生了改变，就会在开发者工具的控制台中输出新值和旧值等信息。

（2）监视 reactive 定义的数据

在子组件 Child. vue 的 setup 函数中，声明变量 goods，并指定初始值，代码如下。

```
let goods = reactive({ name: "冰箱", price: 3999, });
```

然后，将 goods 变量 return 出去，代码如下。

```
return { ..., goods };
```

在子组件 Child. vue 的 setup 函数中，定义函数 changeGoodsName，用于更新 goods 对象中的 name 属性值，代码如下。

```
const changeGoodsName = () => {
      goods.name = "洗衣机";
};
```

将 changeCounter 函数 return 出去，代码如下。

```
return { ..., goods, changeGoodsName };
```

在子组件 Child. vue 的 setup 函数中，定义 watch 函数，用来监视 goods 对象中属性值的变化，代码如下。

```
watch(goods, (newVal, oldVal) => {
      console.log("商品名称改变了", newVal, oldVal);
});
```

在子组件 Child. vue 的 template 模板中，添加一个< button >标签“改变 goods 对象 name 属性值”，并绑定 click 事件。单击“改变 goods 对象 name 属性值”按钮时，会调用自定义函数 changeGoodsName，代码如下。

```
<hr /><button @click = "changeGoodsName">改变 goods 对象 name 属性值</button>
```

启动项目并通过浏览器访问，页面输出如图 11-36 所示。

在图 11-36 中，按下 F12 键，打开开发者工具，单击“改变 goods 对象 name 属性值”按钮。此时，在 Console 中可以看到输出信息，如图 11-37 所示。

图 11-36　watch 监视 reactive 定义的数据示例

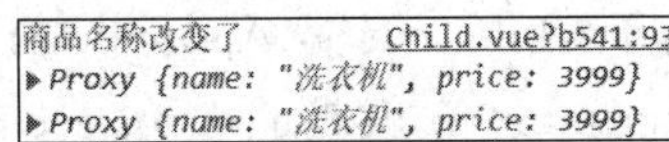

图 11-37　开发者工具 Console 输出信息

由图 11-37 可知，单击“改变 goods 对象 name 属性值”按钮，调用函数 changeGoodsName 修改 goods 对象的 name 属性值，watch 函数监视到 name 属性值发生了改变，就会在开发者工具的控制台中输出信息，但是输出的 newVal 和 oldVal 值都是修改后的新值。

11.14　provide 与 inject

将数据从父组件传递到子组件时，可以使用 props。但是，对于层次比较深的组件，将数据从祖组件传递到孙组件时，使用 props 可能会比较麻烦。这时，可以使用 provide 和 inject。祖组件（父组件）通过 provide 属性来提供数据，孙组件（子组件）则通过 inject 属性来使用这个数据。

在项目 vue3-demo 的 components 目录下，新建一个组件 Grandson. vue。初始时，Grandson. vue 组件内容如下。

```
<template>
  <div>
      <h3>我是 Grandson 组件</h3>
  </div>
</template>
<script>
export default {
    name: "Grandson"
}
</script>
<style>
</style>
```

为了让组件 Grandson. vue 成为 Child. vue 的子组件，在组件 Child. vue 的 script 部分，

通过 import 引入 Grandson. vue，代码如下。

```
import Grandson from "../components/Grandson.vue";
```

在组件 Child. vue 的 script 部分，给 export default 添加一个 components 属性，注册组件 Grandson，代码如下。

```
components: { Grandson },
```

在组件 Child. vue 的 template 模板部分，使用组件 Grandson. vue，代码如下。

```
<hr /><Grandson />
```

由于 Child. vue 是 App. vue 的子组件，而 Grandson. vue 又是 Child. vue 的子组件，因此 Grandson. vue 就成了 App. vue 的孙子组件。

为了演示从组件 App. vue 向 Grandson. vue 传递数据，在组件 App. vue 的 setup 函数中，声明变量 student，代码如下。

```
const student = reactive({
  name: "zhangsan",
  age: 21,
});
```

使用 reactive 函数，需要引入。在组件 App. vue 的 script 部分，通过 import 引入 reactive 函数，代码如下。

```
import { reactive, ref } from "vue";
```

在 return 部分，将变量 student 返回出去，代码如下。

```
return { isShow, showMsg, student };
```

在组件 App. vue 的 setup 函数的 return 前，使用 provide 函数向孙子组件 Grandson. vue 传递数据，代码如下。

```
provide("student", student);
```

使用 provide 函数，需要引入。在组件 App. vue 的 script 部分，通过 import 引入 provide 函数，代码如下。

```
import { provide, reactive, ref } from "vue";
```

在孙子组件 Grandson. vue 中，可以通过 inject 函数来使用这个数据。使用 inject 函数，需要引入。在组件 Grandson. vue 的 script 部分，通过 import 引入 inject 函数，代码如下。

```
import { inject } from "vue";
```

在孙子组件 Grandson.vue 的 script 部分的 export default 里面，先添加一个 setup 函数。在 setup 函数中，使用 inject 函数接收数据，再添加一个 return，将变量 student 返回出去，代码如下。

```
export default {
  name: "Grandson",
  setup() {
    let student = inject("student");
    return {
      student,
    };
  },
};
```

在组件 Grandson.vue 的 template 模板部分，通过文本插值，显示接收的数据，代码如下。

```
<template>
  <div>
    <h3>我是 Grandson 组件</h3>
    <h3>我从 App.vue 接收到的数据：{{ student.name }}、{{ student.age }}</h3>
  </div>
</template>
```

启动项目并通过浏览器访问，页面输出如图 11-38 所示。

图 11-38　provide 与 inject 示例

11.15　Vue 路由

Vue.js 路由可用来实现根据不同的 URL 访问不同的内容，从而实现单页面富应用（SPA）。使用 Vue.js 路由前需要安装 vue-router。

打开项目 vue3-demo 的 package.json 文件，查看 dependencies 字段指定的项目运行所依赖的模块，代码如下。

```
"dependencies": {
    "core - js": "^3.6.5",
    "vue": "^3.0.0"
},
```

上述代码并没有相关 vue-router 信息，即未安装 vue-router，需要自行安装。在 Visual Studio Code 的终端控制台中，输入如下命令。

```
npm install vue - router@4.0.0 - beta.13
```

再次打开 package.json 文件，可以看到 vue-router 的版本信息表明 vue-router 安装成功，代码如下。

```
"ependencies": {
    "core - js": "^3.6.5",
    "vue": "^3.0.0",
    "vue - router": "^4.0.0 - beta.13"
},
```

在项目 src 目录下，创建 route 文件夹，并在 route 文件夹下创建 index.js 文件，如图 11-39 所示。

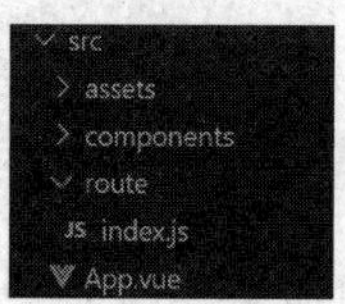

图 11-39 创建 route 文件夹及其下的 index.js 文件

在 index.js 文件中，从 vue-router 中引入 createRouter 和 createWebHistory 这两个函数，代码如下。

```
import { createRouter, createWebHistory } from "vue - router";
```

在项目 vue3-demo 的 components 目录下，创建 Cart.vue 和 List.vue 两个组件。其中，组件 Cart.vue 的内容如下。

```
<template>
  <h1>购物车页</h1>
</template>
<script>
export default {
    name: "Cart"
}
</script>
<style>
</style>
```

组件 List.vue 的内容如下。

```
<template>
  <h1>商品列表页</h1>
</template>
<script>
export default {
    name: "List"
}
</script>
<style>
</style>
```

在 index.js 中，引入 Cart.vue 和 List.vue 两个组件，使用 createWebHistory 函数从哈希模式切换到 history 模式，再使用 createRouter 创建路由器，它接受一个对象，history 属性值设置为 routerHistory，routes 属性值是一个数组，用来指定路由匹配列表，每一个路由映射一个组件，最后导出 router，代码如下。

```
// 引入组件
import Cart from "../components/Cart.vue";
import List from "../components/List.vue";
const routerHistory = createWebHistory();
const router = createRouter({
    history: routerHistory,
    routes: [
        {
            path: "/list",
            name: "商品列表",
            component: List
        },
        {
            path: "/cart",
            name: "购物车",
            component: Cart
        }
    ]
})
export default router
```

修改 main.js，引入 router 并挂载，代码如下。

```
import { createApp } from "vue"
import App from "./App.vue"
// 引入 router
import router from "./route/index"
// 挂载 router
createApp(App).use(router).mount("#app")
```

至此，完成路由的配置。接下来设置跳转，可以使用 vue-router 提供的<router-link>，它会被渲染为一个< a >标签。在组件 App. vue 的 template 模板中，添加两个< router-link >，用作购物车页和商品列表页的链接，添加一个< router-view/>，它可以将当前路由地址所对应的内容显示出来，代码如下。

```
<template>
    <router-link to="/cart">购物车</router-link>      
    <router-link to="/list">商品列表</router-link>
    <router-view />
  <img alt="Vue logo" src="./assets/logo.png" />
  <HelloWorld msg="Welcome to Your Vue.js App" v-if="isShow" />
…
</template>
```

启动项目并通过浏览器访问，页面输出如图 11-40 所示。

图 11-40　Vue 路由示例

在图 11-40 中，单击“购物车”链接，打开购物车页，如图 11-41 所示；单击“商品列表”链接，则打开商品列表页，如图 11-42 所示。

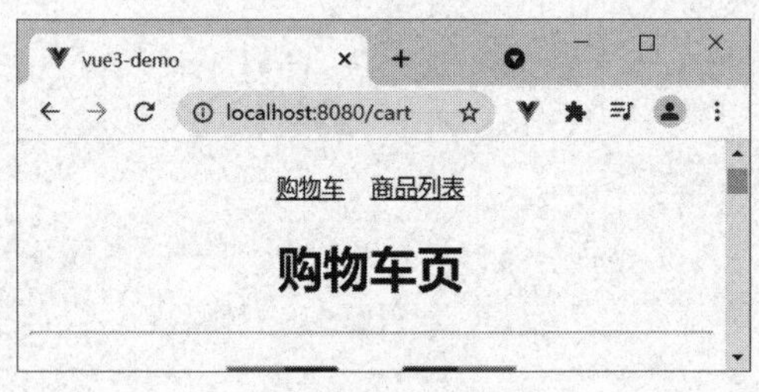

图 11-41　购物车页

图 11-42　商品列表页

11.16　axios 发送请求

axios 是基于 promise 的 http 库，主要用于向后台发起请求，并能对请求进行适当的控制。在创建项目 vue3-demo 时，还没有安装 vue-router，需要自己安装。在 Visual Studio Code 的终端控制台中，输入如下命令。

```
npm install axios --save
```

再次打开 package.json 文件，可以看到 axios 的相关信息，表明 axios 安装成功，代码如下。

```
"dependencies": {
    "axios": "^0.21.1",
    "core-js": "^3.6.5",
    "vue": "^3.0.0",
    "vue-router": "^4.0.0-beta.13"
},
```

在 main.js 文件中，依次引入 axios、设置发送请求的 baseURL，全局注册 axios，代码如下。

```
import { createApp } from 'vue'
import App from './App.vue'
// 引入 router
import router from './route/index'
// 引入 axios
import axios from 'axios'
// 设置发送请求的 baseURL
axios.defaults.baseURL = "https://api.github.com/";
const app = createApp(App)
// 全局注册 axios
app.config.globalProperties.$http = axios
// 挂载 router
app.use(router).mount('#app')
```

为了全局注册 axios，将原来如下代码：

```
createApp(App).use(router).mount('#app')
```

修改为：

```
const app = createApp(App)
app.use(router).mount('#app')
```

为了获取全局注册的 axios，需要使用 getCurrentInstance 函数。在组件 Child.vue 的 script 部分，通过 import 引入 getCurrentInstance 函数，代码如下。

```
import { ..., getCurrentInstance } from "vue";
```

在 setup 函数中，调用 getCurrentInstance 函数获取上下文实例，并从中解构出 proxy，代码如下。

```
const { proxy } = getCurrentInstance();
```

得到 proxy，就可以通过 proxy.$http 使用 axios，向后台服务器发送 get 或 post 请求。

在 setup 函数中，声明变量 userlist，用于保存从服务器获取的用户信息列表，代码如下。

```
const userlist = ref(null);
```

自定义函数 getUsers，通过 axios 向后台服务器发送 get 请求，请求的完整 url 为 https://api.github.com/users，再从服务器的返回结果中解构出 data 属性值，赋值给 res。最后将 res 赋值给变量 userlist，代码如下。

```
const getUsers = async () => {
  const { data: res } = await proxy.$http.get("users");
  userlist.value = res;
};
```

在浏览器地址栏中，输入 https://api.github.com/users，页面显示如图 11-43 所示。

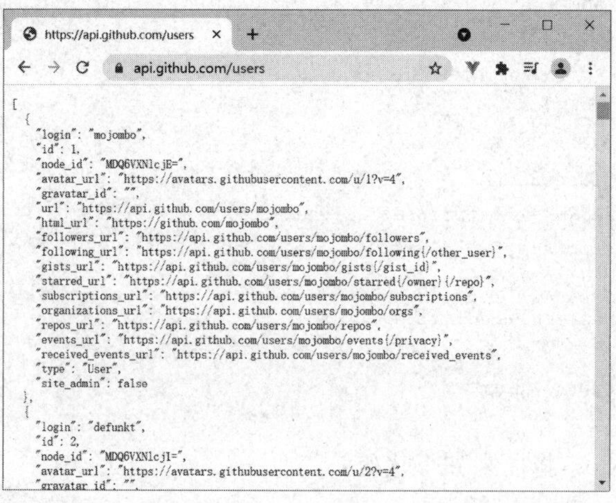

图 11-43　访问 https://api.github.com/users

在 return 中，将变量 userlist 和函数 getUsers 返回出去，供模板中使用，代码如下。

```
return {
  …
  userlist,
  getUsers
};
```

在组件 Child.vue 的 template 模板中，添加<ul>和<li>标签，将变量 userlist 中保存的用户姓名和头像循环显示，代码如下。

```
<ul>
    <li v-for="user in userlist" :key="user.id">
     <h3>{{ user.login }}</h3>
     <img :src="user.avatar_url" />
    </li>
</ul>
```

再添加“加载网络资源”的< button >标签，并绑定 click 事件。当单击“加载网络资源”按钮时，会执行函数 getUsers，代码如下。

```
<button @click = "getUsers">加载网络资源</button>
```

启动项目并通过浏览器访问，页面输出如图 11-44 所示。

图 11-44 axios 示例

在图 11-44 中，单击“加载网络资源”按钮，从指定服务器获取数据并显示，如图 11-45 所示。

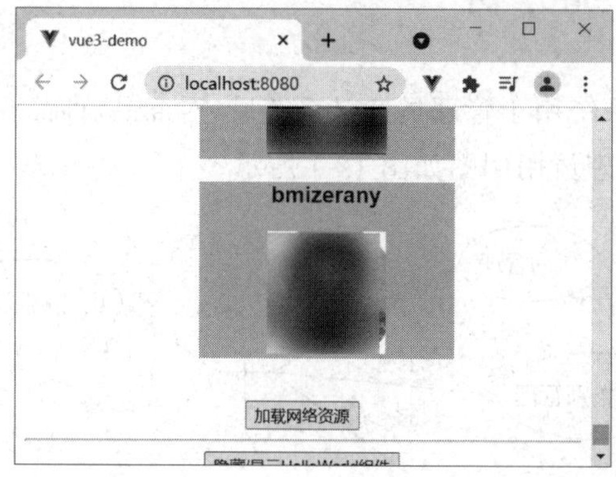

图 11-45 加载网络资源

11.17 小结

本章介绍了当前流行的 Vue.js 前端交互框架，内容包括 Vue 简介、Vue 脚手架、目录结构、初始 setup 和 ref、模板语法、生命周期钩子、reactive 函数、初始 Vue 组件、深入 setup、计算属性、条件渲染、列表渲染、watch 监视、provide 和 inject、Vue 路由、axios 发送请求。掌握这些知识点，不仅可以帮助读者更好地学习本书后面的综合案例，还可以为读者进一步系统地学习 Vue.js 打下扎实的基础。

第12章 电商平台后台管理系统

本章使用 Spring Boot＋Vue＋Element Plus 框架，介绍一个前后端分离的综合案例——电商平台后台管理系统的实现过程。

12.1 需求与系统分析

电商平台后台管理系统用于管理员登录系统后，对商品、商品类别、订单、用户权限进行管理。在这个系统中，管理员用例图如图 12-1 所示。

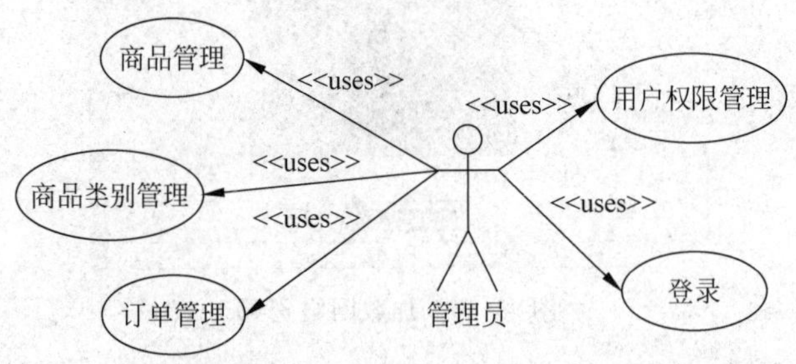

图 12-1 管理员用例图

根据需求分析，管理员拥有如下功能权限。

(1) 商品管理，包括添加商品、修改商品、查询商品。

(2) 商品类别管理，包括添加商品类别、修改商品类别、查询商品类别。

(3) 订单管理，包括创建订单、查询订单、删除订单、查看订单明细。

(4) 用户权限管理，包括前台用户管理、后台用户管理、角色管理。其中，前台用户管理包括查询前台用户、修改前台用户和删除前台用户；后台用户管理包括查询后台用户、修改后台用户、删除后台用户和分配角色；角色管理包括添加角色、修改角色、删除角色和设置权限。

根据上述分析，可以得到系统的功能模块结构，如图 12-2 所示。

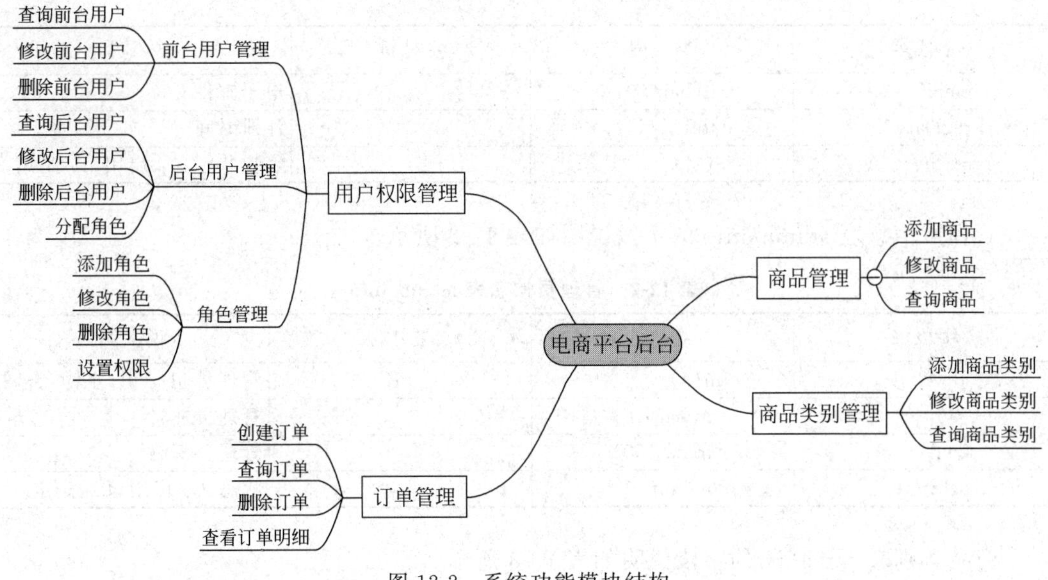

图 12-2　系统功能模块结构

12.2　数据库设计

根据系统需求，创建名称为 eshop 的数据库，创建 10 张数据表，如下所示。

(1) 客户信息表 user_info，用于记录前台客户基本信息。

(2) 管理员信息表 admin_info，用于记录管理员基本信息。

(3) 商品类别表 category，用于记录商品类别。

(4) 商品信息表 goods_info，用于记录商品信息。

(5) 订单信息表 order_info，用于记录订单主要信息。

(6) 订单明细表 order_detail，用于记录订单详细信息。

(7) 系统功能表 functions，用于记录系统功能信息。

(8) 角色表 role，用于记录系统角色信息。

(9) 角色功能表 role_functions，用于记录各个角色拥有的系统功能。

(10) 管理员角色表 admin_role，用于记录各个管理员对应的角色。

其中，客户信息表 user_info 的字段说明如表 12-1 所示。

表 12-1　客户信息表 user_info

字段名	类　型	主外键	说　明
id	int(4)	PK	客户 id 标识，主键，自增
userName	varchar(16)		登录名
password	varchar(16)		登录密码
realName	varchar(8)		真实姓名
sex	varchar(4)		性别
address	varchar(255)		联系地址

续表

字段名	类　型	主外键	说　明
email	varchar(50)		电子邮件
regDate	date		注册时间
status	int(4)		客户状态,0：启用,1：禁用

管理员信息表 admin_info 的字段说明如表 12-2 所示。

表 12-2　管理员信息表 admin_info

字段名	类　型	主外键	说　明
id	int(4)	PK	管理员 id 标识,主键,自增
name	varchar(16)		管理员姓名
pwd	varchar(50)		管理员密码
delState	smallint		状态,0：启用,1：禁用

商品类别表 category 的字段说明如表 12-3 所示。

表 12-3　商品类别表 category

字段名	类　型	主外键	说　明
id	int(4)	PK	类型 id 标识,主键,自增
name	varchar(20)		类别名称

商品信息表 goods_info 字段说明如表 12-4 所示。

表 12-4　商品信息表 goods_info

字段名	类　型	主外键	说　明
id	int(4)	PK	商品 id 标识,主键,自增
code	varchar(16)		商品编号
name	varchar(255)		商品名称
cid	int(4)	FK	商品类别 id,外键
brand	varchar(20)		品牌
pic	varchar(255)		商品图片
num	int(4)		商品数量
price	decimal(10,0)		商品价格
intro	longtext		商品描述
status	int(4)		商品状态,1 表示在售,0 表示不售

订单信息表 order_info 的字段说明如表 12-5 所示。

表 12-5　订单信息表 order_info

字段名	类　型	主外键	说　明
id	int(4)	PK	订单 id 标识,主键,自增
uid	int(4)	FK	客户 id,外键

续表

字段名	类　型	主外键	说　明
orderNo	varchar(20)		订单号
status	varchar(16)		订单状态
ordertime	timestamp		下单时间
orderprice	decimal(8,2)		订单金额

订单明细表 order_detail 的字段说明如表 12-6 所示。

表 12-6　订单明细表 order_detail

字段名	类　型	主外键	说　明
id	int(4)	PK	订单明细 id,主键,自增
oid	int(4)	FK	订单 id,外键
gid	int(4)	FK	商品 id,外键
quantity	int(4)		购买数量

系统功能表 functions 的字段说明如表 12-7 所示。

表 12-7　系统功能表 functions

字段名	类　型	主外键	说　明
id	int(4)	PK	系统功能 id,主键,自增
name	varchar(20)		功能菜单名称
parentid	int(4)		父节点 id
url	varchar(50)		功能页面
isleaf	bit(1)		是否叶子节点
nodeorder	int(4)		节点顺序

角色表 role 的字段说明如表 12-8 所示。

表 12-8　角色表 role

字段名	类　型	主外键	说　明
roleId	int(4)	PK	角色 id,主键,自增
roleName	varchar(50)		角色名称
delState	int(4)		状态,0：启用,1：禁用

角色功能表 role_functions 的字段说明如表 12-9 所示。

表 12-9　角色功能表 role_functions

字段名	类　型	主外键	说　明
rid	int(4)	PK	角色 id,主键
fid	int(4)	PK	功能 id,主键

管理员角色表 admin_role 的字段说明如表 12-10 所示。

表 12-10　管理员角色表 admin_role

字段名	类　型	主外键	说　明
arId	int(4)	PK	管理员角色 id,主键,自增
aid	int(4)	FK	管理员 id,外键
rid	int(4)	FK	角色 id,外键

创建数据表时,还需要设置数据表之间的关联关系,如图 12-3 所示。

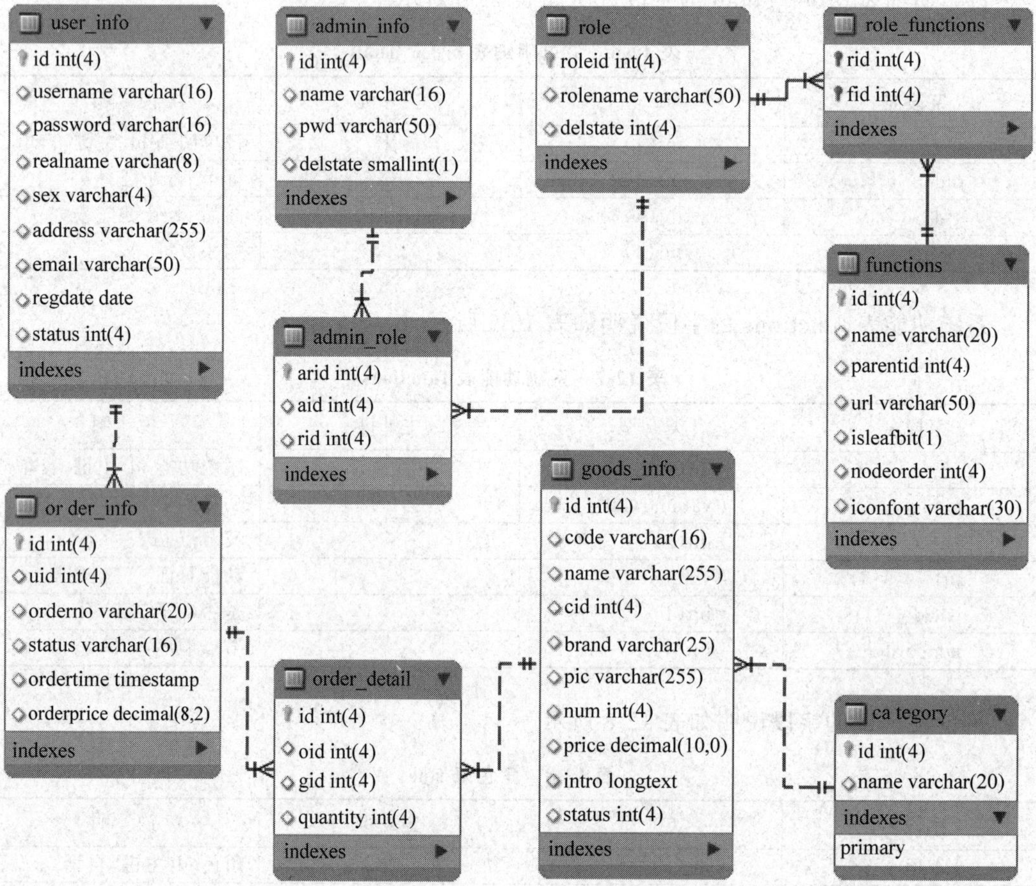

图 12-3　系统数据表之间关系图

12.3　环境搭建

可以参照本书 Spring Boot 整合 MyBatis 的章节,完成电商平台后台管理系统后端程序 eshop 的框架搭建。

12.3.1　后端程序目录结构

后端程序 eshop 的目录结构如图 12-4 所示。

com. my. eshop. controller 包用于存放控制器类，com. my. eshop. service 包用于存放业务逻辑层接口，com. my. eshop. service. impl 包用于存放业务逻辑层接口的实现类，com. my. eshop. dao 包用于存放 Mapper 接口，com. my. eshop. pojo 包用于存放实体类，com. my . eshop. config 包用于存放配置类，com. my. eshop . interceptor 包用于存放拦截器类，com. my. eshop. utils 包用于存放工具栏。在 main/resources/mapper 文件夹中，存放了 Mapper 接口的映射文件。在 main\resources\config 文件夹中，存放了 Mybatis 的配置文件 mybatis-config. xml。

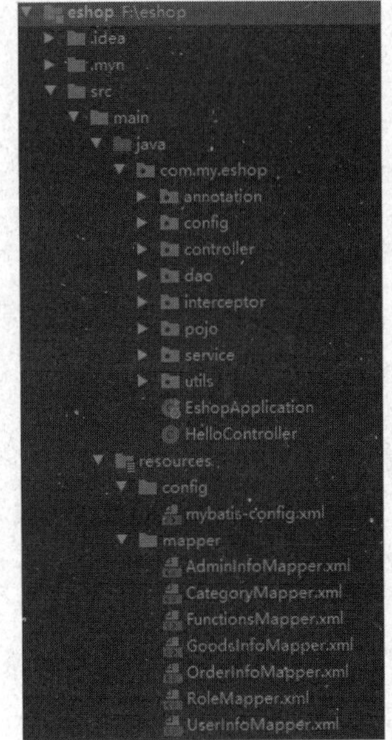

图 12-4　目录结构图

12.3.2　编辑 Spring Boot 配置文件

在 application. properties 文件中，配置端口号如下。

```
server.port = 8888
```

在 application. yml 文件中，配置对 Mybatis 映射文件和配置文件的引用，配置 MySQL 数据源，配置 Redis，代码如下。

```
mybatis:
  mapper - locations: classpath:/mapper/ * .xml
  config - location: classpath:/config/mybatis - config.xml
spring:
  datasource:
    username: root
    password: 123456
    url: jdbc: mysql://127.0.0.1: 3306/eshop?useUnicode = true&
    characterEncoding = utf8&zeroDateTimeBehavior = convertToNull
    driver - class - name: com.mysql.jdbc.Driver
  redis:
    host: 127.0.0.1
```

12.3.3　创建 MyBatis 配置文件

在 MyBatis 配置文件 mybatis-config. xml 中，通过< settings >和< setting >标签，将设置项 mapUnderscoreToCamelCase 的值指定为 true，开启驼峰命名规则映射；通过< typeAliases >和< package >标签，为 SQL 映射文件中的输入和输出参数设置类型别名，< package >标签配置了一个包名，MyBatis 会扫描该包下的所有类，并注册一个别名，别名是类名或者是类名第一个字母小名的类名。代码如下。

```
<?xml version = "1.0" encoding = "UTF - 8"?>
<! DOCTYPE configuration
     PUBLIC " - //mybatis.org//DTD Config 3.0//EN"
     "http://mybatis.org/dtd/mybatis - 3 - config.dtd">
< configuration >
     < settings >
          < setting name = "mapUnderscoreToCamelCase" value = "true"/>
     </settings >
     < typeAliases >
          < package name = "com.my.eshop.pojo"/>
     </typeAliases >
</configuration >
```

12.3.4 集成 JWT 实现 Token 验证

用户登录后，每个请求中都会包含 JWT，用来验证用户身份，并对路由、服务和资源的访问权限进行验证。Spring Boot 集成 JWT 的主要步骤如下。

(1) 引入 JWT 依赖。在 pom.xml 文件中，添加如下内容。

```
< dependency >
     < groupId > com.auth0 </groupId >
     < artifactId > java - jwt </artifactId >
     < version > 3.4.0 </version >
</dependency >
```

(2) 自定义 PassToken 和 UserLoginToken 注解。PassToken 用于跳过验证，代码如下。

```
package com.my.eshop.annotation;
import java.lang.annotation.ElementType;
import java.lang.annotation.Retention;
import java.lang.annotation.RetentionPolicy;
import java.lang.annotation.Target;
@Target({ElementType.METHOD, ElementType.TYPE})
@Retention(RetentionPolicy.RUNTIME)
public @interface PassToken {
     boolean required() default true;
}
```

UserLoginToken 用于标注 Controller 中需要登录才能进行的操作，代码如下。

```
package com.my.eshop.annotation;
import java.lang.annotation.ElementType;
…
@Target({ElementType.METHOD, ElementType.TYPE})
```

```
@Retention(RetentionPolicy.RUNTIME)
public @interface UserLoginToken {
    boolean required() default true;
}
```

(3) 创建 Token 的生成方法。在 com.my.eshop.service 包中，创建接口 TokenService，并添加 getToken()方法，代码如下。

```
package com.my.eshop.service;
import com.my.eshop.pojo.AdminInfo;
public interface TokenService {
    public String getToken(AdminInfo admin);
}
```

在 com.my.eshop.service.impl 包中，创建接口 TokenService 的实现类 TokenServiceImpl，实现 getToken()方法，代码如下。

```
package com.my.eshop.service.impl;
import com.auth0.jwt.JWT;
…
@Service
public class TokenServiceImpl implements TokenService {
    @Override
    public String getToken(AdminInfo admin) {
        String token = "";
        // 将管理员的 id 保存到 token 里面,以 password 作为 token 的密钥
        token = JWT.create().withAudience(String.valueOf(admin.getId())).
sign(Algorithm.HMAC256(admin.getPwd()));
        return token;
    }
}
```

在 getToken()方法中，Algorithm.HMAC256 使用 HS256 生成 Token，密钥是用户的密码。withAudience 用于存入需要保存在 Token 中的信息，这里为管理员的 id。

(4) 创建拦截器获取并验证 Token。在 com.my.eshop.interceptor 包中，创建拦截器类 AuthenticationInterceptor，代码如下。

```
package com.my.eshop.interceptor;
import javax.servlet.http.HttpServletRequest;
…
public class AuthenticationInterceptor implements HandlerInterceptor {
    @Autowired
    AdminInfoService adminInfoService;
    @Override
    public boolean preHandle(HttpServletRequest httpServletRequest,
HttpServletResponse httpServletResponse, Object object)
    throws Exception {
```

```
        // 从 http 请求头中取出token
        String token = httpServletRequest.getHeader("token");
        // 如果不是映射到方法直接通过
        if(!(object instanceof HandlerMethod)){
            return true;
        }
        HandlerMethod handlerMethod = (HandlerMethod)object;
        Method method = handlerMethod.getMethod();
        // 检查是否有 passtoken 注释,有则跳过认证
        if (method.isAnnotationPresent(PassToken.class)) {
            PassToken passToken = method.getAnnotation(PassToken.class);
            if (passToken.required()) {
                return true;
            }
        }
        // 检查有没有需要用户权限的注解
        if (method.isAnnotationPresent(UserLoginToken.class)) {
            UserLoginToken userLoginToken = method.getAnnotation(UserLoginToken.class);
            if (userLoginToken.required()) {
                // 执行认证
                if (token == null) {
                    throw new RuntimeException("无 token,请重新登录");
                }
                // 获取 token 中的用户 id
                String userId;
                try {
                    userId = JWT.decode(token).getAudience().get(0);
                } catch (JWTDecodeException j) {
                    throw new RuntimeException("401");
                }
                 AdminInfo admin = adminInfoService.getAdminById(Integer.parseInt
(userId));
                if (admin == null) {
                    throw new RuntimeException("用户不存在,请重新登录");
                }
                // 验证 token
                JWTVerifier jwtVerifier = JWT.require(Algorithm.HMAC256(admin.getPwd())).
build();
                try {
                    jwtVerifier.verify(token);
                } catch (JWTVerificationException e) {
                    throw new RuntimeException("401");
                }
                // 将验证通过后的用户信息放到请求中
                httpServletRequest.setAttribute("currentUser", admin);
                return true;
            }
        }
```

```
        return true;
    }
    @Override
    public void postHandle(HttpServletRequest httpServletRequest,HttpServletResponse
httpServletResponse, Object o, ModelAndView modelAndView) throws Exception {
    }
    @Override
    public void afterCompletion(HttpServletRequest httpServletRequest,
HttpServletResponse httpServletResponse, Object o, Exception e)
throws Exception {
    }
}
```

在拦截器类 AuthenticationInterceptor 中，实现了 HandlerInterceptor 接口，重写了 preHandle()方法。该方法为预处理回调方法，实现处理器的预处理，第三个参数为响应的处理器，返回值为 true 表示继续流程（如调用下一个拦截器或处理器）或者接着执行 postHandle()方法和 afterCompletion()方法；false 表示流程中断，不会继续调用其他的拦截器或处理器，中断执行。

在 preHandle()方法中，从 http 请求头中取出 Token，先判断是否映射到方法，如果没有映射到方法则返回 true，直接通过，再检查是否有 PassToken 注释，如果有则跳过认证；然后检查有没有需要用户登录的注解，有则需要取出并验证。如果认证通过则可以访问，否则报错。

（5）配置拦截器。在 com. my. eshop. config 包中，创建配置类 InterceptorConfig，代码如下。

```
package com.my.eshop.config;
import com.my.eshop.interceptor.AuthenticationInterceptor;
…
@Configuration
public class InterceptorConfig implements WebMvcConfigurer {
    @Override
    public void addInterceptors(InterceptorRegistry registry) {
        //拦截所有请求,通过判断是否有@UserLoginToken 注解决定是否需要登录
        registry.addInterceptor(authenticationInterceptor())
                .addPathPatterns("/**");
    }
    @Bean
    public AuthenticationInterceptor authenticationInterceptor() {
        return new AuthenticationInterceptor();
    }
}
```

InterceptorConfig 类上标注了注解@Configuration，指明该类是一个配置类并将其作为一个 Spring 的 Bean 添加到 IOC 容器中。

InterceptorConfig 类实现了 WebMvcConfigurer 接口，并重写了 addInterceptors()方

法,用于设置拦截器的过滤路径规则。这里拦截所有请求,通过判断控制器 Controller 类中的方法上是否标注@UserLoginToken 注解,决定是否需要登录。

12.3.5 配置跨域

由于项目前后端分离,所以存在跨域问题。为了解决跨域问题,在 com. my. eshop. config 包中,创建 MyWebMvcConfig 类,代码如下。

```
package com.my.eshop.config;
import org.springframework.beans.factory.annotation.Configurable;
…
@Configuration
public class MyWebMvcConfig implements WebMvcConfigurer{
    @Override
    public void addCorsMappings(CorsRegistry registry) {
        registry.addMapping("/**").allowedHeaders("*").allowedMethods("*")
                .maxAge(1800).allowedOrigins("http://localhost:8080");
    }
}
```

MyWebMvcConfig 类上标注了注解@Configuration,指明该类是一个配置类并将其作为一个 Spring 的 Bean 添加到 IOC 容器中。

MyWebMvcConfig 类实现了 WebMvcConfigurer 接口,并重写了 addCorsMappings() 方法。addMapping 用于设置允许跨域访问的路径,allowedHeaders 用于设置允许的请求头,allowedMethods 用于设置允许的请求方法,* 代表允许所有。allowedOrigins 用于设置允许跨域访问的源,设置为 http://localhost:8080,表示后端程序能接收来自 http://localhost:8080 的请求。

12.4 创建实体类

在 com. my. eshop. pojo 包中,依次创建实体类 UserInfo、AdminInfo、Category、GoodsInfo、OrderInfo、OrderDetail、Functions、Role 和 RoleFunctions。

实体类 UserInfo 用于封装客户信息,代码如下。

```
package com.my.eshop.pojo;
public class UserInfo {
    private int id;
    private String userName;
    private String password;
    private String realName;
    private String sex;
    private String address;
```

```
    private String email;
    private String regDate;
    private int status;
    //省略上述属性的 getter 和 setter 方法
}
```

实体类 AdminInfo 用于封装管理员信息,代码如下。

```
package com.my.eshop.pojo;
import java.util.List;
public class AdminInfo {
    private int id;
    private String name;
    private String pwd;
    private int delState;
    //关联的属性
    private List<Functions> fs;
    private Role role;
    //省略上述属性的 getter 和 setter 方法
}
```

实体类 Category 用于封装商品类别信息,代码如下。

```
package com.my.eshop.pojo;
public class Category {
    private int id;                          //产品类型编号
    private String name;                     //产品类型名称
    //省略上述属性的 getter 和 setter 方法
}
```

实体类 GoodsInfo 用于封装商品信息,代码如下。

```
package com.my.eshop.pojo;
public class GoodsInfo {
    //商品基本信息(部分)
    private int id;                              //商品编号
    private String code;                         //商品编码
    private String name;                         //商品名称
    private int cid;
    private Category category;                   //商品类别
    private String brand;                        //商品品牌
    private String pic;                          //商品小图
    private int num;                             //商品数量
    private double price;                        //商品价格
    private String intro;                        //商品介绍
    private int status;                          //商品状态
    //查询属性
    private double priceFrom;
    private double priceTo;
    //省略上述属性的 getter 和 setter 方法
}
```

实体类 OrderInfo 用于封装订单信息，代码如下。

```
package com.my.eshop.pojo;
import java.util.List;
public class OrderInfo {
    private Integer id;
    private String orderNo;
    private int uid;
    private UserInfo ui;
    private String status;
    private String ordertime;
    private double orderprice;
    private String orderTimeFrom;
    private String orderTimeTo;
    //关联属性
    private List<OrderDetail> orderDetails;
    //省略上述属性的 getter 和 setter 方法
}
```

实体类 OrderDetail 用于封装订单明细信息，代码如下。

```
package com.my.eshop.pojo;
public class OrderDetail {
    private int id;
    private int oid;
    private OrderInfo oi;
    private int gid;
    private String code;
    private GoodsInfo goods;
    private int quantity;
    private double price;
    private double totalprice;
    //省略上述属性的 getter 和 setter 方法
}
```

实体类 Functions 用于封装系统功能信息，代码如下。

```
package com.my.eshop.pojo;
import java.util.HashSet;
import java.util.Set;
public class Functions implements Comparable<Functions> {
    private int id;
    private String name;
    private int parentid;
    private boolean isleaf;
    private String iconfont;
    private String url;
    //关联的属性
    private Set ais = new HashSet();
    //省略上述属性的 getter 和 setter 方法
```

```
    @Override
    public int compareTo(Functions arg0) {
        return ((Integer) this.getId()).compareTo((Integer) (arg0.getId()));
    }
}
```

在实体类 Functions 中，重写了 compareTo(Functions arg0)方法。该方法用于在排序时将两个 Functions 对象的 id 进行比较，根据比较的结果是小于、等于或者大于而返回一个负数、零或者正数。

实体类 Role 用于封装角色信息，代码如下。

```
package com.my.eshop.pojo;
public class Role {
    private int roleId;
    private String roleName;
    private int delState;
    // 省略上述属性的 getter 和 setter 方法
}
```

实体类 RoleFunctions 用于封装角色功能信息，代码如下。

```
package com.my.eshop.pojo;
public class RoleFunctions {
    private int rid;
    private int fid;
    // 省略上述属性的 getter 和 setter 方法
}
```

此外，还创建了辅助类 Pager 和 TreeNode。Pager 类用于封装分页信息，代码如下。

```
package com.my.eshop.pojo;
public class Pager {
    private int curPage;                              // 待显示页
    private int perPageRows;                          // 每页显示的记录数
    private int rowCount;                             // 记录总数
    private int pageCount;                            // 总页数
    // 根据 rowCount 和 perPageRows 计算总页数
    public int getPageCount() {
        return (rowCount + perPageRows - 1) / perPageRows;
    }
    // 分页显示时，获取当前页的第一条记录的索引
    public int getFirstLimitParam() {
        return (this.curPage - 1) * this.perPageRows;
    }
    // 省略其他属性的 getter 和 setter 方法
}
```

TreeNode 类用于封装树形控件的节点信息，代码如下。

```
package com.my.eshop.pojo;
import java.util.List;
public class TreeNode {
    private int id;                              // 节点 id
    private String text;                         // 节点名称
    private int pid;                             // 父节点 id
    private String iconfont;                     // 节点图标字体
    private String url;                          // 跳转路径
    private List<TreeNode> children;             // 包含的子节点
    // 省略其他属性的 getter 和 setter 方法
}
```

12.5 创建 Mapper 接口及映射文件

在 MyBatis 框架中，使用 Mapper 接口和映射文件的方式需要满足以下条件。

(1) 映射文件的 namespace 的值必须是 Mapper 接口的全路径名称。

(2) Mapper 接口中声明的方法名在映射文件中必须有一个 id 值与之对应。

(3) 映射文件的名称必须和 Mapper 接口的名称一致。

在 com.my.eshop.dao 包中，创建接口 UserInfoMapper，声明下列方法。

```
package com.my.eshop.dao;
import com.my.eshop.pojo.AdminInfo;
…
@Repository
public interface UserInfoMapper {
    // 根据状态获取客户列表
    List<UserInfo> findUserInfoByStatus(int status);
    // 根据客户 id 获取客户信息
    UserInfo findUserInfoById(int id);
    // 根据条件查询客户总数
    Integer count(Map<String,Object> params);
    // 分页获取客户信息
    List<UserInfo> findByPage(Map<String,Object> params);
    // 更新客户状态
    int updateStatus(int id);
    // 根据客户姓名获取客户信息
    UserInfo findUserInfoByUserName(String userName);
    // 修改客户信息
    int updateUserInfo(UserInfo userInfo);
}
```

在 main/resources/mapper 文件夹中，创建接口 UserInfoMapper 的映射文件 UserInfoMapper.xml，代码如下。

```
<?xml version = "1.0" encoding = "UTF - 8"?>
<! DOCTYPE mapper PUBLIC " - //mybatis.org//DTD Mapper 3.0//EN" " http://mybatis.org/dtd/
mybatis - 3 - mapper.dtd">
< mapper namespace = "com.my.eshop.dao.UserInfoMapper">
    <!-- 根据状态获取客户列表 -- >
    < select id = "findUserInfoByStatus" parameterType = "map" resultType = "UserInfo">
        SELECT * FROM user_info
        < where >
            < if test = "status > 0">
                status = #{status}
            </if >
        </where >
    </select >
    <!-- 根据客户 id 获取客户信息 -- >
    < select id = "findUserInfoById" parameterType = "int" resultType = "UserInfo">
        SELECT * FROM user_info WHERE id = #{id}
    </select >
    <!-- 根据条件查询客户总数 -- >
    < select id = "count" parameterType = "map" resultType = "int">
        SELECT count( * ) FROM user_info
        < where >
            < if test = "userName != ''">
                userName LIKE '%' #{userName} '%'
            </if >
        </where >
    </select >
    <!-- 分页获取客户信息 -- >
    < select id = "findByPage" parameterType = "map" resultType = "UserInfo">
        SELECT * FROM user_info
        < where >
            < if test = "userName != ''">
                userName LIKE '%' #{userName} '%'
            </if >
        </where >
        limit #{pager.firstLimitParam} , #{pager.perPageRows}
    </select >
    <!-- 更新客户状态 -- >
    < update id = "updateStatus" parameterType = "Integer">
        UPDATE user_info SET status = 1 where id = #{id}
    </update >
    <!-- 根据客户姓名获取客户信息 -- >
    < select id = "findUserInfoByUserName" parameterType = "String" resultType = "UserInfo">
        SELECT * FROM user_info WHERE userName = #{userName}
    </select >
    <!-- 修改客户信息 -- >
    < update id = "updateUserInfo" parameterType = "UserInfo">
        UPDATE user_info SET userName = #{userName}, password = #{password} , realName =
#{realName}, sex = #{sex}, address = #{address} , email = #{email} where id = #{id}
    </update >
</mapper >
```

创建接口 AdminInfoMapper,声明下列方法。

```
package com.my.eshop.dao; import com.my.eshop.pojo.AdminInfo;
import org.apache.ibatis.annotations.Param;
…
@Repository
public interface AdminInfoMapper {
    // 根据姓名和密码查询管理员
    AdminInfo findByNameAndPwd(AdminInfo adminInfo);
    // 根据姓名查询管理员
    AdminInfo findByName(@Param("name") String name);
    // 根据 id 查询管理员
    AdminInfo findAdminById(@Param("id") int id);
    // 查询所有管理员列表
    List<AdminInfo> findAll();
    // 根据条件查询管理员总数
    Integer count(Map<String,Object> params);
    // 分页获取管理员列表
    List<AdminInfo> findByPage(Map<String,Object> params);
    // 插入管理员
    int insertAdminInfo(AdminInfo adminInfo);
    // 修改管理员
    int updateAdminInfo(AdminInfo adminInfo);
    // 更新管理员状态
    int updateDelState(int id);
}
```

创建接口 AdminInfoMapper 的映射文件 AdminInfoMapper.xml,代码如下。

```
<?xml version="1.0" encoding="UTF-8"?>
<!DOCTYPE mapper PUBLIC "-//mybatis.org//DTD Mapper 3.0//EN" "http://mybatis.org/dtd/
mybatis-3-mapper.dtd">
<mapper namespace="com.my.eshop.dao.AdminInfoMapper">
    <!-- 查询语句查询结果映射 -->
    <resultMap type="AdminInfo" id="adminInfoMap">
        <id property="id" column="id"/>
        <result property="name" column="name"/>
        <result property="pwd" column="pwd"/>
        <result property="delState" column="delState"/>
        <!-- 一对一关联映射 -->
        <association property="role" column="id" select="com.my.eshop.dao.RoleMapper.
findRoleByAdminId"
                     javaType="com.my.eshop.pojo.Role"/>
    </resultMap>
    <!--根据姓名和密码查询管理员 -->
    <select id="findByNameAndPwd" parameterType="AdminInfo"
            resultType="com.my.eshop.pojo.AdminInfo">
    select * from admin_info where name = #{name} and pwd = #{pwd}
   </select>
```

```
    <!-- 根据姓名查询管理员 -->
    <select id="findByName" resultType="AdminInfo">
      SELECT * FROM admin_info
      where
      name = #{name}
    </select>
    <!-- 根据 id 查询管理员 -->
    <select id="findAdminById" resultType="AdminInfo">
        SELECT * FROM admin_info where id = #{id}
    </select>
    <!-- 查询所有管理员列表 -->
    <select id="findAll" resultType="AdminInfo">
        SELECT * FROM admin_info
    </select>
    <!-- 根据条件查询管理员总数 -->
    <select id="count" parameterType="map" resultType="int">
        SELECT count(*) FROM admin_info
        <where>
            <if test="name != ''">
                name LIKE '%' #{name} '%'
            </if>
        </where>
    </select>
    <!-- 分页获取管理员列表 -->
    <select id="findByPage" parameterType="map" resultMap="adminInfoMap">
        SELECT * FROM admin_info
        <where>
            <if test="name != ''">
                name LIKE '%' #{name} '%'
            </if>
        </where>
        limit #{pager.firstLimitParam} , #{pager.perPageRows}
    </select>
    <!-- 插入管理员 -->
    <insert id="insertAdminInfo" parameterType="AdminInfo" useGeneratedKeys="true"
keyProperty="id">
            insert into admin_info(name,pwd) values(#{name},#{pwd})
    </insert>
    <!-- 修改管理员 -->
    <update id="updateAdminInfo" parameterType="AdminInfo">
        UPDATE admin_info SET name = #{name}, pwd = #{pwd} where id = #{id}
    </update>
    <!-- 更新管理员状态 -->
    <update id="updateDelState" parameterType="Integer">
        UPDATE admin_info SET delState = 1 where id = #{id}
    </update>
</mapper>
```

创建接口 CategoryMapper，声明下列方法。

```
package com.my.eshop.dao;
import com.my.eshop.pojo.Category;
import org.springframework.stereotype.Repository;
import java.util.List;
@Repository
public interface CategoryMapper {
    // 根据类别名获取类别列表
    List<Category> findCategories(String name);
    // 插入类别
    int insertCategory(String name);
    // 根据类别名获取类别对象
    Category findByName(String name);
    // 根据类别 id 获取类别对象
    Category findById(int id);
    // 修改类别
    int updateCategory(Category category);
}
```

创建接口 CategoryMapper 的映射文件 CategoryMapper.xml，代码如下。

```
<?xml version="1.0" encoding="UTF-8"?>
<!DOCTYPE mapper PUBLIC "-//mybatis.org//DTD Mapper 3.0//EN" "http://mybatis.org/dtd/
mybatis-3-mapper.dtd">
<mapper namespace="com.my.eshop.dao.CategoryMapper">
    <!-- 根据类别名获取类别列表 -->
    <select id="findCategories" parameterType="String" resultType="Category">
        SELECT * FROM category
        <where>
            <if test="name != ''">
                name like '%${name}%'
            </if>
        </where>
    </select>
    <!-- 插入类别 -->
    <insert id="insertCategory" parameterType="String">
            insert into category(name) values(#{name})
    </insert>
    <!-- 根据类别名获取类别对象 -->
    <select id="findByName" parameterType="String" resultType="Category">
      SELECT * FROM category
      where
      name = #{name}
    </select>
    <!-- 根据类别 id 获取类别对象 -->
    <select id="findById" parameterType="Integer" resultType="Category">
        SELECT * FROM category where id = #{id}
    </select>
    <!-- 修改类别 -->
    <update id="updateCategory" parameterType="Category">
```

```
        UPDATE category SET name = #{name} where id = #{id}
    </update>
</mapper>
```

创建接口 GoodsInfoMapper，声明下列方法。

```
package com.my.eshop.dao;
import com.my.eshop.pojo.GoodsInfo;
…
@Repository
public interface GoodsInfoMapper {
    // 根据状态获取商品列表
    List<GoodsInfo> findGoodsInfoByStatus(int status);
    // 获取满足条件的商品总数
    Integer count(Map<String,Object> params);
    // 分页获取商品列表
    List<GoodsInfo> findByPage(Map<String,Object> params);
    // 插入商品
    int insertGoodsInfo(GoodsInfo goodsInfo);
    // 根据 id 获取商品对象
    GoodsInfo findById(int id);
    // 修改商品
    int updateGoodsInfo(GoodsInfo goodsInfo);
}
```

创建接口 GoodsInfoMapper 的映射文件 GoodsInfoMapper.xml，代码如下。

```
<?xml version = "1.0" encoding = "UTF-8"?>
<!DOCTYPE mapper PUBLIC "-//mybatis.org//DTD Mapper 3.0//EN" "http://mybatis.org/dtd/
mybatis-3-mapper.dtd">
<mapper namespace = "com.my.eshop.dao.GoodsInfoMapper">
    <!-- 查询语句查询结果映射 -->
    <resultMap type = "GoodsInfo" id = "goodsInfoMap">
        <!-- 多对一关联映射 -->
        <association property = "category" column = "cid" select = "com.my.eshop.dao.
CategoryMapper.findById" javaType = "Category"/>
    </resultMap>
    <!-- 根据状态获取商品列表 -->
    <select id = "findGoodsInfoByStatus" parameterType = "map" resultType = "com.my.eshop.
pojo.GoodsInfo">
        SELECT * FROM goods_info where status = #{status}
    </select>
    <!-- 获取满足条件的商品总数 -->
    <select id = "count" parameterType = "map" resultType = "int">
        SELECT count(*) FROM goods_info where status = 1
        <if test = "name != ''"> and name like '%${name}%'
        </if>
    </select>
    <!-- 分页获取商品列表 -->
```

```
    <select id="findByPage" parameterType="map" resultMap="goodsInfoMap">
        SELECT * FROM goods_info where status = 1
        <if test="name != ''">
            and name like '%${name}%'
        </if>
        limit #{pager.firstLimitParam} , #{pager.perPageRows}
    </select>
    <!-- 插入商品 -->
    <insert id="insertGoodsInfo" parameterType="com.my.eshop.pojo.GoodsInfo">
        insert into goods_info(code,name,cid,brand,num,price,intro)
        values(#{code},#{name},#{cid},#{brand},#{num},#{price},#{intro})
    </insert>
    <!-- 修改商品 -->
    <insert id="updateGoodsInfo" parameterType="com.my.eshop.pojo.GoodsInfo">
        update goods_info set code = #{code}, name = #{name}, cid = #{cid}, brand =
#{brand}, num = #{num}, price = #{price}, intro = #{intro} where id = #{id}
    </insert>
    <!-- 修改商品 -->
    <select id="findById" parameterType="Integer" resultType="GoodsInfo">
        SELECT * FROM goods_info where id = #{id}
    </select>
</mapper>
```

创建接口 OrderInfoMapper，声明下列方法。

```
package com.my.eshop.dao;
import com.my.eshop.pojo.AdminInfo;
…
@Repository
public interface OrderInfoMapper {
    // 插入订单
    int insertOrderInfo(OrderInfo oi);
    // 插入订单明细
    int insertOrderDetail(OrderDetail od);
    // 获取满足条件的订单总数
    Integer count(Map<String, Object> params);
    // 分页获取订单列表
    List<OrderInfo> findByPage(Map<String, Object> params);
    // 根据订单 id 删除订单明细
    int deleteOrderDetailByOid(int oid);
    // 根据订单 id 删除订单
    int deleteOrderInfoById(int id);
    // 根据订单 id 获取订单明细列表
    List<OrderDetail> findOrderDetailByOid(int oid);
}
```

创建接口 OrderInfoMapper 的映射文件 OrderInfoMapper.xml，代码如下。

```
<?xml version = "1.0" encoding = "UTF - 8"?>
<! DOCTYPE mapper PUBLIC " - //mybatis. org//DTD Mapper 3. 0//EN" " http://mybatis. org/dtd/
mybatis - 3 - mapper. dtd">
< mapper namespace = "com. my. eshop. dao. OrderInfoMapper">
    <!-- 查询语句查询结果映射 -- >
    < resultMap type = "OrderInfo" id = "orderInfoMap">
        < id property = "id" column = "id"/>
        < result property = "orderNo" column = "orderNo"/>
        < result property = "status" column = "status"/>
        < result property = "ordertime" column = "ordertime"/>
        < result property = "orderprice" column = "orderprice"/>
        <!-- 多对一关联映射 -- >
          < association property = " ui " column = " uid " select = " com. my. eshop. dao.
UserInfoMapper. findUserInfoById" javaType = "UserInfo"/>
    </resultMap >
    <!-- 查询语句查询结果映射 -- >
    < resultMap type = "OrderDetail" id = "orderDetailMap">
        < id property = "id" column = "id"/>
        < id property = "oid" column = "oid"/>
        < id property = "gid" column = "gid"/>
        < id property = "quantity" column = "quantity"/>
        <!-- 关联映射 -- >
          < association property = " goods " column = " gid " select = " com. my. eshop. dao.
GoodsInfoMapper. findById"  javaType = "GoodsInfo"/>
    </resultMap >
    <!-- 插入订单 -- >
     < insert id = " insertOrderInfo " parameterType = " com. my. eshop. pojo. OrderInfo "
useGeneratedKeys = "true" keyProperty = "id" keyColumn = "id">
        insert into order_info(uid, orderNo, status, ordertime, orderprice)
        values( # {uid}, # {orderNo}, # {status}, # {ordertime}, # {orderprice})
    </insert >
    <!-- 插入订单明细 -- >
    < insert id = "insertOrderDetail" parameterType = "com. my. eshop. pojo. OrderDetail">
        insert into order_detail(oid, gid, quantity) values( # {oid}, # {gid}, # {quantity})
    </insert >
    <!-- 获取满足条件的订单总数 -- >
    < select id = "count" parameterType = "map" resultType = "int">
        SELECT count( * ) FROM order_info
        < where >
            < if test = "orderNo != ''">
                orderNo = # {orderNo}
            </if >
            < if test = "uid > 0">
                and uid = # {uid}
            </if >
            < if test = "status != ''">
                and status = # {status}
            </if >
        </where >
    </select >
```

```
    <!-- 分页获取订单列表 -->
    <select id="findByPage" parameterType="map" resultMap="orderInfoMap">
        SELECT * FROM order_info
        <where>
            <if test="orderNo != ''">
                orderNo = #{orderNo}
            </if>
            <if test="uid > 0">
                and uid = #{uid}
            </if>
            <if test="status != ''">
                and status = #{status}
            </if>
        </where>
        limit #{pager.firstLimitParam} , #{pager.perPageRows}
    </select>
    <!-- 根据订单 id 删除订单子表 -->
    <delete id="deleteOrderDetailByOid" parameterType="int">
        delete from order_detail where oid = #{oid}
    </delete>
    <!-- 根据订单 id 删除订单主表 -->
    <delete id="deleteOrderInfoById" parameterType="int">
        delete from order_info where id = #{id}
    </delete>
    <!-- 根据订单 id 获取订单明细 -->
    <select id="findOrderDetailByOid" parameterType="int" resultMap="orderDetailMap">
        select * from order_detail where oid = #{oid}
    </select>
</mapper>
```

创建接口 FunctionsMapper，声明下列方法。

```
package com.my.eshop.dao;
import com.my.eshop.pojo.Functions;
import org.springframework.stereotype.Repository;
import java.util.List;
@Repository
public interface FunctionsMapper {
    // 根据 id 获取系统功能对象
    Functions findFunctionsById(int id);
    // 获取系统功能列表
    List<Functions> findAllFunctions();
    // 查询管理员可以操作的功能模块编号集合
    List<String> findFunctionIdsByAid(int aid);
}
```

创建接口 FunctionsMapper 的映射文件 FunctionsMapper.xml，代码如下。

```
<?xml version = "1.0" encoding = "UTF - 8"?>
<! DOCTYPE mapper PUBLIC " - //mybatis. org//DTD Mapper 3. 0//EN" "http://mybatis. org/dtd/
mybatis - 3 - mapper. dtd">
< mapper namespace = "com. my. eshop. dao. FunctionsMapper">
    <!-- 查询管理员可以操作的功能模块编号集合 -- >
    < select id = "findFunctionIdsByAid" parameterType = "Integer" resultType = "java. lang.
String">
        select distinct fid from role_functions
        where rid in (select rid from admin_role where aid = #{aid})
    </select >
    <!-- 获取系统功能列表 -- >
    < select id = "findAllFunctions" resultType = "com. my. eshop. pojo. Functions">
        SELECT * FROM functions
    </select >
    <!-- 根据 id 获取系统功能对象 -- >
    < select id = "findFunctionsById" parameterType = "Integer" resultType = "com. my. eshop.
pojo. Functions">
        SELECT * FROM functions Where id = #{id}
    </select >
</mapper >
```

创建接口 RoleMapper，声明下列方法。

```
package com. my. eshop. dao;
import com. my. eshop. pojo. Category;
…
@Repository
public interface RoleMapper {
    // 获取所有角色列表
    List < Role > findAllRole();
    // 根据状态获取角色列表
    List < Role > findRoleByDelState( int delState);
    // 根据角色 id 获取对应的功能集合
    List < Integer > findFunctionIdsByRid( int roleId);
    // 删除 roleid 下的 functions
    public int deleteFunctionsByRoleId( int roleId);
    // 插入角色功能记录
    public int insertFunctionsByRoleId(Map < String, Object > params);
    // 根据管理员 id 获取相应的角色
    public Role findRoleByAdminId( int aid);
    // 更新管理员的角色
    public int updateRoleByAdminId(Map < String, Object > params);
    // 插入角色
    int insertRole(String roleName);
    // 修改角色
    int updateRole(Role role);
    // 根据名称获取角色对象
    Role findByRoleName(String roleName);
    // 根据角色 id 获取角色
    Role findByRoleId( int roleId);
    // 修改角色状态
    int updateDelState( int roleId);
```

```
    // 插入管理员角色
    int insertAdminRole(Map < String, Object > params);
}
```

创建接口 RoleMapper 的映射文件 RoleMapper.xml，代码如下。

```
<?xml version = "1.0" encoding = "UTF - 8"?>
<! DOCTYPE mapper PUBLIC " - //mybatis.org//DTD Mapper 3.0//EN" "http://mybatis.org/dtd/mybatis - 3 - mapper.dtd">
< mapper namespace = "com.my.eshop.dao.RoleMapper">
    <!-- 获取所有角色列表 -->
    < select id = "findAllRole" resultType = "com.my.eshop.pojo.Role">
        SELECT * FROM role
    </select >
    <!-- 根据状态获取角色列表 -->
    < select id = "findRoleByDelState" parameterType = "Integer" resultType = "com.my.eshop.pojo.Role">
        SELECT * FROM role where delState = #{delState}
    </select >
    <!-- 根据角色 id 获取对应的功能 id 集合 -->
    < select id = "findFunctionIdsByRid" parameterType = "Integer" resultType = "Integer">
        select fid from role_functions
        where rid = #{roleId}
    </select >
    <!-- 删除 roleid 下的 functions -->
    < delete id = "deleteFunctionsByRoleId" parameterType = "Integer">
        delete from role_functions
        where rid = #{roleId}
    </delete >
    <!-- 插入角色功能记录 -->
    < insert id = "insertFunctionsByRoleId">
        insert into role_functions(rid, fid) values(#{rid}, #{fid})
    </insert >
    <!-- 根据管理员 id 获取相应的角色 -->
    < select id = "findRoleByAdminId" parameterType = "Integer" resultType = "com.my.eshop.pojo.Role">
        select role.* from admin_role ar, role
        where ar.rid = role.roleId and aid = #{adminId}
    </select >
    <!-- 更新管理员的角色 -->
    < update id = "updateRoleByAdminId">
        UPDATE admin_role
        set rid = #{roleId}
        where aid = #{adminId}
    </update >
    <!-- 添加管理员角色 -->
    < insert id = "insertAdminRole" parameterType = "map">
        insert into admin_role (aid, rid) values(#{adminId}, #{roleId})
    </insert >
```

```
    <!-- 插入角色 -->
    <insert id="insertRole" parameterType="String">
        insert into role(roleName) values(#{roleName})
    </insert>
    <!-- 修改角色 -->
    <update id="updateRole" parameterType="Role">
        UPDATE role SET roleName = #{roleName} where roleId = #{roleId}
    </update>
    <!-- 根据名称获取角色对象 -->
    <select id="findByRoleName" parameterType="String" resultType="Role">
      SELECT * FROM role
      where
      roleName = #{roleName}
    </select>
    <!-- 根据角色 id 获取角色 -->
    <select id="findByRoleId" parameterType="Integer" resultType="Role">
        SELECT * FROM role where roleId = #{roleId}
    </select>
    <!-- 修改角色状态 -->
    <update id="updateDelState" parameterType="Integer">
        UPDATE role SET delState = 1 where roleId = #{roleId}
    </update>
</mapper>
```

12.6　创建 Service 接口及实现类

在 com.my.eshop.service 包中，创建接口 UserInfoService，声明下列方法。

```
package com.my.eshop.service;
import com.my.eshop.pojo.UserInfo;
import java.util.List;
import java.util.Map;
public interface UserInfoService {
    // 根据状态获取客户列表
    List<UserInfo> getUserInfoByStatus(int status);
    // 根据条件查询客户总数
    Integer count(Map<String, Object> params);
    // 分页获取客户信息
    List<UserInfo> getUserInfos(Map<String, Object> params);
    // 更新客户状态
    int updateStatus(int id);
    // 根据客户 id 获取客户信息
    UserInfo getUserInfoById(int id);
    // 根据客户姓名获取客户信息
    UserInfo getByUserName(String userName);
    // 修改客户信息
    int editUserInfo(UserInfo userInfo);
}
```

在 com. my. eshop. service. impl 包中,创建接口 UserInfoService 的实现类 UserInfoServiceImpl,实现接口 UserInfoService 中声明的方法,代码如下。

```
package com.my.eshop.service.impl;
import com.my.eshop.dao.UserInfoMapper;
…
@Service
@Transactional(propagation = Propagation.REQUIRED)
public class UserInfoServiceImpl implements UserInfoService {
    @Autowired
    private UserInfoMapper userInfoMapper;
    //根据状态获取客户列表
    @Override
    public List<UserInfo> getUserInfoByStatus(int status) {
        return userInfoMapper.findUserInfoByStatus(status);
    }
    //根据条件查询客户总数
    @Override
    public Integer count(Map<String, Object> params) {
        return userInfoMapper.count(params);
    }
    //分页获取客户信息
    @Override
    public List<UserInfo> getUserInfos(Map<String, Object> params) {
        return userInfoMapper.findByPage(params);
    }
    //更新客户状态
    @Override
    public int updateStatus(int id) {
        return userInfoMapper.updateStatus(id);
    }
    //根据客户 id 获取客户信息
    @Override
    public UserInfo getUserInfoById(int id) {
        return userInfoMapper.findUserInfoById(id);
    }
    //根据客户姓名获取客户信息
    @Override
    public UserInfo getByUserName(String userName) {
        return userInfoMapper.findUserInfoByUserName(userName);
    }
    //修改客户信息
    @Override
    public int editUserInfo(UserInfo userInfo) {
        return userInfoMapper.updateUserInfo(userInfo);
    }
}
```

创建接口 AdminInfoService,声明下列方法。

```
package com.my.eshop.service;
import com.my.eshop.pojo.AdminInfo;
…
public interface AdminInfoService {
    // 登录验证
    public AdminInfo login(AdminInfo ai);
    // 根据姓名查询管理员
    public AdminInfo getByName(AdminInfo ai);
    // 根据 id 查询管理员
    AdminInfo getAdminById(int id);
    // 查询所有管理员列表
    List<AdminInfo> getAll();
    // 根据条件查询管理员总数
    Integer count(Map<String, Object> params);
    // 分页获取管理员列表
    List<AdminInfo> getAdminInfo(Map<String, Object> params);
    // 根据管理员 id 获取 Functions 列表
    public List<Functions> getFunctionsByAdminId(int adminId);
    // 插入管理员
    int saveAdminInfo(AdminInfo adminInfo);
    // 修改管理员
    int editAdminInfo(AdminInfo adminInfo);
    // 更新管理员状态
    int updateDelState(int id);
}
```

创建接口 AdminInfoService 的实现类 AdminInfoServiceImpl，实现接口 AdminInfoService 中声明的方法，代码如下。

```
package com.my.eshop.service.impl;
import com.my.eshop.dao.AdminInfoMapper;
…
@Service
public class AdminInfoServiceImpl implements AdminInfoService {
    @Autowired
    private AdminInfoMapper adminInfoMapper;
    @Autowired
    private FunctionsMapper functionsMapper;
    @Autowired
    private RoleMapper roleMapper;
    // 根据姓名和密码查询管理员
    @Override
    public AdminInfo login(AdminInfo ai) {
        return adminInfoMapper.findByNameAndPwd(ai);
    }
    // 根据姓名查询管理员
    @Override
    public AdminInfo getByName(AdminInfo ai) {
        return adminInfoMapper.findByName(ai.getName());
```

```
    }
    // 根据 id 查询管理员
    @Override
    public AdminInfo getAdminById(int id) {
        return adminInfoMapper.findAdminById(id);
    }
    // 查询所有管理员列表
    @Override
    public List<AdminInfo> getAll() {
        return adminInfoMapper.findAll();
    }
    // 根据条件查询管理员总数
    @Override
    public Integer count(Map<String, Object> params) {
        return adminInfoMapper.count(params);
    }
    // 分页获取管理员列表
    @Override
    public List<AdminInfo> getAdminInfo(Map<String, Object> params) {
        return adminInfoMapper.findByPage(params);
    }
    // 根据管理员 id 获取 Functions 列表
    @Override
    public List<Functions> getFunctionsByAdminId(int adminId) {
        // 查询该管理员可以操作的功能模块编号集合
        List<String> fids = functionsMapper.findFunctionIdsByAid(adminId);
        List<Functions> functionsList = new ArrayList<>();
        if ((fids != null) && (fids.size() > 0)) {
            for (int i = 0; i < fids.size(); i++) {
                Functions functions = functionsMapper.findFunctionsById(Integer.parseInt
(fids.get(i).toString()));
                functionsList.add(functions);
            }
        }
        return functionsList;
    }
    // 插入管理员
    @Override
    public int saveAdminInfo(AdminInfo adminInfo) {
        int result1 = 1, result2 = 1;
        result1 = adminInfoMapper.insertAdminInfo(adminInfo);
        Map<String, Object> params = new HashMap<String, Object>();
        params.put("adminId", adminInfo.getId());
        params.put("roleId", 0);
        result2 = roleMapper.insertAdminRole(params);
        return result1 * result2;
    }
    // 修改管理员
    @Override
    public int editAdminInfo(AdminInfo adminInfo) {
```

```
        return adminInfoMapper.updateAdminInfo(adminInfo);
    }
    //更新管理员状态
    @Override
    public int updateDelState(int id) {
        return adminInfoMapper.updateDelState(id);
    }
}
```

创建接口 CategoryService,声明下列方法。

```
package com.my.eshop.service;
import com.my.eshop.pojo.Category;
import java.util.List;
public interface CategoryService {
    //根据类别名获取类别列表
    List<Category> getCategories(String name);
    //根据类别名获取类别对象
    Category getByName(String name);
    //根据类别 id 获取类别对象
    Category getById(int id);
    //插入类别
    int addCategory(String name);
    //修改类别
    int editCategory(Category category);
}
```

创建接口 CategoryService 的实现类 CategoryServiceImpl,实现接口 CategoryService 中声明的方法,代码如下。

```
package com.my.eshop.service.impl;
import com.my.eshop.dao.CategoryMapper;
…
@Service
@Transactional(propagation = Propagation.REQUIRED)
public class CategoryServiceImpl implements CategoryService {
    @Autowired
    private CategoryMapper categoryMapper;
    //根据类别名获取类别列表
    @Override
    public List<Category> getCategories(String name) {
        return categoryMapper.findCategories(name);
    }
    //插入类别
    @Override
    public int addCategory(String name) {
        return categoryMapper.insertCategory(name);
    }
    //根据类别名获取类别对象
```

```
    @Override
    public Category getByName(String name) {
        return categoryMapper.findByName(name);
    }
    //根据类别 id 获取类别对象
    @Override
    public Category getById(int id) {
        return categoryMapper.findById(id);
    }
    //修改类别
    @Override
    public int editCategory(Category category) {
        return categoryMapper.updateCategory(category);
    }
}
```

创建接口 FunctionsService，声明下列方法。

```
package com.my.eshop.service;
import com.my.eshop.pojo.Functions;
import java.util.List;
public interface FunctionsService {
    //获取系统功能列表
    List<Functions> getAllFunctions();
}
```

创建接口 FunctionsService 的实现类 FunctionsServiceImpl，实现接口 FunctionsService 中声明的方法，代码如下。

```
package com.my.eshop.service.impl;
import com.my.eshop.dao.FunctionsMapper;
…
@Service
public class FunctionsServiceImpl implements FunctionsService {
    @Autowired
    private FunctionsMapper functionsMapper;
    //获取系统功能列表
    @Override
    public List<Functions> getAllFunctions() {
        return functionsMapper.findAllFunctions();
    }
}
```

创建接口 GoodsInfoService，声明下列方法。

```
package com.my.eshop.service;
import com.my.eshop.pojo.GoodsInfo;
import java.util.List;
```

```
import java.util.Map;
public interface GoodsInfoService {
    //获取有效的商品列表
    List<GoodsInfo> getValidGoodstInfo();
    //获取满足条件的商品总数
    Integer count(Map<String,Object> params);
    //分页获取商品列表
    List<GoodsInfo> getGoodsInfo(Map<String,Object> params);
    //插入商品
    int addGoodsInfo(GoodsInfo goodsInfo);
    //根据商品 id 获取商品对象
    GoodsInfo getById(int id);
    //修改商品
    int editGoodsInfo(GoodsInfo goodsInfo);
}
```

创建接口 GoodsInfoService 的实现类 GoodsInfoServiceImpl，实现接口 GoodsInfoService 中声明的方法，代码如下。

```
package com.my.eshop.service.impl;
import com.my.eshop.dao.GoodsInfoMapper;
…
@Service
public class GoodsInfoServiceImpl implements GoodsInfoService {
    @Autowired
    private GoodsInfoMapper goodsInfoMapper;
    //获取有效的商品列表
    @Override
    public List<GoodsInfo> getValidGoodstInfo() {
        return goodsInfoMapper.findGoodsInfoByStatus(1);
    }
    //获取满足条件的商品总数
    @Override
    public Integer count(Map<String, Object> params) {
        return goodsInfoMapper.count(params);
    }
    //分页获取商品列表
    @Override
    public List<GoodsInfo> getGoodsInfo(Map<String, Object> params) {
        return goodsInfoMapper.findByPage(params);
    }
    //插入商品
    @Override
    public int addGoodsInfo(GoodsInfo goodsInfo) {
        return goodsInfoMapper.insertGoodsInfo(goodsInfo);
    }
    //根据商品 id 获取商品对象
    @Override
    public GoodsInfo getById(int id) {
```

```
        return goodsInfoMapper.findById(id);
    }
    //修改商品
    @Override
    public int editGoodsInfo(GoodsInfo goodsInfo) {
        return goodsInfoMapper.updateGoodsInfo(goodsInfo);
    }
}
```

创建接口 OrderInfoService，声明下列方法。

```
package com.my.eshop.service;
import com.my.eshop.pojo.OrderDetail;
…
public interface OrderInfoService {
    //添加订单
    public int addOrder(OrderInfo oi);
    //获取满足条件的订单总数
    Integer count(Map<String, Object> params);
    //分页获取订单列表
    List<OrderInfo> getOrderInfo(Map<String, Object> params);
    //根据订单id删除订单明细
    int deleteOrderDetailByOid(int oid);
    //根据订单id删除订单
    int deleteOrderInfoById(int id);
    //根据订单id,删除订单记录和订单明细记录
    int deleteOrderById(int id);
    //根据订单id获取订单明细列表
    List<OrderDetail> getOrderDetailByOid(int oid);
}
```

创建接口 OrderInfoService 的实现类 OrderInfoServiceImpl，实现接口 OrderInfoService 中声明的方法，代码如下。

```
package com.my.eshop.service.impl;
import com.my.eshop.dao.OrderInfoMapper;
…
@Service
public class OrderInfoServiceImpl implements OrderInfoService {
    @Autowired
    private OrderInfoMapper orderInfoMapper;
    //添加订单
    @Override
    public int addOrder(OrderInfo oi) {
        int result1 = 1, result2 = 1;
        //保存订单信息
        result1 = orderInfoMapper.insertOrderInfo(oi);
        for (OrderDetail od : oi.getOrderDetails()) {
            od.setOid(oi.getId());
```

```
            //保存订单明细
            result2 * = orderInfoMapper.insertOrderDetail(od);
        }
        return result1 * result2;
    }
    //获取满足条件的订单总数
    @Override
    public Integer count(Map<String, Object> params) {
        return orderInfoMapper.count(params);
    }
    //分页获取订单列表
    @Override
    public List<OrderInfo> getOrderInfo(Map<String, Object> params) {
        return orderInfoMapper.findByPage(params);
    }
    //根据订单 id 删除订单明细
    @Override
    public int deleteOrderDetailByOid(int oid) {
        return orderInfoMapper.deleteOrderDetailByOid(oid);
    }
    //根据订单 id 删除订单
    @Override
    public int deleteOrderInfoById(int id) {
        return orderInfoMapper.deleteOrderInfoById(id);
    }
    //根据订单 id,删除订单记录和订单明细记录
    @Override
    public int deleteOrderById(int id) {
        int result1 = 1;
        int result2 = 2;
        result1 = orderInfoMapper.deleteOrderDetailByOid(id);
        result2 = orderInfoMapper.deleteOrderInfoById(id);
        return result1 * result2;
    }
    //根据订单 id 获取订单明细列表
    @Override
    public List<OrderDetail> getOrderDetailByOid(int oid) {
        return orderInfoMapper.findOrderDetailByOid(oid);
    }
}
```

创建接口 RoleService,声明下列方法。

```
package com.my.eshop.service;
import com.my.eshop.pojo.Role;
…
public interface RoleService {
    //获取所有角色列表
    List<Role> getAllRole();
```

```
    // 获取有效角色列表
    List<Role> getValidRole();
    // 根据角色 id 获取对应的功能集合
    List<Integer> getFunctionIdsByRid(int roleId);
    // 根据 roleid 绑定 Functions
    @Transactional
    public int bindFunctionsByRoleId(int roleId, String fids);
    // 根据管理员 id 获取相应的角色
    public Role getRoleByAdminId(int aid);
    // 更新管理员的角色
    public int updateRoleByAdminId(Map<String, Object> params);
    // 添加角色
    int addRole(String roleName);
    // 修改角色
    int editRole(Role role);
    // 根据名称获取角色对象
    Role getByRoleName(String roleName);
    // 根据角色 id 获取角色
    Role getByRoleId(int roleId);
    // 修改角色状态
    int updateDelState(int roleId);
}
```

创建接口 RoleService 的实现类 RoleServiceImpl，实现接口 RoleService 中声明的方法，代码如下。

```
package com.my.eshop.service.impl;
import com.my.eshop.dao.FunctionsMapper;
…
@Service
public class RoleServiceImpl implements RoleService {
    @Autowired
    private RoleMapper roleMapper;
    @Autowired
    private FunctionsMapper functionsMapper;
    // 获取所有角色列表
    @Override
    public List<Role> getAllRole() {
        return roleMapper.findAllRole();
    }
    // 获取有效角色列表
    @Override
    public List<Role> getValidRole() {
        return roleMapper.findRoleByDelState(0);
    }
    // 根据角色 id 获取对应的功能集合
    @Override
    public List<Integer> getFunctionIdsByRid(int roleId) {
        List<Integer> result = new ArrayList<>();
```

```
        List<Integer> fids = roleMapper.findFunctionIdsByRid(roleId);
        if ((fids != null) && (fids.size() > 0)) {
            for (int i = 0; i < fids.size(); i++) {
                Functions functions = functionsMapper.findFunctionsById(Integer.parseInt
(fids.get(i).toString()));
                if (functions.isIsleaf()) {
                    result.add(functions.getId());
                }
            }
        }
        return result;
    }
    // 根据 roleid 绑定 Functions
    @Override
    public int bindFunctionsByRoleId(int roleId, String fids) {
        String[] fidStrings = fids.split(",");
        int result1 = 1, result2 = 1;
        List<Integer> ids = roleMapper.findFunctionIdsByRid(roleId);
        if (ids.size() > 0) {
            result1 = roleMapper.deleteFunctionsByRoleId(roleId);
        }
        Map<String, Object> params = new HashMap<>();
        params.put("rid", roleId);
        params.put("fid", 1);
        result2 *= roleMapper.insertFunctionsByRoleId(params);
        for (int i = 0; i < fidStrings.length; i++) {
            params = new HashMap<>();
            params.put("rid", roleId);
            params.put("fid", fidStrings[i]);
            result2 *= roleMapper.insertFunctionsByRoleId(params);
        }
        return result1 * result2;
    }
    // 根据管理员 id 获取相应的角色
    @Override
    public Role getRoleByAdminId(int aid) {
        return roleMapper.findRoleByAdminId(aid);
    }
    // 更新管理员的角色
    @Override
    public int updateRoleByAdminId(Map<String, Object> params) {
        return roleMapper.updateRoleByAdminId(params);
    }
    // 添加角色
    @Override
    public int addRole(String roleName) {
        return roleMapper.insertRole(roleName);
    }
    // 修改角色
    @Override
```

```
    public int editRole(Role role) {
        return roleMapper.updateRole(role);
    }
    // 根据名称获取角色对象
    @Override
    public Role getByRoleName(String roleName) {
        return roleMapper.findByRoleName(roleName);
    }
    // 根据角色 id 获取角色
    @Override
    public Role getByRoleId(int roleId) {
        return roleMapper.findByRoleId(roleId);
    }
    // 修改角色状态
    @Override
    public int updateDelState(int roleId) {
        return roleMapper.updateDelState(roleId);
    }
}
```

12.7 创建 Controller 控制器类

在 com.my.eshop.controller 包中，创建控制器类 AdminInfoController，代码如下。

```
package com.my.eshop.controller;
import com.my.eshop.annotation.UserLoginToken;
…
@RestController
public class AdminInfoController {
    @Autowired
    private AdminInfoService adminInfoService;
    @Autowired
    private TokenService tokenService;
    @Autowired
    private RoleService roleService;
    // 用户名和密码合法性校验
    @PostMapping("/eshop/login")
    public Map<String,Object> login (@RequestBody AdminInfo adminInfo){
        Map<String,Object> result = new HashMap<>();
        AdminInfo ai = adminInfoService.getByName(adminInfo);
        if(ai!= null){
            if (!ai.getPwd().equals(adminInfo.getPwd())){
                result.put("code",2);
                result.put("msg","登录失败,密码错误");
            }else {
                String token = tokenService.getToken(ai);
                result.put("code",0);
```

```
                result.put("token", token);
                result.put("admin", ai);
            }
        }else {
            result.put("code",1);
            result.put("msg","登录失败,用户不存在");
        }
        return result;
    }
    //获取登录用户功能菜单列表
    @UserLoginToken
    @GetMapping("/eshop/menus")
    public Map<String,Object> getMenus (HttpServletRequest
httpServletRequest){
        Map<String,Object> result = new HashMap<>();
        List<TreeNode> nodes = new ArrayList<TreeNode>();
        AdminInfo adminInfo =
(AdminInfo) httpServletRequest.getAttribute("currentUser");
        List<Functions> functionsList =
adminInfoService.getFunctionsByAdminId(adminInfo.getId());
        //对 List<Functions>类型的 Functions 对象集合排序
        Collections.sort(functionsList);
        //将排序后的 Functions 对象集合转换到 List<TreeNode>类型的列表 nodes
        for (Functions functions : functionsList) {
            TreeNode treeNode = new TreeNode();
            treeNode.setIconfont(functions.getIconfont());
            treeNode.setId(functions.getId());
            treeNode.setPid(functions.getParentid());
            treeNode.setText(functions.getName());
            treeNode.setUrl(functions.getUrl());
            nodes.add(treeNode);
        }
        //调用自定义的工具类 JsonFactory 的 buildtree 方法,
        //为 nodes 列表中的各个 TreeNode 元素中的
        //children 属性赋值(该节点包含的子节点)
        List<TreeNode> treeNodes = JsonFactory.buildtree(nodes, 0);
        result.put("code",0);
        result.put("msg","获取菜单成功");
        result.put("data",treeNodes);
        return result;
    }
    //分页获取管理员列表
    @UserLoginToken
    @GetMapping("/eshop/admininfos")
    public Map<String,Object> getAdmininfos (Integer curPage, Integer pageSize, AdminInfo
adminInfo){
        //创建对象 result,保存查询结果数据
        Map<String,Object> result = new HashMap<>();
        //创建对象 params,封装查询条件
        Map<String, Object> params = new HashMap<String, Object>();
```

```
        params.put("name",adminInfo.getName());
        //根据查询条件,获取管理员记录数
        int totalCount = adminInfoService.count(params);
        //创建分页类对象
        Pager pager = new Pager();
        pager.setCurPage(curPage);
        pager.setPerPageRows(pageSize);
        params.put("pager", pager);
        //根据查询条件,分页获取管理员列表
        List<AdminInfo> adminInfos = adminInfoService.getAdminInfo(params);
        System.out.println(adminInfos.size());
        if(adminInfos.size()>0){
            result.put("code",0);
            result.put("page",curPage);
            result.put("total",totalCount);
            result.put("adminInfos", adminInfos);
            result.put("msg","获取管理员列表成功");
        }else{
            result.put("code",1);
            result.put("msg","没有管理员记录");
        }
        return result;
    }
    //分配角色
    @UserLoginToken
    @PutMapping("/eshop/admininfos/{adminId}/role/{roleId}")
    public Map<String, Object> saveRole(@PathVariable("adminId") Integer adminId,
@PathVariable("roleId") Integer roleId){
        Map<String,Object> result = new HashMap<>();
        Map<String, Object> params = new HashMap<String, Object>();
        params.put("roleId",roleId);
        params.put("adminId",adminId);
        if(roleService.updateRoleByAdminId(params)>0){
            result.put("code",0);
            result.put("msg","分配角色成功!");
        }else {
            result.put("code",1);
            result.put("msg","分配角色失败!");
        }
        return result;
    }
    //添加管理员
    @UserLoginToken
    @PostMapping("/eshop/admininfos")
    public Map<String,Object> addAdminInfo(@RequestBody AdminInfo adminInfo){
        Map<String,Object> result = new HashMap<>();
        int flag = adminInfoService.saveAdminInfo(adminInfo);
        if(flag>0){
            result.put("code",0);
            result.put("msg","添加成功");
```

```
        }else {
            result.put("code",1);
            result.put("msg","添加失败");
        }
        return result;
    }
    //判断管理员名是否存在
    @UserLoginToken
    @GetMapping("/eshop/admininfos/name/{name}")
    public Map<String,Object> isExistAdminInfoName(@PathVariable String name){
        Map<String,Object> result = new HashMap<>();
        AdminInfo ai = new AdminInfo();
        ai.setName(name);
        AdminInfo adminInfo = adminInfoService.getByName(ai);
        if(adminInfo!= null){
            result.put("code",1);
            result.put("msg","管理员名称已存在");
        }else {
            result.put("code",0);
        }
        return result;
    }
    //判断管理员名和 id 是否存在
    @UserLoginToken
    @GetMapping("/eshop/admininfos/name/{name}/id/{id}")
    public Map<String, Object> isExistAdminInfoName_Id(@PathVariable String name,
@PathVariable int id){
        Map<String,Object> result = new HashMap<>();
        AdminInfo ai = new AdminInfo();
        ai.setName(name);
        AdminInfo adminInfo = adminInfoService.getByName(ai);
        if(adminInfo!= null && adminInfo.getId()!= id ){
            result.put("code",1);
            result.put("msg","管理员名称已存在");
        }else {
            result.put("code",0);
        }
        return result;
    }
    //根据管理员 id 获取管理员信息
    @UserLoginToken
    @GetMapping("/eshop/admininfos/id/{id}")
    public Map<String,Object> getAdminInfoById(@PathVariable int id){
        Map<String,Object> result = new HashMap<>();
        AdminInfo adminInfo = adminInfoService.getAdminById(id);
        if(adminInfo!= null){
            result.put("code",0);
            result.put("adminInfo",adminInfo);
        }else {
            result.put("code",1);
```

```
            result.put("msg","获取管理员失败");
        }
        return result;
    }
    // 修改管理员信息
    @UserLoginToken
    @PutMapping("/eshop/admininfos")
    public Map<String,Object> editAdminInfo(@RequestBody AdminInfo adminInfo){
        Map<String,Object> result = new HashMap<>();
        int flag = adminInfoService.editAdminInfo(adminInfo);
        if(flag>0){
            result.put("code",0);
            result.put("msg","修改成功");
        }else {
            result.put("code",1);
            result.put("msg","修改失败");
        }
        return result;
    }
    // 禁用管理员
    @UserLoginToken
    @PutMapping("/eshop/admininfos/id/{id}/delState")
    public Map<String,Object> editRole(@PathVariable int id){
        Map<String,Object> result = new HashMap<>();
        int flag = adminInfoService.updateDelState(id);
        if(flag>0){
            result.put("code",0);
            result.put("msg","修改成功");
        }else {
            result.put("code",1);
            result.put("msg","修改失败");
        }
        return result;
    }
}
```

创建控制器类 UserInfoController，代码如下。

```
package com.my.eshop.controller;
import com.my.eshop.annotation.UserLoginToken;
…
@RestController
public class UserInfoController {
    @Autowired
    private UserInfoService userInfoService;
    // 获取客户列表
    @UserLoginToken
    @GetMapping("/eshop/customers")
    public Map<String, Object> getCustomers(int status) {
```

```
        Map<String, Object> result = new HashMap<>();
        List<UserInfo> userInfos = userInfoService.getUserInfoByStatus(status);
        if (userInfos.size() > 0) {
            result.put("code", 0);
            result.put("msg", "获取客户列表成功");
            result.put("data", userInfos);
        } else {
            result.put("code", 1);
            result.put("msg", "获取客户列表失败");
        }
        return result;
    }
    //分页获取客户列表
    @UserLoginToken
    @GetMapping("/eshop/userinfos")
    public Map<String, Object> getUserInfos(Integer curPage, Integer pageSize, UserInfo
userInfo) {
        //创建对象 result,保存查询结果数据
        Map<String, Object> result = new HashMap<>();
        //创建对象 params,封装查询条件
        Map<String, Object> params = new HashMap<String, Object>();
        params.put("userName", userInfo.getUserName());
        //根据查询条件,获取用户记录数
        int totalCount = userInfoService.count(params);
        //创建分页类对象
        Pager pager = new Pager();
        pager.setCurPage(curPage);
        pager.setPerPageRows(pageSize);
        params.put("pager", pager);
        //根据查询条件,分页获取用户列表
        List<UserInfo> userInfos = userInfoService.getUserInfos(params);
        //System.out.println(userInfos.size());
        if (userInfos.size() > 0) {
            result.put("code", 0);
            result.put("page", curPage);
            result.put("total", totalCount);
            result.put("userInfos", userInfos);
            result.put("msg", "获取用户列表成功");
        } else {
            result.put("code", 1);
            result.put("msg", "没有用户记录");
        }
        return result;
    }
    //禁用用户
    @UserLoginToken
    @PutMapping("/eshop/userinfos/id/{id}/status")
    public Map<String, Object> editRole(@PathVariable int id) {
        Map<String, Object> result = new HashMap<>();
```

```
        int flag = userInfoService.updateStatus(id);
        if (flag > 0) {
            result.put("code", 0);
            result.put("msg", "禁用成功");
        } else {
            result.put("code", 1);
            result.put("msg", "禁用失败");
        }
        return result;
    }
    // 根据客户 id 获取客户信息
    @UserLoginToken
    @GetMapping("/eshop/userinfos/id/{id}")
    public Map<String, Object> getUserInfoById(@PathVariable int id) {
        Map<String, Object> result = new HashMap<>();
        UserInfo userInfo = userInfoService.getUserInfoById(id);
        if (userInfo != null) {
            result.put("code", 0);
            result.put("userInfo", userInfo);
        } else {
            result.put("code", 1);
            result.put("msg", "获取用户失败");
        }
        return result;
    }
    // 判断客户名称和 id 是否存在
    @UserLoginToken
    @GetMapping("/eshop/userinfos/userName/{userName}/id/{id}")
    public Map<String, Object> isExistAdminInfoName_Id(@PathVariable String userName,
@PathVariable int id) {
        Map<String, Object> result = new HashMap<>();
        UserInfo userInfo = userInfoService.getByUserName(userName);
        if (userInfo != null && userInfo.getId() != id) {
            result.put("code", 1);
            result.put("msg", "用户名称已存在");
        } else {
            result.put("code", 0);
        }
        return result;
    }
    // 修改
    @UserLoginToken
    @PutMapping("/eshop/userinfos")
    public Map<String, Object> editUserInfo(@RequestBody UserInfo userInfo) {
        Map<String, Object> result = new HashMap<>();
        int flag = userInfoService.editUserInfo(userInfo);
        if (flag > 0) {
            result.put("code", 0);
            result.put("msg", "修改成功");
```

```
        } else {
            result.put("code", 1);
            result.put("msg", "修改失败");
        }
        return result;
    }
}
```

创建控制器类 CategoryController，代码如下。

```
package com.my.eshop.controller;
import com.my.eshop.annotation.UserLoginToken;
…
@RestController
public class CategoryController {
    @Autowired
    private CategoryService categoryService;
    // 根据类别名称获取类别列表
    @UserLoginToken
    @GetMapping("/eshop/categories")
    public Map<String, Object> getCategories(String name) {
        Map<String, Object> result = new HashMap<>();
        List<Category> categories = categoryService.getCategories(name);
        if (categories.size() > 0) {
            result.put("code", 0);
            result.put("msg", "获取商品类别列表成功");
            result.put("categories", categories);
        } else {
            result.put("code", 1);
            result.put("msg", "获取商品类别列表失败");
        }
        return result;
    }
    // 判断类别名称是否存在
    @UserLoginToken
    @GetMapping("/eshop/categories/name/{name}")
    public Map<String, Object> isExistCategoryName(@PathVariable String name) {
        Map<String, Object> result = new HashMap<>();
        Category category = categoryService.getByName(name);
        if (category != null) {
            result.put("code", 1);
            result.put("msg", "商品类别名称已存在");
        } else {
            result.put("code", 0);
        }
        return result;
    }
    // 判断类别名称和 id 是否存在
```

```
    @UserLoginToken
    @GetMapping("/eshop/categories/name/{name}/id/{id}")
    public Map<String, Object> isExistCategoryName_Id(@PathVariable String name,
@PathVariable int id) {
        Map<String, Object> result = new HashMap<>();
        Category category = categoryService.getByName(name);
        if (category != null && category.getId() != id) {
            result.put("code", 1);
            result.put("msg", "商品类别名称已存在");
        } else {
            result.put("code", 0);
        }
        return result;
    }
    //根据类别id获取类别信息
    @UserLoginToken
    @GetMapping("/eshop/categories/id/{id}")
    public Map<String, Object> getCategoryById(@PathVariable int id) {
        Map<String, Object> result = new HashMap<>();
        Category category = categoryService.getById(id);
        if (category != null) {
            result.put("code", 0);
            result.put("category", category);
        } else {
            result.put("code", 1);
            result.put("msg", "获取商品类别失败");
        }
        return result;
    }
    //添加类别
    @UserLoginToken
    @PostMapping("/eshop/categories")
    public Map<String, Object> addCategory(@RequestBody Category category) {
        Map<String, Object> result = new HashMap<>();
        int flag = categoryService.addCategory(category.getName());
        if (flag > 0) {
            result.put("code", 0);
            result.put("msg", "添加成功");
        } else {
            result.put("code", 1);
            result.put("msg", "添加失败");
        }
        return result;
    }
    //修改类别
    @UserLoginToken
    @PutMapping("/eshop/categories")
    public Map<String, Object> editCategory(@RequestBody Category category) {
        Map<String, Object> result = new HashMap<>();
```

```
        int flag = categoryService.editCategory(category);
        if (flag > 0) {
            result.put("code", 0);
            result.put("msg", "修改成功");
        } else {
            result.put("code", 1);
            result.put("msg", "修改失败");
        }
        return result;
    }
}
```

创建控制器类 FunctionsController，代码如下。

```
package com.my.eshop.controller;
import com.my.eshop.annotation.UserLoginToken;
…
@RestController
public class FunctionsController {
    @Autowired
    private FunctionsService functionsService;
    @Autowired
    private AdminInfoService adminInfoService;
    //获取所有功能菜单列表
    @UserLoginToken
    @GetMapping("/eshop/functions")
    public Map<String,Object> getFunctions (HttpServletRequest httpServletRequest){
        Map<String,Object> result = new HashMap<>();
        List<TreeNode> nodes = new ArrayList<TreeNode>();
        List<Functions> functionsList = functionsService.getAllFunctions();
        //对 List<Functions>类型的 Functions 对象集合排序
        Collections.sort(functionsList);
        //将排序后的 Functions 对象集合转换到 List<TreeNode>类型的列表 nodes
        for (Functions functions : functionsList) {
            TreeNode treeNode = new TreeNode();
            treeNode.setIconfont(functions.getIconfont());
            treeNode.setId(functions.getId());
            treeNode.setPid(functions.getParentid());
            treeNode.setText(functions.getName());
            treeNode.setUrl(functions.getUrl());
            nodes.add(treeNode);
        }
        //调用自定义的工具类 JsonFactory 的 buildtree 方法,
        //为 nodes 列表中的各个 TreeNode 元素中的
        //children 属性赋值(该节点包含的子节点)
        List<TreeNode> treeNodes = JsonFactory.buildtree(nodes, 0);
        result.put("code",0);
        result.put("msg","获取所有功能菜单成功");
        result.put("data",treeNodes);
        return result;
    }
}
```

创建控制器类 RoleController，代码如下。

```
package com.my.eshop.controller;
import com.my.eshop.annotation.UserLoginToken;
…
@RestController
public class RoleController {
    @Autowired
    private RoleService roleService;
    //获取所有角色列表
    @UserLoginToken
    @GetMapping("/eshop/roles")
    public Map<String,Object> getAllRoles (HttpServletRequest httpServletRequest){
        Map<String,Object> result = new HashMap<>();
        List<Role> roles = roleService.getAllRole();
        if(roles.size()> 0){
            result.put("code",0);
            result.put("msg","获取所有角色列表成功");
            result.put("data",roles);
        }else {
            result.put("code",1);
            result.put("msg","获取所有角色列表失败");
        }
        return result;
    }
    //获取未禁用角色列表
    @UserLoginToken
    @GetMapping("/eshop/roles/valid")
    public Map<String,Object> getValidRoles (HttpServletRequest httpServletRequest){
        Map<String,Object> result = new HashMap<>();
        List<Role> roles = roleService.getValidRole();
        if(roles.size()> 0){
            result.put("code",0);
            result.put("msg","获取可用角色列表成功");
            result.put("data",roles);
        }else {
            result.put("code",1);
            result.put("msg","获取可用角色列表失败");
        }
        return result;
    }
    //根据角色 id 获取叶子节点功能菜单 id 列表
    @UserLoginToken
    @GetMapping("/eshop/getLeafFunctionsByRid")
    public List<Integer> getLeafFunctionsByRid (Integer roleId){
        List<Integer> result = roleService.getFunctionIdsByRid(roleId);
        return result;
    }
    //保存权限
    @UserLoginToken
```

```
    @PostMapping("/eshop/roles/{roleId}")
    public Map < String, Object > savePowers (@ PathVariable ( " roleId") Integer roleId,
@RequestBody Map params){
        Map<String,Object> result = new HashMap<>();
        if(roleService.bindFunctionsByRoleId(roleId, params.get("fids").toString())>0){
            result.put("code",0);
            result.put("msg","设置权限成功!");
        }else {
            result.put("code",1);
            result.put("msg","设置权限失败!");
        }
        return result;
    }
    //根据 roleId 获取对象
    @UserLoginToken
    @GetMapping("/eshop/roles/roleId/{roleId}")
    public Map<String,Object> getRoleByRoleId(@PathVariable("roleId") int roleId) {
        Map<String,Object> result = new HashMap<>();
        Role role = roleService.getByRoleId(roleId);
        if(role!= null){
            result.put("code",0);
            result.put("msg","获取角色成功!");
            result.put("data",role);
        }else{
            result.put("code",1);
            result.put("msg","获取角色失败!");
        }
        return result;
    }
    //添加角色
    @UserLoginToken
    @PostMapping("/eshop/roles")
    public Map<String,Object> addRole(@RequestBody Role role){
        Map<String,Object> result = new HashMap<>();
        int flag = roleService.addRole(role.getRoleName());
        if(flag>0){
            result.put("code",0);
            result.put("msg","添加成功");
        }else {
            result.put("code",1);
            result.put("msg","添加失败");
        }
        return result;
    }
    //修改角色
    @UserLoginToken
    @PutMapping("/eshop/roles")
    public Map<String,Object> editRole(@RequestBody Role role){
        Map<String,Object> result = new HashMap<>();
```

```
        int flag = roleService.editRole(role);
        if(flag>0){
            result.put("code",0);
            result.put("msg","修改成功");
        }else {
            result.put("code",1);
            result.put("msg","修改失败");
        }
        return result;
    }
    // 判断角色名是否存在
    @UserLoginToken
    @GetMapping("/eshop/roles/roleName/{roleName}")
    public Map<String,Object> isExistRoleName(@PathVariable String roleName){
        Map<String,Object> result = new HashMap<>();
        Role role = roleService.getByRoleName(roleName);
        if(role!= null){
            result.put("code",1);
            result.put("msg","角色名称已存在");
        }else {
            result.put("code",0);
        }
        return result;
    }
    // 判断角色名和角色 id 是否存在
    @UserLoginToken
    @GetMapping("/eshop/roles/roleName/{roleName}/roleId/{roleId}")
    public Map<String, Object> isExistRoleName_RoleId(@PathVariable String roleName,
@PathVariable int roleId){
        Map<String,Object> result = new HashMap<>();
        Role role = roleService.getByRoleName(roleName);
        if(role!= null && role.getRoleId()!= roleId ){
            result.put("code",1);
            result.put("msg","角色名称已存在");
        }else {
            result.put("code",0);
        }
        return result;
    }
    // 禁用角色
    @UserLoginToken
    @PutMapping("/eshop/roles/roleId/{roleId}/delState")
    public Map<String,Object> editRole(@PathVariable int roleId){
        Map<String,Object> result = new HashMap<>();
        int flag = roleService.updateDelState(roleId);
        if(flag>0){
            result.put("code",0);
            result.put("msg","修改成功");
        }else {
```

```
            result.put("code",1);
            result.put("msg","修改失败");
        }
        return result;
    }
}
```

创建控制器类 GoodsInfoController，代码如下。

```
package com.my.eshop.controller;
import com.my.eshop.annotation.UserLoginToken;
…
@RestController
public class GoodsInfoController {
    @Autowired
    private GoodsInfoService goodsInfoService;
    //分页获取商品列表
    @UserLoginToken
    @GetMapping("/eshop/goods")
    public Map<String, Object> getGoodsInfos(Integer curPage, Integer pageSize, GoodsInfo
goodsInfo) {
        //创建对象 result,保存查询结果数据
        Map<String, Object> result = new HashMap<>();
        //创建对象 params,封装查询条件
        Map<String, Object> params = new HashMap<String, Object>();
        params.put("name", goodsInfo.getName());
        //根据查询条件,获取商品记录数
        int totalCount = goodsInfoService.count(params);
        //创建分页类对象
        Pager pager = new Pager();
        pager.setCurPage(curPage);
        pager.setPerPageRows(pageSize);
        params.put("pager", pager);
        //根据查询条件,分页获取商品列表
        List<GoodsInfo> goodsInfos = goodsInfoService.getGoodsInfo(params);
        System.out.println(goodsInfos.size());
        if (goodsInfos.size() > 0) {
            result.put("code", 0);
            result.put("page", curPage);
            result.put("total", totalCount);
            result.put("goodsInfos", goodsInfos);
            result.put("msg", "获取商品列表成功");
        } else {
            result.put("code", 1);
            result.put("msg", "没有商品记录");
        }
        return result;
    }
    //获取商品列表
```

```
@UserLoginToken
@GetMapping("/eshop/goods/status")
public Map<String, Object> getValidGoodstInfo() {
    Map<String, Object> result = new HashMap<>();
    List<GoodsInfo> goodsInfos = goodsInfoService.getValidGoodstInfo();
    if (goodsInfos.size() > 0) {
        result.put("code", 0);
        result.put("msg", "获取商品列表成功");
        result.put("data", goodsInfos);
    } else {
        result.put("code", 1);
        result.put("msg", "获取商品列表失败");
    }
    return result;
}
// 根据商品 id 获取商品信息
@UserLoginToken
@GetMapping("/eshop/goods/id/{id}")
public Map<String, Object> getGoodsInfoById(@PathVariable int id) {
    Map<String, Object> result = new HashMap<>();
    GoodsInfo goodsInfo = goodsInfoService.getById(id);
    if (goodsInfo != null) {
        result.put("code", 0);
        result.put("goodsInfo", goodsInfo);
    } else {
        result.put("code", 1);
        result.put("msg", "获取商品失败");
    }
    return result;
}
// 添加商品
@UserLoginToken
@PostMapping("/eshop/goods")
public Map<String, Object> addGoods(@RequestBody GoodsInfo goodsInfo) {
    Map<String, Object> result = new HashMap<>();
    int flag = goodsInfoService.addGoodsInfo(goodsInfo);
    if (flag > 0) {
        result.put("code", 0);
        result.put("msg", "添加成功");
    } else {
        result.put("code", 1);
        result.put("msg", "添加失败");
    }
    return result;
}
// 修改商品
@UserLoginToken
@PutMapping("/eshop/goods")
```

```
    public Map<String, Object> editGoods(@RequestBody GoodsInfo goodsInfo) {
        Map<String, Object> result = new HashMap<>();
        int flag = goodsInfoService.editGoodsInfo(goodsInfo);
        if (flag > 0) {
            result.put("code", 0);
            result.put("msg", "修改成功");
        } else {
            result.put("code", 1);
            result.put("msg", "修改失败");
        }
        return result;
    }
}
```

创建控制器类 OrderInfoController,代码如下。

```
package com.my.eshop.controller;
import com.my.eshop.annotation.UserLoginToken;
…
@RestController
public class OrderInfoController {
    @Autowired
    private OrderInfoService orderInfoService;
    //提交订单
    @UserLoginToken
    @PostMapping("/eshop/commitOrder")
    public Map<String, Object> commitOrder(@RequestBody OrderInfo orderInfo) throws
ParseException {
        Map<String, Object> result = new HashMap<>();
        String str1 = "";
        String ordertime = orderInfo.getOrdertime();
        ordertime = ordertime.replace("Z", " UTC");
        SimpleDateFormat format = new SimpleDateFormat("yyyy-MM-dd'T'HH:mm:ss.SSS Z");
        SimpleDateFormat defaultFormat = new SimpleDateFormat("yyyy-MM-dd HH:mm:ss");
        try {
            Date time = format.parse(ordertime);
            str1 = defaultFormat.format(time);
            // System.out.println("str1 is " + str1);
        } catch (Exception e) {
            e.printStackTrace();
        }
        orderInfo.setOrdertime(str1);
        int result1, result2 = 1;
        try {
            //保存订单
            if (orderInfoService.addOrder(orderInfo) > 0) {
                result.put("code", 0);
                result.put("msg", "创建订单成功!");
```

```
            } else {
                result.put("code", 1);
                result.put("msg", "创建订单失败!");
            }
        } catch (Exception e) {
            result.put("code", 1);
            result.put("msg", "创建订单失败!");
        }
        return result;
    }
    //分页获取订单列表
    @UserLoginToken
    @GetMapping("/eshop/orderinfos")
    public Map<String, Object> getOrderinfos(Integer curPage, Integer pageSize, OrderInfo
orderInfo) {
        //创建对象result,保存查询结果数据
        Map<String, Object> result = new HashMap<>();
        //创建对象params,封装查询条件
        Map<String, Object> params = new HashMap<String, Object>();
        params.put("orderNo", orderInfo.getOrderNo());
        params.put("uid", orderInfo.getUid());
        params.put("status", orderInfo.getStatus());
        //根据查询条件,获取订单记录数
        int totalCount = orderInfoService.count(params);
        //创建分页类对象
        Pager pager = new Pager();
        pager.setCurPage(curPage);
        pager.setPerPageRows(pageSize);
        params.put("pager", pager);
        //根据查询条件,分页获取订单列表
        List<OrderInfo> orderInfos = orderInfoService.getOrderInfo(params);
        System.out.println(orderInfos.size());
        if (orderInfos.size() > 0) {
            result.put("code", 0);
            result.put("page", curPage);
            result.put("total", totalCount);
            result.put("orderInfos", orderInfos);
            result.put("msg", "获取订单列表成功");
        } else {
            result.put("code", 1);
            result.put("msg", "没有订单记录");
        }
        return result;
    }
    //删除订单
    @UserLoginToken
    @DeleteMapping("/eshop/orderinfos/{id}")
    public Map<String, Object> deleteOrderInfo(@PathVariable("id") String id) {
```

```
        Map<String, Object> result = new HashMap<>();
        if (orderInfoService.deleteOrderById(Integer.parseInt(id)) > 0) {
            result.put("code", 0);
            result.put("msg", "删除订单成功!");
        } else {
            result.put("code", 1);
            result.put("msg", "删除订单失败!");
        }
        return result;
    }
    //根据订单 id 获取订单明细列表
    @UserLoginToken
    @RequestMapping("/eshop/orderinfos/oid/{oid}")
    public Map<String, Object> getOrderDetailByOid(@PathVariable("oid") int oid) {
        Map<String, Object> result = new HashMap<>();
        List<OrderDetail> orderDetails = orderInfoService.getOrderDetailByOid(oid);
        if (orderDetails != null && orderDetails.size() > 0) {
            result.put("code", 0);
            result.put("orderDetails", orderDetails);
        } else {
            result.put("code", 1);
            result.put("msg", "获取订单明细数据失败");
        }
        return result;
    }
}
```

创建控制器类 SequenceController,代码如下。

```
package com.my.eshop.controller;
import com.my.eshop.annotation.UserLoginToken;
…
@RestController
public class SequenceController {
    @Autowired
    private StringRedisTemplate stringRedisTemplate;
    //使用 redis 生成订单编号
    @UserLoginToken
    @GetMapping("/eshop/getSequence")
    public Map<String, Object> getSequence(String prefix) {
        //创建对象 result,保存查询结果数据
        Map<String, Object> result = new HashMap<>();
        //时间戳,后面拼接流水号
        String datetime = new SimpleDateFormat("yyyyMMdd").format(new Date());
        //Redis key 的前缀
        String key = MessageFormat.format("{0}:{1}:{2}", "sys", prefix, datetime);
```

```
        //查询 key 是否存在，不存在返回 1，存在则自增加 1
        Long autoID = stringRedisTemplate.opsForValue().increment(key, 1);
        //设置 key 过期时间，保证每天的流水号从 1 开始
        if (autoID == 1) {
            stringRedisTemplate.expire(key, 86400, TimeUnit.SECONDS);
        }
        //id 为 6 位,不够的在左边补 0
        String value = StringUtils.leftPad(String.valueOf(autoID), 4, "0");
        //然后把时间戳和优化后的 ID 拼接
        String code = MessageFormat.format("{0}{1}{2}", prefix, datetime, value);
        result.put("code", 0);
        result.put("data", code);
        result.put("msg", "获取流水号成功");
        return result;
    }
}
```

12.8 前端程序目录结构

在电商平台后台管理系统中，采用前后端分离的方式，前端使用 Vue+Element Plus 框架。通过 Vue3 的脚手架创建前端程序 eshop-vue3，目录结构如图 12-5 所示。

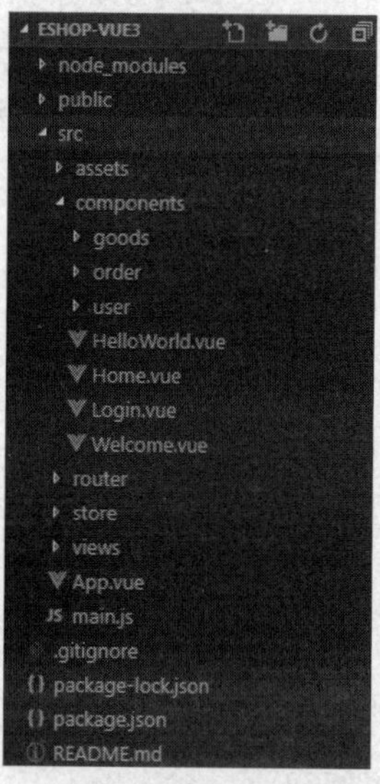

图 12-5 前端程序 eshop-vue3 的目录结构

在图 12-5 中，main.js 是项目的入口文件，用于初始化 Vue 实例，并引入需要的插件和各种公共组件。项目中所有的页面都会加载 main.js，代码如下。

```
import { createApp } from "vue";
import App from "./App.vue";
```

```
// 导入 router 文件夹下 index.js
import router from "./router";
// 导入 store 文件夹下 index.js
import store from "./store";
// 引入 Element Plus 组件库
import ElementPlus from "element-plus";
import "element-plus/lib/theme-chalk/index.css";
// 导入全局样式表
import "./assets/css/global.css";
// 引入 axios
import axios from "axios";
// 给 axios 组件设置全局 api 地址
axios.defaults.baseURL = "http://localhost:8888/eshop/";
// 请求拦截器
axios.interceptors.request.use(config => {
  // axios 通过 headers 传参,设置请求头 token
  config.headers.token = window.sessionStorage.getItem("token");
  return config;
});
const app = createApp(App);
// 挂载 axios
app.config.globalProperties.$axios = axios
app.use(store).use(router).use(ElementPlus).mount("#app");
```

在 main.js 文件中，依次导入 router 文件夹下的 index.js、store 文件夹下的 index.js、引入 Element Plus 组件库、导入全局样式表、引入 axios，给 axios 组件设置全局 api 地址、设置请求拦截器，axios 通过 headers 传参，设置请求头 token、挂载 axios。

router 文件夹下的 index.js 是项目的路由文件，用于控制页面跳转，代码如下。

```
import { createRouter, createWebHistory } from "vue-router";
import Home from "../components/Home.vue";
import Welcome from "../components/Welcome.vue";
import Login from "../components/Login.vue";
import Category from "../components/goods/Category.vue";
import GoodsList from "../components/goods/GoodsList.vue";
import OrderList from "../components/order/OrderList.vue";
import CreateOrder from "../components/order/CreateOrder.vue";
import Backusers from "../components/user/Backusers.vue";
import Frontusers from "../components/user/Frontusers.vue";
import Role from "../components/user/Role.vue";
const routes = [
  { path: "/", redirect: "/login" },
  { path: "/login", component: Login },
  { path: "/home", component: Home, redirect: "/welcome",
    children: [ { path: "/welcome", component: Welcome },
      { path: "/backusers", component: Backusers },
      { path: "/frontusers", component: Frontusers },
      { path: "/roles", component: Role },
```

```
        { path: "/categorys", component: Category },
        { path: "/goods",component: GoodsList },
        { path: "/createorder", component: CreateOrder },
        { path: "/orders", component: OrderList } ]
    }
];
const router = createRouter({
  history: createWebHistory(process.env.BASE_URL),
  routes
});
export default router;
```

App.vue 是项目的主组件或根组件，是页面入口文件，整个应用只有一个，所有页面都在 App.vue 下进行切换，内容如下。

```
<template>
   <router-view/>
</template>
<style>
</style>
```

src/components 是公共组件目录，存放用户创建的功能组件。

12.9 登录与管理首页面

实现系统登录的组件为 Login.vue，登录页面效果如图 12-6 所示。

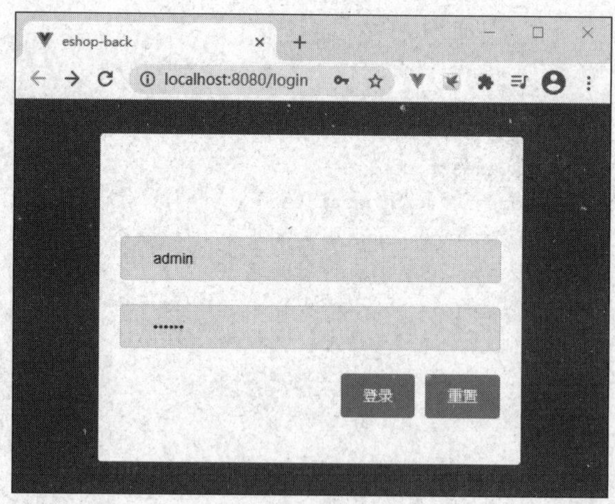

图 12-6 系统登录页

在 Login.vue 组件中，表示 html 结构的 template 部分代码如下。

```
<template>
  <div class="login-container">
```

```
    <div class = "login - box">
      <el - form ref = "loginFormRef" :model = "loginForm"
        :rules = "loginFormRules" label - width = "0px" class = "login - form">
        <!-- 用户名 -->
        <el - form - item prop = "name">
          <el - input v - model = "loginForm.name" prefix - icon = "iconfont icon - user"
placeholder = "请输入用户名"></el - input>
        </el - form - item>
        <el - form - item prop = "pwd">
          <el - input type = "password" v - model = "loginForm.pwd"
            prefix - icon = "iconfont icon - mima" placeholder = "请输入密码">
          </el - input>
        </el - form - item>
        <!-- 按钮 -->
        <el - form - item class = "login - btns">
          <el - button type = "primary" @click = "login">登录</el - button>
          <el - button type = "info" @click = "resetLoginForm()">重置</el - button>
        </el - form - item>
      </el - form>
    </div>
  </div>
</template>
```

在 template 模板中，< el-form >代表一个表单，rules 属性用于设置表单验证规则，绑定的是一个对象数组，这里指定为响应式变量 loginFormRules；ref 属性表示被引用时的名称和标识，这里指定为响应式变量 loginFormRef；model 属性用来指定表单使用的数据，这里指定为响应式变量 loginForm。

每个< el-form-item >是一个表单项，< el-form-item >的 prop 属性取值必须是< el-form >的 model 的直接属性，否则会导致校验失败。

< el-input >是输入框，使用 v-model 指令来实现与 loginForm 对象中属性的双向数据绑定。

< el-button >是按钮，这里定义了登录和重置两个按钮，并给这两个按钮绑定 click 事件。单击登录按钮时，会执行自定义函数 login。单击重置按钮时，会执行自定义函数 resetLoginForm。

在 Login.vue 组件中，表示 js 逻辑的 script 部分代码如下。

```
<script>
import { ref, reactive, getCurrentInstance } from "vue";
export default {
  setup() {
    const { ctx } = getCurrentInstance();
    const loginForm = reactive({ name: "", pwd: "" });
    const loginFormRef = ref(null);
    const loginFormRules = reactive({
      name: [{ required: true, message: "请输入用户名", trigger: "blur" }],
```

```
      pwd: [{ required: true, message: "请输入密码", trigger: "blur" }],
    });
    const login = () => {
      loginFormRef.value.validate(async (valid) => {
        if (!valid) return;
        const { data: res } = await ctx.$axios.post("login", {
          name: loginForm.name,
          pwd: loginForm.pwd,
        });
        if (res.code !== 0) return ctx.$message.error("登录失败");
        ctx.$message.success("登录成功");
        window.sessionStorage.setItem("token", res.token);
        ctx.$router.push("/home");
      });
    };
    const resetLoginForm = () => {
      loginFormRef.value.resetFields();
    };
    return {loginFormRef, loginForm, loginFormRules, login, resetLoginForm, };
  },
};
</script>
```

在 script 部分，通过 import 引入 ref、reactive 和 getCurrentInstance 这三个函数。调用 getCurrentInstance()方法获取上下文实例，并从中解构出 ctx，通过 ctx.$axios 可以使用 axios 向后台服务器发送 get 和 post 请求。

使用 reactive 函数定义响应式变量 loginForm，<el-form>表单的 model 属性动态绑定该变量。使用 ref 函数定义变量 loginFormRef，el-form 表单的 ref 属性指定该变量。使用 reactive 函数定义 loginFormRules 对象，对 loginForm 中的属性 name 和 pwd 定义了校验规格。

在登录页面中，输入用户名(如 admin)和密码(如 123456)，单击“登录”按钮，会执行自定义函数 login。在 login 函数中，会判断用户名和密码的非空校验是否通过。如果通过校验，则使用 axios 发送一个 post 请求到后端程序 eshop，该请求的 url 为 login，其完整路径为 http://localhost:8888/eshop/login，这个请求将由后端程序 eshop 的控制器类 AdminInfoController 中的 login()方法处理，对用户名和密码进行合法性验证。AdminInfoController 类的 login()方法返回结果进行解构后，赋值给 res。如果 res.code 等于 0，表示登录成功，先通过 window.sessionStorage 存储 token 令牌，再将 Vue 页面跳转到 home，该路径对应于 src/ components 目录下的 Home.Vue 文件；否则通过消息框给出登录失败提示。

登录成功后，进入管理首页面，如图 12-7 所示。

管理首页面功能是在组件 Home.vue 中实现的，该组件位于 components 目录下。在组件 Home.vue 中，表示 html 结构的 template 部分代码如下。

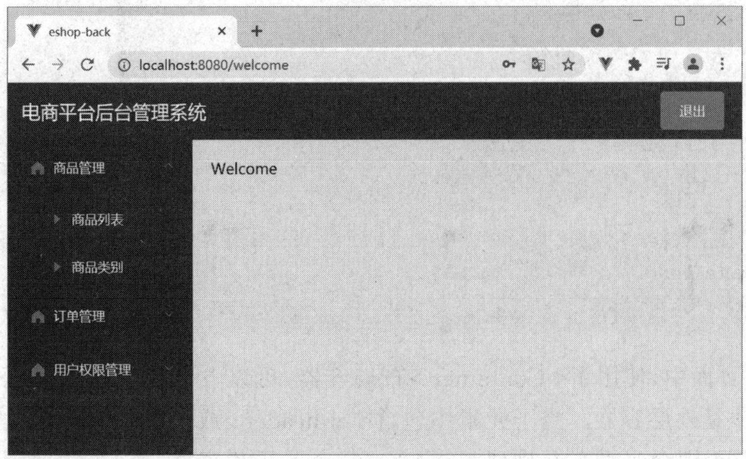

图 12-7　管理首页面

```
<template>
  <el-container class="home-container">
    <!-- 头部 -->
    <el-header class="el-header">
      <div>
        <img src="../assets/logo.png" width="35px" height="35px" alt>
        <span>电商平台后台管理系统</span>
      </div>
      <el-button type="info" @click="logout">退出</el-button>
    </el-header>
    <!-- 主体 -->
    <el-container>
      <!-- 菜单 -->
      <el-aside class="el-aside" width="200px">
        <el-menu v-if="menus" background-color="#333744"
          text-color="#fff" active-text-color="#ffd04b"
          unique-opened router :default-active="currentActiveMeunItem">
          <el-submenu :index="item.id + ''" v-for="item in menus"
                      :key="item.id">
            <template #title>
              <i class="el-icon-s-home"></i>
              <span>{{item.text}}</span>
            </template>
            <el-menu-item :index="'/' + subItem.url"
              v-for="subItem in item.children"
              :key="subItem.id"
              @click="savaCurrentActiveMeunItem('/' + subItem.url)">
              <i class="el-icon-caret-right"></i>
              <span slot="title">{{subItem.text}}</span>
            </el-menu-item>
          </el-submenu>
        </el-menu>
```

```
        <ul id = "array - rendering">
        </ul>
      </el - aside>
      <!-- 操作区 -->
      <el - main class = "el - main">
        <router - view></router - view>
      </el - main>
    </el - container>
  </el - container>
</template>
```

在管理首页面中，使用了< Container >布局容器，可以方便快速搭建页面的基本结构。< el-container >是外层容器。当子元素中包含< el-header >或< el-footer >时，全部子元素会垂直上下排列，否则会水平左右排列。< el-header >是顶栏容器，< el-aside >是侧边栏容器，< el-main >是主要区域容器，< el-footer >是底栏容器。在组件 Home. vue 的 style 部分，定义样式，将布局呈现出如图 12-7 所示的效果。

< el-menu >标签用来构建下拉菜单，使用 v-if 指令，判断响应式变量 menus 是否为空，如果不为空则渲染下拉菜单，否则不渲染。< el-menu >标签的 default-active 属性指定当前激活菜单的 index，这里动态绑定响应式变量 currentActiveMeunItem。unique-opened 属性用来指定是否只保持一个子菜单的展开，这里设置只保持一个展开。router 属性用来指定是否使用 vue-router 的模式，启用该模式会在激活导航时以 index 作为 path 进行路由跳转，这里设置为启动 vue-router 的模式。

< el-menu >包含< el-submenu >子标签，用于构建子菜单，通过 v-for 循环生成子菜单。< el-submenu >里面包含< template ＃title >，用来显示子菜单的标题和图标。

< el-submenu >还可以包含< el-menu-item >标签，用于构建子菜单下的菜单项，通过 v-for 循环生成各个子菜单下的菜单项，每个菜单项绑定 click 事件，单击菜单项时，执行自定义函数 savaCurrentActiveMeunItem，保存当前菜单项。

< el-main >标签里包含< router-view >，作为一个容器，用来渲染 vue-router 指定的组件。

在组件 Home. vue 中，script 部分表示 js 逻辑，代码如下。

```
<script>
import { ref, onMounted, getCurrentInstance, nextTick } from "vue";
export default {
  setup() {
    const { ctx } = getCurrentInstance();
    const menus = ref(null);
    const currentActiveMeunItem = ref(null);
    currentActiveMeunItem.value = window.sessionStorage.getItem(
      "currentActiveMeunItem"
    );
    const logout = () => {
      window.sessionStorage.clear();
      ctx.$router.push("/login");
```

```
    };
    const getMenus = async () => {
      const { data: res } = await ctx. $axios.get("menus");
      menus.value = res.data[0].children;
    };
    const savaCurrentActiveMeunItem = (activeMeunItem) => {
      window.sessionStorage.setItem("currentActiveMeunItem",
activeMeunItem);
      currentActiveMeunItem.value = activeMeunItem;
    };
    onMounted(() => {
      getMenus();
    });
    return {getMenus, menus,currentActiveMeunItem,logout,
          savaCurrentActiveMeunItem};
  },
};
</script>
```

在 Vue 生命周期钩子函数 onMounted 中，调用自定义方法 getMenus()，使用 axios 向后台程序 eshop 发送一个 get 请求，该请求的 url 为 menus，其完整路径为 http://localhost:8888/eshop/menus，该请求将由后台程序 eshop 的控制器类 AdminInfoController 中的 getMenus()方法处理，获取登录用户的功能菜单列表。getMenus()方法返回结果进行解构后，赋值给 res，并将 res.data[0].children 赋值给 ref 函数定义的响应式变量 menus，作为渲染< el-menu >下拉菜单的数据源。

为了保存单击的菜单项，定义一个响应式变量 currentActiveMeunItem，每次单击菜单项时，执行自定义函数 savaCurrentActiveMeunItem，函数的参数 activeMeunItem 代表单击的菜单项。在函数 savaCurrentActiveMeunItem 中，将单击的菜单项保存到变量 currentActiveMeunItem，同时保存到本地存储 sessionStorage 中。sessionStorage 用于临时保存同一窗口(或标签页)的数据，在关闭窗口(或标签页)后，会删除这些数据。在组件 Home.vue 生命周期函数 setup 执行时，都会先从 sessionStorage 中获取所保存的单击菜单项，并赋值给变量 currentActiveMeunItem。

自定义函数 logout 用于退出登录，在函数 logout 中，先清除本地存储 sessionStorage 中的数据，再使用 router.push 方法导航到登录组件 Login.vue。

12.10　商品管理

商品管理包括商品列表和商品类别两部分，商品列表模块包括商品列表分页显示与查询、添加商品和修改商品功能。商品类别模块包括商品类别分页显示与查询、添加商品类别和修改商品类别功能。

12.10.1　商品列表

在管理首页面中，单击商品管理下的商品列表菜单，打开商品列表页，如图 12-8 所示。

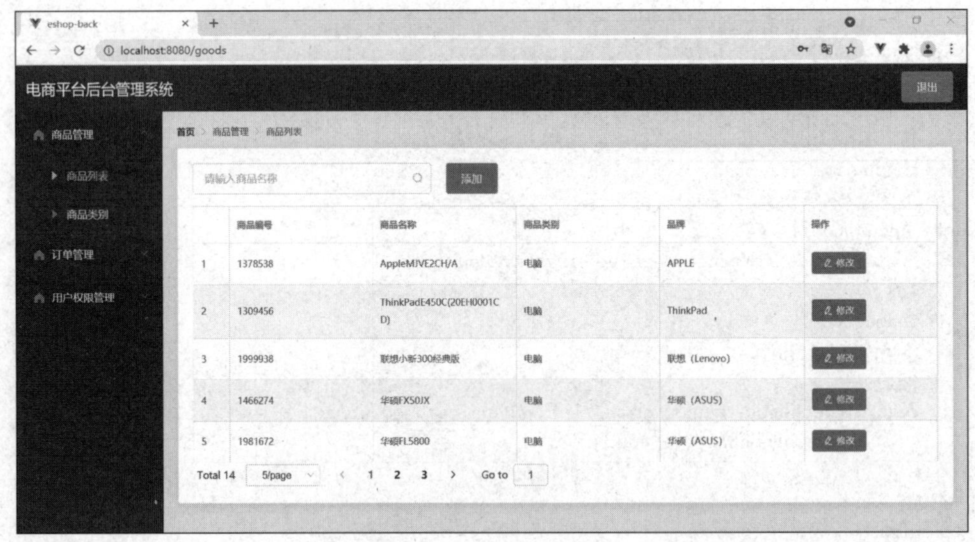

图 12-8 商品列表页

在商品列表页中，可以分页显示商品、查询商品、添加商品和修改商品。商品列表页的所有功能在组件 GoodsList. vue 中实现，该组件位于 components/goods 目录下。

(1) 页面导航。在组件 GoodsList. vue 的 template 模板部分，添加面包屑导航，代码如下。

```
<el-breadcrumb separator-class="el-icon-arrow-right">
  <el-breadcrumb-item :to="{ path: '/home' }">首页</el-breadcrumb-item>
  <el-breadcrumb-item>商品管理</el-breadcrumb-item>
  <el-breadcrumb-item>商品列表</el-breadcrumb-item>
</el-breadcrumb>
```

<el-breadcrumb>是面包屑导航组件，用来显示当前页面路径，实现快速返回之前的页面。<el-breadcrumb-item>表示从首页开始的每一级。通过<el-breadcrumb>标签的 separator-class 属性，可以设置图标分隔符。通过<el-breadcrumb-item>标签的 to 属性，可以设置链接的目标路由。

(2) 商品分页显示与查询。在组件 GoodsList. vue 的 template 模板部分，添加商品查询表单、商品列表、分页条等内容，相关代码如下。

```
<el-card>
  <!-- 查询表单区 -->
  <el-row :gutter="20">
    <el-col :span="8">
     <el-input placeholder="请输入商品名称" suffix-icon="el-icon-search"
        v-model="queryInfo.name" clearable @clear="getGoods"@change="getGoods(-1)">
     </el-input>
    </el-col>
    <el-col :span="4">
```

```
      <el-button type="primary" @click="showAddGoodsInfoDialog"
          >添加</el-button>
      </el-col>
    </el-row>
    <!-- 商品列表 -->
    <el-table :data="goodslist" border stripe>
      <el-table-column type="index"></el-table-column>
      <el-table-column label="商品编号" prop="code"></el-table-column>
      <el-table-column label="商品名称" prop="name"></el-table-column>
      <el-table-column label="商品类别" prop="category.name">
      </el-table-column>
      <el-table-column label="品牌" prop="brand"></el-table-column>
      <el-table-column label="操作" width="180px">
        <template v-slot="scope">
          <el-button type="primary" icon="el-icon-edit" size="mini"
            @click="openEditGoodsDialog(scope.row.id)">修改</el-button>
        </template>
      </el-table-column>
    </el-table>
    <!-- 分页条 -->
    <div v-if="goodslist != null && goodslist.length > 0">
      <el-pagination
       @size-change="handleSizeChange"
       @current-change="handleCurrentChange"
       :current-page="queryInfo.curPage"
       :page-sizes="[5, 10, 20]"
       :page-size="queryInfo.pageSize"
       layout="total, sizes, prev, pager, next, jumper"
       :total="total"></el-pagination>
    </div>
  </el-card>
```

<el-card>是卡片容器，用来将信息聚合在卡片容器中显示。在卡片容器中，自上而下依次放置查询表单、商品列表、分页条。

查询表单通过<el-row>布局，<el-row>用来标识一行，通过:gutter属性设置各列之间的间隔。<el-col>用来标识每行内的一列，通过:span属性设置每列占整行的宽度比例。在查询表单中，设置一行两列。在第一列中，添加一个输入框标签<el-input>，用于输入要查询的商品名称。在<el-input>标签中，通过placeholder属性指定输入框占位文本，通过suffix-icon属性指定输入框尾部图标，通过clearable属性指定输入框内容是否可清空，通过v-model属性将输入框内容绑定到响应式对象queryInfo的name属性。同时，还给商品名称输入框绑定clear和change两个事件，clear事件会在单击由clearable属性生成的清空按钮时触发，这时会调用自定义函数getGoods，重新查询所有商品列表。change事件会在输入框失去焦点或用户按下回车键时触发，这时会执行自定义函数getGoods(-1)，查询商品列表。在第二列中，添加一个"添加"按钮标签<el-button>，给该按钮绑定click事件。单击按钮时，会执行自定义函数showAddGoodsInfoDialog，打开添加商品对话框。

商品列表使用<el-table>标签，该标签可用于展示多条结构类似的数据。在<el-table>

标签中，通过 data 属性动态绑定响应式变量 goodslist，变量 goodslist 用于存放<el-table>要显示的数据。<el-table>标签包含<el-table-column>子标签，作为表格列。在<el-table-column>标签中，label 属性用于定义表格的列名，prop 属性用于指定列内容的字段名，type 属性用于指定对应列的类型，当 index 设置为 index 时，则显示该行的索引（从 1 开始）。在列名为“操作”这一列中，使用自定义列模板。在<el-table-column>标签内，添加一个<template>标签，使用 v-slot 指令绑定一个作用域插槽，插槽名为 scope。这样，通过 scope. row. id 就可以从数据源中，获得每一行记录的 id 字段值。在<template>标签内再添加一个“修改”按钮标签<el-button>，并绑定 click 事件。单击该按钮，会执行自定义函数 openEditGoodsDialog，用于打开修改商品对话框，函数 openEditGoodsDialog 的参数为该行商品的 id 号。

分页条使用<el-pagination>标签，在<el-pagination>标签中，current-page 属性用于指定当前页码，这里动态绑定到响应式变量 queryInfo 中的 curPage 属性。page-size 属性用于指定每页显示记录条数，这里动态绑定到响应式变量 queryInfo 中的 pageSize 属性。total 属性用于指定总记录条数，这里动态绑定到响应式变量 total。page-sizes 属性用于每页显示个数选择器的选项设置。此外，<el-pagination>绑定 size-change 和 current-change 两个事件。size-change 事件在 pageSize 改变时触发，此时会执行自定义函数 handleSizeChange。current-change 事件在 currentPage 改变时触发，此时会执行自定义函数 handleCurrentChange。layout 属性用于设置组件布局，可选值包括 sizes，prev，pager，next，jumper，->，total，slot，默认值为 prev，pager，next，jumper，->，total，选项之间用逗号分隔。

在组件 GoodsList. vue 的 script 部分，与商品分页显示和查询功能相关的代码如下。

```
const { ctx } = getCurrentInstance();
// 商品列表
const goodslist = ref(null);
// 商品总数
const total = ref(0);
// 封装查询参数
const queryInfo = reactive({ name: "", curPage: 1, pageSize: 5, });
// 联网获取商品列表
const getGoods = async (flag) => {
  if (flag == 0 || flag == -1) {
    queryInfo.curPage = 1;
  }
  // 分页获取管理员列表
  const { data: res } = await ctx.$axios.get("goods", {
    params: {
      name: queryInfo.name,
      curPage: queryInfo.curPage,
      pageSize: queryInfo.pageSize,
    },
  });
```

```
    if (res.code != 0) {
      goodslist.value = [];
      return ctx.$message.error("没有商品记录!");
    }
    goodslist.value = res.goodsInfos;
    total.value = res.total;
  };
  // pageSize 改变事件处理函数
  const handleSizeChange = (pageSize) => {
    queryInfo.pageSize = pageSize;
    getGoods();
  };
  // page 改变事件处理函数
  const handleCurrentChange = (curPage) => {
    queryInfo.curPage = curPage;
    getGoods();
  };
  onMounted(() => {
    getGoods();
  });
  return { ..., queryInfo, goodslist,total, getGoods, handleSizeChange,
    handleCurrentChange, catelist };
```

在上述代码中,依次定义变量 goodslist 保存商品列表,total 保存商品总数,queryInfo 封装查询参数。

添加自定义函数 getGoods,用来向后台服务器发送请求,获取商品列表。在函数 getGoods 中,使用 axios 发送 get 请求,请求地址为 http://localhost:8888/eshop/goods,这个请求将由后端程序 eshop 的控制器类 GoodsInfoController 中的 getGoodsInfos()方法处理,分页获取商品列表。返回结果解构后,赋值给 res。如果 res.code 等于 0,表示成功获取商品列表,则将商品列表赋值给响应式变量 goodslist,将商品总数赋值给响应式变量 total。否则,通过消息框提示“没有商品记录”。

添加自定义函数 handleSizeChange,用来处理分页条绑定的 size-change 事件,该事件在分页条上每页显示个数选择器的选项发生变化时触发。添加自定义函数 handleCurrentChange,用来处理分页条绑定的 current-change 事件,该事件在分页条上当前页页码发生变化时触发。

在 Vue 生命周期钩子函数 onMounted 中,调用自定义函数 getGoods。最后,将定义的变量、添加的自定义函数通过 return 返回出去,供 template 模板使用。

(3) 添加商品。在图 12-8 所示的商品列表页中,单击“添加”按钮,打开“添加商品信息”对话框,如图 12-9 所示。

在组件 GoodsList.vue 的 template 模板部分,与“添加商品信息”对话框相关的代码如下。

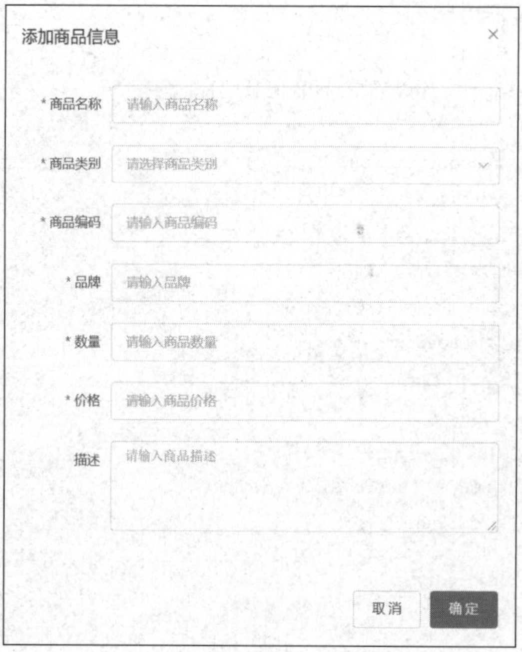

图 12-9 “添加商品信息”对话框

```
<el-dialog
  title="添加商品信息" v-model="addGoodsInfoDialogVisible" width="40%"
  @close="resetAddGoodsInfoForm()">
  <el-form ref="addGoodsInfoFormRef" :model="addGoodsInfoForm"
    :rules="addGoodsInfoFormRules" label-width="100px" class="login-form">
    <!-- 商品类别名称 -->
    <el-form-item prop="name" label="商品名称">
      <el-inputv-model="addGoodsInfoForm.name"
        placeholder="请输入商品名称"></el-input>
    </el-form-item>
    <!-- 商品类别 -->
    <el-form-item prop="cid" label="商品类别">
      <el-select
        style="width: 100%"
        v-model="addGoodsInfoForm.cid"
        placeholder="请选择商品类别">
        <el-option v-for="item in catelist" :key="item.id"
          :label="item.name" :value="item.id"></el-option>
      </el-select>
    </el-form-item>
    <!-- 商品编码 -->
    <el-form-item prop="code" label="商品编码">
      <el-inputv-model="addGoodsInfoForm.code"
        placeholder="请输入商品编码"></el-input>
    </el-form-item>
    <!-- 品牌 -->
```

```
    <el-form-item prop="brand" label="品牌">
      <el-input v-model="addGoodsInfoForm.brand"
        placeholder="请输入品牌"></el-input>
    </el-form-item>
    <!-- 数量 -->
    <el-form-item prop="num" label="数量">
      <el-input type="number" min="1" v-model="addGoodsInfoForm.num"
        placeholder="请输入商品数量"></el-input>
    </el-form-item>
    <!-- 价格 -->
    <el-form-item prop="price" label="价格">
      <el-input type="number" min="0" v-model="addGoodsInfoForm.price"
        placeholder="请输入商品价格"></el-input>
    </el-form-item>
    <!-- 描述 -->
    <el-form-item prop="intro" label="描述">
      <el-input type="textarea" :rows="4" v-model="addGoodsInfoForm.intro"
        placeholder="请输入商品描述"></el-input>
    </el-form-item>
  </el-form>
  <template #footer>
    <span class="dialog-footer">
      <el-button @click="addGoodsInfoDialogVisible = false">取 消
      </el-button>
      <el-button type="primary" @click="addGoodsInfo()">确 定</el-button>
    </span>
  </template>
</el-dialog>
```

<el-dialog>是对话框标签，width 属性用于指定对话框的宽度，通过 v-model 与响应式变量 addGoodsInfoDialogVisible 实现双向绑定，用于控制对话框是否显示。给<el-dialog>绑定 close 事件，该事件在对话框关闭时触发，此时会调用自定义函数 resetAddGoodsInfoForm，用来重置添加商品信息表单。在<el-dialog>标签中，添加表单标签<el-form>。在<el-form>中，添加多个<el-form-item>子标签。在这些<el-form-item>标签内，依次添加用于商品名称输入的<el-input>标签、商品类别选择的<el-select>标签、商品编码、品牌、商品数量、商品价格和商品描述输入的<el-input>标签。

<el-select>通过下拉菜单展示并选择内容，通过 v-model 与响应式变量 addGoodsInfoForm 中的属性 cid 实现双向绑定。在<el-select>标签内，添加<el-option>标签。在<el-option>标签上，通过 v-for 指令循环生成下拉菜单的选项，数据源为响应式变量 catelist。label 属性用于指定选项的显示文本，value 属性用于指定选项中的返回值。

在<el-dialog>标签内，添加<template #footer>，用于定义一个具名为 footer 的 slot。在<template>标签内，放置“取消”和“确定”两个按钮，并绑定 click 事件。单击“确定”按钮时，会执行自定义函数 addGoodsInfo，向服务器发送请求，添加商品信息。

在组件 GoodsList.vue 的 script 部分，与添加商品信息功能相关的代码如下。

```
// 控制添加商品对话框是否可见
const addGoodsInfoDialogVisible = ref(false);
// 绑定添加商品表单数据
const addGoodsInfoForm = reactive({
  name: "", cid: "", code: "", brand: "", num: "", price: "", intro: "",
});
// 指定添加商品表单被引用时的名称和标识
const addGoodsInfoFormRef = ref(null);
// 设置添加商品表单验证规则
const addGoodsInfoFormRules = reactive({
  name: [{ required: true, message: "请输入商品类别名称", trigger: "blur" }],
  code: [{ required: true, message: "请输入商品编码", trigger: "blur" }],
  cid: [{ required: true, message: "请选择商品类别", trigger: "change" }],
  brand: [{ required: true, message: "请输入品牌", trigger: "blur" }],
  num: [{ required: true, message: "请输入数量", trigger: "blur" }],
  price: [{ required: true, message: "请输入价格", trigger: "blur" }],
});
// 商品类别列表
const catelist = ref(null);
// 打开添加商品信息对话框
const showAddGoodsInfoDialog = async () => {
  console.log("打开添加商品信息对话框");
  const { data: res } = await ctx.$axios.get("categories");
  catelist.value = res.categories;
  await nextTick();
  addGoodsInfoDialogVisible.value = true;
};
// 重置添加商品信息表单
const resetAddGoodsInfoForm = () => {};
// 联网添加商品
const addGoodsInfo = () => {
  addGoodsInfoFormRef.value.validate(async (valid) => {
    if (!valid) return;
    var name = addGoodsInfoForm.name;
    var code = addGoodsInfoForm.code;
    var cid = addGoodsInfoForm.cid;
    var brand = addGoodsInfoForm.brand;
    var num = addGoodsInfoForm.num;
    var price = addGoodsInfoForm.price;
    var intro = addGoodsInfoForm.intro;
    const { data: res } = await ctx.$axios.post('goods', {
      name: name, code: code, cid: cid, brand: brand, num: num,
      price: price, intro: intro, });
    if (res.code !== 0) return ctx.$message.error("添加失败!");
    ctx.$message.success("添加成功!");
    getGoods();
    addGoodsInfoDialogVisible.value = false;
  });
};
return { ..., showAddGoodsInfoDialog, addGoodsInfoDialogVisible,
```

```
    resetAddGoodsInfoForm, addGoodsInfoFormRef,addGoodsInfoForm,
    addGoodsInfoFormRules,addGoodsInfo };
```

在上述代码中，变量 addGoodsInfoDialogVisible 用于控制添加商品对话框是否可见，变量 addGoodsInfoForm 用于绑定添加商品表单数据，变量 addGoodsInfoFormRef 用于指定添加商品表单被引用时的名称和标识，变量 addGoodsInfoFormRules 用于设置添加商品表单验证规则，变量 catelist 用于保存商品类别列表。

showAddGoodsInfoDialog 函数用于打开添加商品信息对话框，resetAddGoodsInfoForm 函数用于重置添加商品信息表单，addGoodsInfo 函数用于向后台程序发送请求，添加商品信息。最后，将这些变量和函数 return 出去，供 template 模板使用。

(4) 修改商品。在图 12-8 所示的商品列表页中，每行记录后面都有一个“修改”按钮。单击“修改”按钮，可以打开“修改商品信息”对话框，如图 12-10 所示。

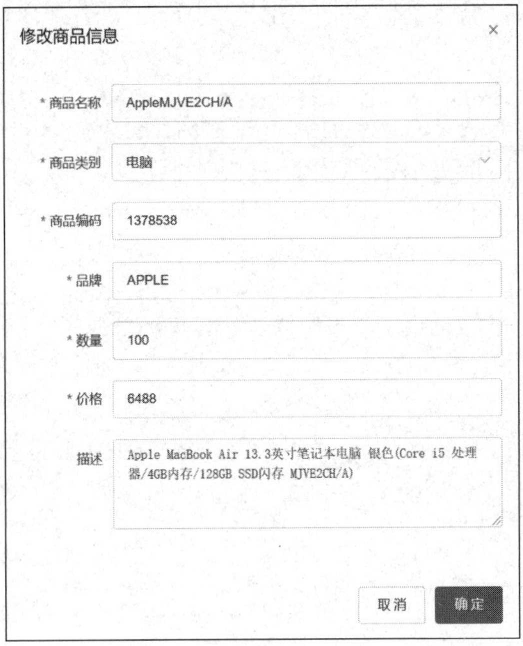

图 12-10　“修改商品信息”对话框

在组件 GoodsList.vue 的 template 模板部分，与“修改商品信息”对话框相关的代码如下。

```
<el-dialog
  title="修改商品信息" v-model="editGoodsInfoDialogVisible"
  width="40%" @close="resetAddGoodsInfoForm()">
  <el-form
    ref="editGoodsInfoFormRef" :model="editGoodsInfoForm"
    :rules="editGoodsInfoFormRules"label-width="100px" class="login-form">
    <!-- 商品类别名称 -->
```

```
    <el-form-item prop="name" label="商品名称">
      <el-inputv-model="editGoodsInfoForm.name"
        placeholder="请输入商品名称"></el-input>
    </el-form-item>
    <!-- 商品类别 -->
    <el-form-item prop="cid" label="商品类别">
      <el-select style="width: 100%" v-model="editGoodsInfoForm.cid"
        placeholder="请选择商品类别">
        <el-option v-for="item in catelist" :key="item.id"
          :label="item.name" :value="item.id"></el-option>
      </el-select>
    </el-form-item>
    <!-- 商品编码 -->
    <el-form-item prop="code" label="商品编码">
      <el-input v-model="editGoodsInfoForm.code"
        placeholder="请输入商品编码"></el-input>
    </el-form-item>
    <!-- 品牌 -->
    <el-form-item prop="brand" label="品牌">
      <el-inputv-model="editGoodsInfoForm.brand"
        placeholder="请输入品牌"></el-input>
    </el-form-item>
    <!-- 数量 -->
    <el-form-item prop="num" label="数量">
      <el-inputtype="number" min="1" v-model="editGoodsInfoForm.num"
        placeholder="请输入商品数量"></el-input>
    </el-form-item>
    <!-- 价格 -->
    <el-form-item prop="price" label="价格">
      <el-inputtype="number" min="0" v-model="editGoodsInfoForm.price"
        placeholder="请输入商品价格"></el-input>
    </el-form-item>
    <!-- 描述 -->
    <el-form-item prop="intro" label="描述">
      <el-inputtype="textarea":rows="4" v-model="editGoodsInfoForm.intro"
        placeholder="请输入商品描述"></el-input>
    </el-form-item>
  </el-form>
  <template #footer>
    <span class="dialog-footer">
      <el-button @click="editGoodsInfoDialogVisible = false">取消
      </el-button>
      <el-button type="primary" @click="editGoodsInfo()">确定</el-button>
    </span>
  </template>
</el-dialog>
```

在“修改商品信息”对话框中，使用的标签与“添加商品信息”对话框类似，由于篇幅，此处不再赘述。

在组件 GoodsList.vue 的 script 部分，与修改商品信息功能相关的代码如下。

```
// 控制修改商品对话框是否可见
const editGoodsInfoDialogVisible = ref(false);
// 绑定修改商品表单数据
const editGoodsInfoForm = reactive({
  id: "", name: "", cid: "", code: "", brand: "", num: "", price: "",
  intro: "", });
// 指定修改商品表单被引用时的名称和标识
const editGoodsInfoFormRef = ref(null);
// 设置修改商品表单验证规则
const editGoodsInfoFormRules = reactive({
  name: [{ required: true, message: "请输入商品类别名称", trigger: "blur" }],
  code: [{ required: true, message: "请输入商品编码", trigger: "blur" }],
  cid: [{ required: true, message: "请选择商品类别", trigger: "change" }],
  brand: [{ required: true, message: "请输入品牌", trigger: "blur" }],
  num: [{ required: true, message: "请输入数量", trigger: "blur" }],
  price: [{ required: true, message: "请输入价格", trigger: "blur" }],
});
// 打开修改对话框
const openEditGoodsDialog = async (id) => {
  const { data: res1 } = await ctx.$axios.get("categories");
  catelist.value = res1.categories;
  await nextTick();
  editGoodsInfoDialogVisible.value = true;
  const { data: res2 } = await ctx.$ctx.$axios.get(`goods/id/${id}`);
  if (res2.code != 0) return ctx.$message.error("获取数据失败!");
  editGoodsInfoForm.id = res2.goodsInfo.id;
  editGoodsInfoForm.name = res2.goodsInfo.name;
  editGoodsInfoForm.code = res2.goodsInfo.code;
  editGoodsInfoForm.cid = res2.goodsInfo.cid;
  editGoodsInfoForm.brand = res2.goodsInfo.brand;
  editGoodsInfoForm.num = res2.goodsInfo.num;
  editGoodsInfoForm.price = res2.goodsInfo.price;
  editGoodsInfoForm.intro = res2.goodsInfo.intro;
};
// 联网修改商品
const editGoodsInfo = () => {
  editGoodsInfoFormRef.value.validate(async (valid) => {
    if (!valid) return;
    var id = editGoodsInfoForm.id;
    var name = editGoodsInfoForm.name;
    var code = editGoodsInfoForm.code;
    var cid = editGoodsInfoForm.cid;
    var brand = editGoodsInfoForm.brand;
    var num = editGoodsInfoForm.num;
    var price = editGoodsInfoForm.price;
    var intro = editGoodsInfoForm.intro;
    const { data: res } = await ctx.$axios.put(`goods`, {
     id: id, name: name, code: code, cid: cid, brand: brand, num: num,
```

```
        price: price, intro: intro,});
      if (res.code !== 0) return ctx. $message.error("修改失败!");
      ctx. $message.success("修改成功!");
      getGoods();
      editGoodsInfoDialogVisible.value = false;
    });
  };
  return { ..., openEditGoodsDialog,editGoodsInfoDialogVisible,
  editGoodsInfoForm, editGoodsInfoFormRef,editGoodsInfoFormRules,
  editGoodsInfo };
```

在上述代码中,变量 editGoodsInfoDialogVisible 用于控制修改商品对话框是否可见,变量 editGoodsInfoForm 用于绑定修改商品表单数据,变量 editGoodsInfoFormRef 用于指定修改商品表单被引用时的名称和标识,变量 editGoodsInfoFormRules 用于设置修改商品表单验证规则,变量 openEditGoodsDialog 用于打开修改对话框。editGoodsInfo 函数用于向后台程序发送请求,修改商品信息。最后,将这些变量和函数 return 出去,供 template 模板使用。

12.10.2 商品类别

在管理首页面中,单击商品管理下的商品类别菜单,打开商品类别页。在商品类别页中,可以显示与查询类别、添加和修改类别,如图 12-11 所示。

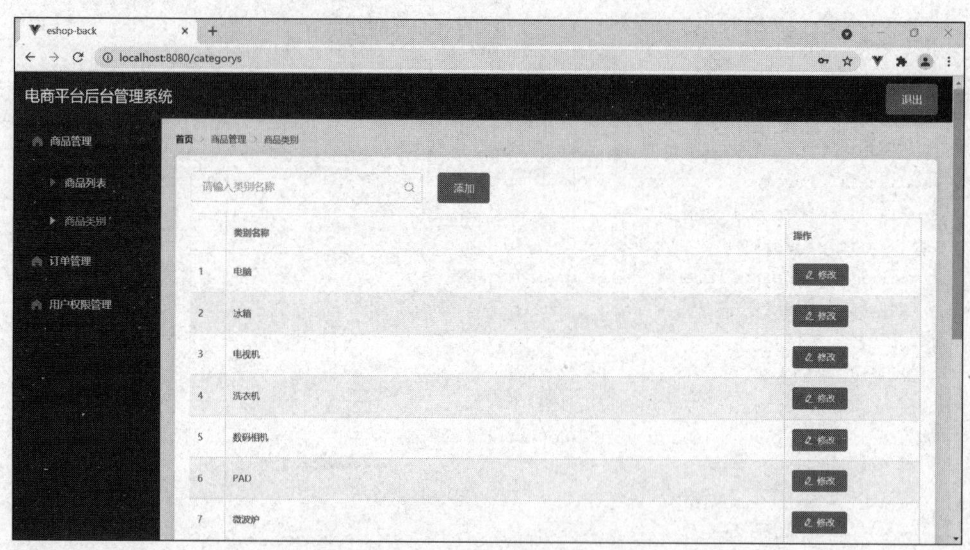

图 12-11 商品类别页

商品类别页的所有功能在组件 Category. vue 中实现,该组件位于 components/goods 目录下。

(1) 页面导航。在组件 Category. vue 的 template 模板部分,添加面包屑导航,代码如下。

```
<el-breadcrumb separator-class="el-icon-arrow-right">
  <el-breadcrumb-item :to="{ path: '/home' }">首页</el-breadcrumb-item>
  <el-breadcrumb-item>商品管理</el-breadcrumb-item>
  <el-breadcrumb-item>商品类别</el-breadcrumb-item>
</el-breadcrumb>
```

(2) 商品类别显示与查询。在组件 Category.vue 的 template 模板部分，添加类别查询表单、类别列表，相关代码如下。

```
<el-card>
  <el-row :gutter="20">
    <el-col :span="8">
      <el-inputplaceholder="请输入类别名称" suffix-icon="el-icon-search"
        v-model="queryInfo.name" clearable @clear="getCategories"
        @change="getCategories">
      </el-input>
    </el-col>
    <el-col :span="4">
      <el-button type="primary" @click="showAddCategoryDialog">添加
      </el-button>
    </el-col>
  </el-row>
  <!-- 类别列表 -->
  <el-table :data="catelist" border stripe>
    <el-table-column type="index"></el-table-column>
    <el-table-column label="类别名称" prop="name"></el-table-column>
    <el-table-column label="操作" width="180px">
      <template v-slot="scope">
        <el-button type="primary" icon="el-icon-edit" size="mini"
          @click="openEditCateDialog(scope.row.id)">修改</el-button>
      </template>
    </el-table-column>
  </el-table>
</el-card>
```

在上述代码中，使用了<el-card>和<el-table>等标签，在介绍商品列表功能模块时，对这些标签已做介绍，此处不再赘述。

在组件 Category.vue 的 script 部分，与类别显示和查询功能相关的代码如下。

```
const { ctx } = getCurrentInstance();
const catelist = ref(null);
const queryInfo = reactive({ name: "",});
const getCategories = async () => {
  // 分页获取类别列表
  const { data: res } = await ctx.$axios.get("categories", {
    params: {
      name: queryInfo.name,
```

```
      },
    });
    if (res.code != 0) {
      catelist.value = [];
      return ctx.$message.error("没有商品类别记录!");
    }
    catelist.value = res.categories;
  };
  onMounted(() => {
    getCategories();
  });
  return { queryInfo, catelist, getCategories, ...};
```

在上述代码中,响应式变量 catelist 用于保存类别列表,queryInfo 用于封装查询参数。

自定义函数 getCategories 用来向后台服务器发送请求获取类别列表,在函数 getCategories 中,使用 axios 发送 get 请求,请求地址为 http://localhost:8888/eshop/categories,并传递参数 name,这个请求将由后端程序 eshop 的控制器类 CategoryController 中的 getCategories()方法处理,返回结果解构后,赋值给 res。如果 res.code 等于 0,表示成功获取类别列表,则将类别列表赋值给变量 catelist。否则,通过消息框提示"没有商品类别记录"。

在 Vue 生命周期钩子函数 onMounted 中,调用自定义函数 getCategories。

最后,将定义的变量、添加的自定义函数通过 return 返回出去,供 template 模板使用。

(3) 添加商品类别。在图 12-11 所示的商品类别页中,单击"添加"按钮,可以打开"添加商品类别"对话框,如图 12-12 所示。

图 12-12 "添加商品类别"对话框

在组件 Category.vue 的 template 模板部分,与"添加商品类别"对话框相关的代码如下。

```
<!-- 添加商品类别对话框 -->
<el-dialog
  title="添加商品类别" v-model="addCategoryDialogVisible" width="40%"
  @close="resetAddCategoryForm()">
  <el-form
    ref="addCategoryFormRef" :model="addCategoryForm"
    :rules="addCategoryFormRules" label-width="10px" class="login-form">
    <!-- 商品类别名称 -->
```

```
      <el-form-item prop="name">
        <el-input v-model="addCategoryForm.name"
          prefix-icon="iconfont icon-user"
          placeholder="请输入商品类别名称"></el-input>
      </el-form-item>
    </el-form>
    <template #footer>
      <span class="dialog-footer">
        <el-button @click="addCategoryDialogVisible = false">取消</el-button>
        <el-button type="primary" @click="addCategory()">确定</el-button>
      </span>
    </template>
  </el-dialog>
```

在上述代码中，使用了<el-dialog>和<el-form>等标签。在介绍商品列表功能模块时，对这些标签已做介绍，此处不再赘述。

在组件 Category.vue 的 script 部分，与添加商品类别功能相关的代码如下。

```
const addCategoryForm = reactive({ name: "", });
const addCategoryFormRef = ref(null);
const addCategoryFormRules = reactive({
  name: [{ required: true,message: "请输入商品类别名称",trigger: "blur" }] });
const addCategoryDialogVisible = ref(false);
// 显示添加商品类别对话框
const showAddCategoryDialog = async () => {
  addCategoryDialogVisible.value = true;
  await nextTick();
};
// 重置添加商品类别对话框中的表单
const resetAddCategoryForm = () => {
  addCategoryFormRef.value.resetFields();
};
// 添加商品类别
const addCategory = () => {
  addCategoryFormRef.value.validate(async (valid) => {
    if (!valid) return;
    var name = addCategoryForm.name;
    // 判断商品类别名称是否已存在
    const { data: res1 } = await ctx.$axios.get(`categories/name/${name}`);
    if (res1.code !== 0) return ctx.$message.error("该商品类别名称已存在!");
    const { data: res2 } = await ctx.$axios.post(`categories`, {
      name: name,
    });
    if (res2.code !== 0) return ctx.$message.error("添加失败!");
    ctx.$message.success("添加成功!");
    getCategories();
    addCategoryDialogVisible.value = false;
  });
};
```

```
return { ..., showAddCategoryDialog, addCategoryDialogVisible,
addCategoryFormRef, addCategoryForm, addCategoryFormRules,
resetAddCategoryForm, addCategory };
```

在上述代码中，变量 addCategoryForm 用于绑定添加商品类别表单数据，变量 addCategoryFormRef 用于指定添加商品类别表单被引用时的名称和标识，变量 addCategoryFormRules 用于设置添加商品类别表单验证规则，变量 addCategoryDialogVisible 用于控制添加商品类别对话框是否可见。

自定义函数 showAddCategoryDialog 用于打开添加商品类别对话框，自定义函数 resetAddCategoryForm 用于重置添加商品类别表单，自定义函数 addCategory 用于向后台程序发送请求，添加商品类别。最后，将这些变量和函数 return 出去，供 template 模板使用。

(4) 修改商品类别。在图 12-11 所示的商品类别页中，每行记录后面都有一个"修改"按钮。单击"修改"按钮，可以打开"修改商品类别"对话框，如图 12-13 所示。

图 12-13 "修改商品类别"对话框

在组件 Category.vue 的 template 模板部分，与"修改商品类别"对话框相关的代码如下。

```
<!-- 修改商品类别对话框 -->
<el-dialog
  title="修改商品类别" v-model="editCategoryDialogVisible" width="50%">
  <el-form
    :model="editCategoryForm" :rules="editCategoryFormRules"
    ref="editCategoryFormRef" label-width="90px">
    <el-form-item label="商品类别" prop="name">
      <el-input v-model="editCategoryForm.name"></el-input>
    </el-form-item>
  </el-form>
  <template #footer>
    <span class="dialog-footer">
      <el-button @click="editCategoryDialogVisible = false">取消
      </el-button>
      <el-button type="primary" @click="editCategory()">确定</el-button>
    </span>
  </template>
</el-dialog>
```

在上述代码中，使用了<el-dialog>和<el-form>等标签。在介绍商品列表功能模块时，对这些标签已做介绍，此处不再赘述。

在组件 Category.vue 的 script 部分，与修改商品类别功能相关的代码如下。

```
const editCategoryForm = reactive({ name: "", id: "", });
const editCategoryFormRef = ref(null);
const editCategoryFormRules = reactive({
  name: [{ required: true, message: "请输入商品类别名称", trigger: "blur" }],
});
const editCategoryDialogVisible = ref(false);
// 打开修改对话框
const openEditCateDialog = async (id) => {
  editCategoryDialogVisible.value = true;
  const { data: res } = await ctx.$axios.get(`categories/id/${id}`);
  await nextTick();
  if (res.code != 0) return ctx.$message.error("获取数据失败!");
  editCategoryForm.name = res.category.name;
  editCategoryForm.id = res.category.id;
};
// 修改商品类别
const editCategory = () => {
  editCategoryFormRef.value.validate(async (valid) => {
    if (!valid) return;
    var name = editCategoryForm.name;
    var id = editCategoryForm.id;
    // 判断商品类别名称是否已存在
    const { data: res1 } = await ctx.$axios.get(
      `categories/name/${name}/id/${id}`
    );
    if (res1.code !== 0)
      return context.root.$message.error("该商品类别名称已存在!");
    const { data: res2 } = await ctx.$axios.put(`categories`, {
      name: name,
      id: id,
    });
    if (res2.code !== 0) return ctx.$message.error("修改失败!");
    ctx.$message.success("修改成功!");
    getCategories();
    editCategoryDialogVisible.value = false;
  });
};
return { ..., editCategoryForm, editCategoryFormRules, editCategoryFormRef, penEditCateDialog,
editCategoryDialogVisible, editCategory };
```

在上述代码中，变量 editCategoryForm 用于绑定修改商品类别表单数据，变量 editCategoryFormRef 用于指定修改商品类别表单被引用时的名称和标识，变量 editCategoryFormRules 用于设置修改商品类别表单验证规则，变量 editCategoryDialogVisible 用于控制修改商品类别对话框是否可见。

自定义函数 openEditCateDialog 用于打开修改商品类别对话框，自定义函数 editCategory

用于向后台程序发送请求，修改商品类别。最后，将这些变量和函数 return 出去，供 template 模板使用。

12.11 订单管理

订单管理包括订单列表和创建订单两部分，订单列表模块包括订单列表分页显示与查询、删除订单功能。

12.11.1 订单列表

在管理首页面中，单击订单管理下的订单列表菜单，打开订单列表页。在订单列表页中，可以分页显示订单、查询订单明细、删除订单，如图 12-14 所示。

图 12-14 订单列表页

订单列表页的所有功能在组件 OrderList. vue 中实现，该组件位于 components/order 目录下。

(1) 页面导航。在组件 OrderList. vue 的 template 模板部分，添加面包屑导航，代码如下。

```
<el - breadcrumb separator - class = "el - icon - arrow - right">
  <el - breadcrumb - item : to = "{ path: '/home' }">首页</el - breadcrumb - item >
  <el - breadcrumb - item >订单管理</el - breadcrumb - item >
  <el - breadcrumb - item >订单列表</el - breadcrumb - item >
</el - breadcrumb >
```

(2) 订单分页显示与查询。在组件 OrderList. vue 的 template 模板部分，添加订单查询表单、订单列表、分页条等内容，相关代码如下。

```
<el-card>
  <el-row :gutter="20">
    <el-col :span="5">
      <el-inputclearable placeholder="请输入订单编号"
        v-model="queryInfo.orderNo"></el-input>
    </el-col>
    <el-col :span="5">
      <el-select v-if="customers" clearablev-model="queryInfo.uid"
        placeholder="请选择客户名称">
        <el-option v-for="item in customers" :key="item.id"
          :label="item.userName" :value="item.id"></el-option>
      </el-select>
    </el-col>
    <el-col :span="5">
      <el-select clearablev-model="queryInfo.orderStatus"
        placeholder="请选择订单状态">
        <el-option v-for="item in orderStatusList" :key="item.value" :label="item.
label" :value="item.value"></el-option>
      </el-select>
    </el-col>
    <el-col :span="4">
      <el-button type="primary" @click="getOrderinfos(0)">查找</el-button>
    </el-col>
  </el-row>
  <!-- 订单列表 -->
  <el-table@expand-change="handleExpandChange": data="orderlist"
    Border stripe ref="tableRef">
    <el-table-column type="expand">
      <template v-slot="scope">
        <!-- 展开行表格 -->
        <el-table :data="scope.row.orderdetails" border stripe>
          <el-table-column type="index"></el-table-column>
          <el-table-column label="商品名称"prop="goods.name">
          </el-table-column>
          <el-table-column label="订单数量"prop="quantity">
          </el-table-column>
        </el-table>
      </template>
    </el-table-column>
    <el-table-column type="index"></el-table-column>
    <el-table-column label="订单号" prop="orderNo"></el-table-column>
    <el-table-column label="客户名称" prop="ui.realName"></el-table-column>
    <el-table-column label="订单状态" prop="status"></el-table-column>
    <el-table-column label="订单日期">
      <template v-slot="scope">{{
        timeFormat(scope.row.ordertime)
      }}</template>
    </el-table-column>
```

```
    <el-table-column label="订单金额" prop="orderprice"></el-table-column>
    <el-table-column label="操作" width="180px">
      <template v-slot="scope">
        <el-button type="danger" icon="el-icon-delete"
          size="mini" @click="handleDeleteOrder(scope.row.id)">
        </el-button>
      </template>
    </el-table-column>
  </el-table>
  <!-- 分页 -->
  <div v-if="orderlist != null && orderlist.length > 0">
    <el-pagination
      @size-change="handleSizeChange"
      @current-change="handleCurrentChange"
      :current-page="queryInfo.curPage"
      :page-sizes="[1, 2, 5, 10]"
      :page-size="queryInfo.pageSize"
      layout="total, sizes, prev, pager, next, jumper"
      :total="total"></el-pagination>
  </div>
</el-card>
```

在上述代码中，使用了< el-card >、< el-row >、< el-col >、< el-table >、< el-table-column >、< el-input >、< el-select >、< el-option >、< el-pagination >等标签，在介绍商品列表功能模块时，对这些标签已做介绍。只是，在此处的< el-table >标签内，使用 Table 控件的展开行功能，用来显示订单明细。要开启展开行功能，可以通过设置 type="expand"和 slot。这样，el-table-column 的模板会被渲染成展开行的内容，展开行可访问的属性与使用自定义列模板时的 slot 相同。

在组件 OrderList.vue 的 script 部分，与订单分页显示与查询功能相关的代码如下。

```
const { ctx } = getCurrentInstance();
// 存放订单列表
const orderlist = ref(null);
// 存放订单明细
const orderdetaillist = ref(null);
// 订单总数
const total = ref(0);
// 封装订单查询参数
const queryInfo = reactive({ orderNo: "", uid: "", orderStatus: "",
  curPage: 1, pageSize: 5, });
// 客户列表
const customers = ref(null);
// 存放订单状态
const orderStatusList = ref(null);
// 初始订单状态
orderStatusList.value = [ {value: "未付款",label: "未付款", },
  {value: "已付款",label: "已付款", },{value: "待发货",label: "待发货", },
```

```
    {value: "已发货",label: "已发货", }, {value: "已完成",label: "已完成", },];
  // 向后台程序发送 get 请求,获取订单列表
  const getOrderinfos = async (flag) => {
    // 订单客户
    var uid;
    if (queryInfo.uid === "") {
      uid = 0;
    } else {
      uid = queryInfo.uid;
    }
    if (flag == 0) {
      queryInfo.curPage = 1;
    }
    // 分页获取订单列表
    const { data: res } = await ctx.$axios.get("orderinfos", {
      params: {
        uid: uid,orderNo: queryInfo.orderNo,status: queryInfo.orderStatus,
        curPage: queryInfo.curPage, pageSize: queryInfo.pageSize,},
    });
    if (res.code != 0) {
      orderlist.value = [];
      return ctx.$message.error("没有订单记录!");
    }
    total.value = res.total;
    res.orderInfos.map((item) => {
      item.orderdetails = [];
    });
    orderlist.value = res.orderInfos;
  };
  // pageSize 改变事件处理函数
  const handleSizeChange = (pageSize) => {
    queryInfo.pageSize = pageSize;
    getOrderinfos();
  };
  // page 改变事件处理函数
  const handleCurrentChange = (curPage) => {
    queryInfo.curPage = curPage;
    getOrderinfos();
  };
  // 向后台程序发送 get 请求,获取有效客户
  const getValidateCustomers = async () => {
    const { data: res } = await ctx.$axios.get("customers", {
      params: { status: 0, },
    });
    customers.value = res.data;
  };
  // 处理<el-table>标签绑定的 expand-change 事件
  const handleExpandChange = async (row) => {
    var id = row.id;
    const { data: res } = await ctx.$axios.get(`orderinfos/oid/${id}`);
```

```
    orderlist.value.forEach((temp, index) => {
      // 找到当前选中的行,把动态获取到的数据赋值进去
      if (temp.id == row.id) {
        orderlist.value[index].orderdetails = res.orderDetails;
      }
    });
  };
  onMounted(() => {
    getOrderinfos();
    getValidateCustomers();
  });
  return { orderlist,queryInfo,getOrderinfos,total,handleSizeChange,
    handleCurrentChange,customers,orderStatusList,orderdetaillist,
    handleExpandChange };
```

在上述代码中，变量 orderlist 用于存放订单列表，orderdetaillist 用于存放订单明细，total 用于存放订单总数，queryInfo 用于封装订单查询参数，customers 用于存放客户列表，orderStatusList 用于存放订单状态。

添加自定义函数 getOrderinfos，向后台程序发送 get 请求，获取订单列表。在 getOrderinfos 函数中，使用 axios 发送 get 请求，请求地址为 http://localhost:8888/eshop/orderinfos，这个请求将由后端程序 eshop 的控制器类 OrderInfoController 中的 getOrderinfos()方法处理，分页获取订单列表。返回结果解构后，赋值给 res。如果 res.code 等于 0，表示成功获取订单列表，则将订单列表赋值给响应式变量 orderlist，将订单总数赋值给响应式变量 total。否则，通过消息框提示没有订单记录。

添加自定义函数 getValidateCustomers，向后台程序发送 get 请求，获取有效客户列表，用来作为查询表单客户下拉列表数据源。在 getValidateCustomers 函数中，使用 axios 发送 get 请求，请求地址为 http://localhost:8888/eshop/customers，这个请求将由后端程序 eshop 的控制器类 UserInfoController 中的 getCustomers()方法处理，并传递参数 status，获取有效客户列表。返回结果解构后，赋值给 res，再将客户列表赋值给响应式变量 customers。

添加自定义函数 handleSizeChange，用来处理分页条绑定的 size-change 事件，该事件在分页条上每页显示个数选择器的选项发生变化时触发。

添加自定义函数 handleCurrentChange，用来处理分页条绑定的 current-change 事件，该事件在分页条上当前页页码发生变化时触发。

添加自定义函数 handleExpandChange，用来处理< el-table >标签绑定的 expand-change 事件，该事件在用户对某一行展开或关闭时触发。在 handleExpandChange 函数中，先获取当前展开行的订单 id，再使用 axios 发送 get 请求，请求地址为 http://localhost:8888/eshop/orderinfos/oid/${id}，这个请求将由后端程序 eshop 的控制器类 OrderInfoController 中的 getOrderDetailByOid 方法处理，根据订单 id 获取订单明细列表。返回结果解构后，赋值给 res。接着遍历订单列表对象 orderlist，找到当前选中的行，将从 res 中取出 data(订单明细数据)赋值给当前订单对象的 orderdetails 属性。

在 Vue 生命周期钩子函数 onMounted 中，调用自定义函数 getOrderinfos 和

getValidateCustomers。最后,将定义的变量、添加的自定义函数通过 return 返回出去,供 template 模板使用。

(3) 删除订单。在图 12-14 所示的订单列表页中,每行记录后面都有一个"删除"按钮,用于从数据库中删除订单及明细数据。单击"删除"按钮,会执行自定义函数 handleDeleteOrder,参数是订单 id,代码如下。

```
const handleDeleteOrder = async (id) =>{
  const { data: res } = await ctx.$axios.delete(`orderinfos/${id}`);
  if (res.code != 0) return ctx.$message.error(res.msg);
  getOrderinfos();
  return ctx.$message.success(res.msg);
};
```

在 handleDeleteOrder 函数中,使用 axios 发送 delete 请求,请求地址为 http://localhost:8888/eshop/orderinfos/${id},这个请求将由后端程序 eshop 的控制器类 OrderInfoController 中的 deleteOrderInfo 方法处理,从数据库中删除指定订单及明细记录。

12.11.2 创建订单

在管理首页面中,单击订单管理下的创建订单菜单,打开创建订单页,如图 12-15 所示。

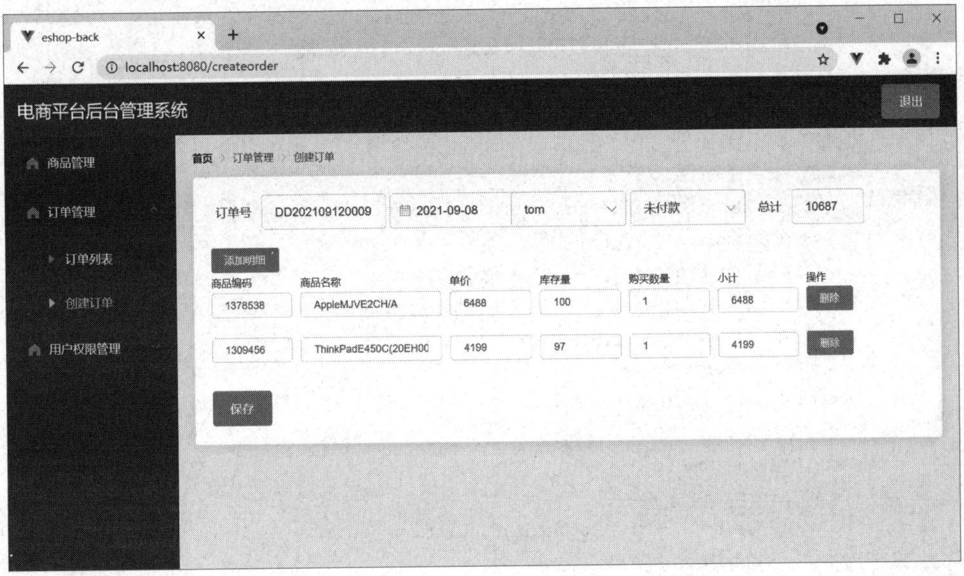

图 12-15 创建订单页

在创建订单页中,订单信息由订单号、订单日期、订单客户、订单状态和总计等构成。其中,订单号通过 Redis 生成,订单日期、订单客户和订单状态通过下拉菜单进行选择,总计根据订单明细自动计算。在图 12-15 中,单击"添加明细"按钮,添加一个空白行,可以输入订单明细数据。单击"删除"按钮,可以删除这条明细。

创建订单页的所有功能在组件 CreateOrder.vue 中实现,该组件位于 components/

order 目录下。

(1) 页面导航。在组件 CreateOrder.vue 的 template 模板部分,添加面包屑导航,代码如下。

```
<el-breadcrumb separator-class="el-icon-arrow-right">
    <el-breadcrumb-item :to="{ path: '/home' }">首页</el-breadcrumb-item>
    <el-breadcrumb-item>订单管理</el-breadcrumb-item>
    <el-breadcrumb-item>创建订单</el-breadcrumb-item>
</el-breadcrumb>
```

(2) 订单主表信息录入布局。在组件 CreateOrder.vue 的 template 模板部分,订单主表信息录入布局代码如下。

```
<el-card>
    <el-form ref="form1">
      <el-row :gutter="5">
        <el-col :span="6">
          <el-form-item label="订单号" label-width="60px">
            <el-input v-model="orderNo" readonly type="text"></el-input>
          </el-form-item>
        </el-col>
        <el-col :span="4">
          <el-form-item>
            <el-date-picker v-model="orderTime" type="date"
              placeholder="订单日期"></el-date-picker>
          </el-form-item>
        </el-col>
        <el-col :span="4">
          <el-form-item>
            <el-select v-if="customers" v-model="customerId"
              placeholder="客户名称">
              <el-option v-for="item in customers"
                :key="item.id" :label="item.userName" :value="item.id">
              </el-option>
            </el-select>
          </el-form-item>
        </el-col>
        <el-col :span="4">
          <el-form-item>
            <el-select v-if="orderStatusList" v-model="orderStatus"
              placeholder="订单状态">
              <el-option v-for="item in orderStatusList"
                :key="item.value" :label="item.label" :value="item.value">
              </el-option>
            </el-select>
          </el-form-item>
        </el-col>
```

```
        <el-col :span="4">
          <el-form-item label="总计" label-width="50px">
            <el-input v-model="totalPrice" readonly type="text"></el-input>
          </el-form-item>
        </el-col>
      </el-row>
    </el-form>
    <!-- 此处省略订单明细部分布局 -->
</el-card>
```

在上述代码中,定义了“订单号”输入框,与响应式变量 orderNo 双向绑定。定义了“订单日期”日期选择器,与响应式变量 orderTime 双向绑定。定义了“客户名称”下拉菜单,数据来自变量 customers。定义了“订单状态”下拉菜单,数据来自变量 orderStatusList。定义了“总计”输入框,与响应式变量 totalPrice 双向绑定。

(3) 订单明细信息录入布局。在组件 CreateOrder.vue 的 template 模板部分,订单明细信息录入布局代码如下。

```
<el-card>
    <!-- 此处省略订单主表信息录入布局 -->
    <!-- 订单明细信息录入布局 -->
    <br />
    <el-row>
      <el-button type="primary" size="mini" @click="handleAdd()">添加明细
</el-button>
    </el-row>
    <el-row>
      <el-col :span="3">商品编码</el-col>
      <el-col :span="5">商品名称</el-col>
      <el-col :span="3">单价</el-col>
      <el-col :span="3">库存量</el-col>
      <el-col :span="3">购买数量</el-col>
      <el-col :span="3">小计</el-col>
      <el-col :span="4">操作</el-col>
    </el-row>
    <el-form ref="form3" :model="orderTableData" :inline="true" size="mini">
      <el-row v-for="(item, index) in orderTableData" :key="index">
        <el-col :span="3">
          <el-form-item :prop="'.' + index + '.goodsCode'"
            :rules="{ required: true,message: '商品编码为空',
              trigger: 'blur', }">
            <el-input @click="selectGoods(index)" v-model="item.goodsCode"
              placeholder="商品编码"/>
          </el-form-item>
        </el-col>
        <el-col :span="5">
          <el-form-item class="el-form-item" label-width="0">
            <el-input readonly v-model="item.goodsName" />
          </el-form-item>
```

```
        </el-col>
        <el-col :span="3">
          <el-form-item class="el-form-item" label-width="0">
            <el-input readonly v-model="item.price" />
          </el-form-item>
        </el-col>
        <el-col :span="3">
          <el-form-item class="el-form-item" label-width="0">
            <el-input readonly v-model="item.num" />
          </el-form-item>
        </el-col>
        <el-col :span="3">
          <el-form-item
            :prop="'.' + index + '.quantity'"
            :rules="{ required: true, message: '购买数量为空',
              trigger: 'blur', }"
            class="el-form-item"
            label-width="0">
            <el-input
              @keydown.native="quantityInputLimit"
              @change="quantityChange(item, index)"
              type="number" v-model="item.quantity"/>
          </el-form-item>
        </el-col>
        <el-col :span="3">
          <el-form-item class="el-form-item" label-width="0">
            <el-input
              @change="subtotalChange(item, index)"
              readonly v-model="item.subtotal"/>
          </el-form-item>
        </el-col>
        <el-col :span="4">
          <template class="el-form-item" label-width="0">
            <el-button size="mini" type="danger" @click="handleDelete(index)">
删除</el-button>
          </template>
        </el-col>
      </el-row>
      <el-row class="saveorder-btn">
        <el-button type="primary" @click="submitHandler">保存</el-button>
      </el-row>
    </el-form>
</el-card>
```

在上述代码中，定义了“添加明细”按钮，单击该按钮，会执行自定义函数 handleAdd，用于生成空行，添加订单明细。在<el-form>标签上，通过 model 属性动态绑定表单使用的数据，通过 inline 属性将表单域变为行内的表单域。在<el-form>标签内，添加了<el-row>标签。在<el-row>标签上，通过 v-for 指令循环生成订单明细行。

(4) 商品选择对话框布局。在订单明细行“商品编码”输入框中单击时,打开“选择商品”对话框,如图 12-16 所示。

选择商品

商品id	商品编号	商品名称	单价	库存量
1	1378538	AppleMJVE2CH/A	6488	100
2	1309456	ThinkPadE450C(20EH0001CD)	4199	97
3	1999938	联想小新300经典版	4399	99
4	1466274	华硕FX50JX	4799	100
5	1981672	华硕FL5800	4999	100
6	1904696	联想G50-70M	3499	12
7	751624	美的BCD-206TM(E)	1298	100

图 12-16 “选择商品”对话框

在组件 CreateOrder.vue 的 template 模板部分,选择商品对话框布局代码如下。

```
<el-dialog title="选择商品" v-model="selectGoodsDialogVisible" width="50%">
    <el-table :data="goodsList" border stripe @row-dblclick="dataBackFillGoods">
      <el-table-column prop="id" label="商品 id"></el-table-column>
      <el-table-column prop="code" label="商品编号"></el-table-column>
      <el-table-column prop="name" label="商品名称"></el-table-column>
      <el-table-column prop="price" label="单价"></el-table-column>
      <el-table-column prop="num" label="库存量"></el-table-column>
    </el-table>
    <template #footer>
      <span class="dialog-footer">
        <el-button @click="selectGoodsDialogVisible = false">取消
        </el-button>
        <el-button type="primary" @click="selectGoodsDialogVisible = false">
确定</el-button>
      </span>
    </template>
</el-dialog>
```

在< el-dialog >标签内,添加了一个< el-table >标签,通过 data 属性动态绑定响应式变量 goodsList,作为数据源。双击表格行,会调用自定义方法 dataBackFillGoods(),将双击行的商品信息填充到订单明细行相应列。

至此，完成了创建订单页的布局。接下来，将介绍创建订单功能的 js 逻辑实现。在组件 CreateOrder.vue 的 script 部分，先引入 ref、reactive 等函数，代码如下。

```
import { ref, reactive, onMounted, getCurrentInstance } from "vue";
```

在 setup 函数中，定义一些响应式变量，代码如下。

```
const { ctx } = getCurrentInstance();                    // 获取当前组件实例
const orderNo = ref("");                                 // 订单号
const orderTime = ref("");                               // 订单时间
const customerId = ref("");                              // 订单客户 id
const totalPrice = ref(0);                               // 订单总金额
const orderStatus = ref("");                             // 订单状态
// 订单明细表单引用时的名称和标识
const form3 = ref(null);
const customers = ref(null);                             // 客户列表
customers.value = [{}];                                  // 初始化客户列表
const orderStatusList = ref(null);                       // 订单状态列表
// 初始化订单状态列表
orderStatusList.value = [ {value: "未付款",label: "未付款", },
  {value: "已付款",label: "已付款", },
  {value: "待发货",label: "待发货", },
  {value: "已发货",label: "已发货", },
  {value: "已完成",label: "已完成", },];
// 控制商品选择对话框是否可见
const selectGoodsDialogVisible = ref(false);
const goodsList = ref(null);                             // 定义商品列表
goodsList.value = [];                                    // 初始化商品列表
// 单击订单明细行商品编码输入框时,当前行索引
const currentIndex = ref(0);
// 订单明细数据列表
const orderTableData = ref(null);
// 初始化订单明细数据列表
orderTableData.value = [];
// 订单明细行商品编码输入框非空校验规则
const rules = reactive({
  goodsCode: [
    { required: true, message: "请输入商品编号", trigger: "blur" },
  ],
});
```

在上述代码中，变量 orderNo 用于保存订单号，变量 orderTime 用于保存订单时间，变量 customerId 用于保存订单客户，变量 totalPrice 用于保存订单总金额，变量 orderStatus 用于保存订单状态，变量 form3 用于保存订单明细表单引用时的名称和标识，变量 customers 用于保存客户列表，变量 orderStatusList 用于保存订单状态列表，变量 selectGoodsDialogVisible 用于控制选择商品对话框是否可见，变量 goodsList 用于保存商品列表，变量 currentIndex 用于保存单击订单明细行商品编码输入框时当前行的索引，变量

orderTableData 用于保存订单明细数据列表，变量 rules 用于保存订单明细行商品编码输入框非空校验规则。

在 setup 函数中，自定义一些函数，代码如下。

```
// 向后台服务器发送 get 请求，获取商品列表
const getGoodsInfoList = async () => {
  const { data: res } = await ctx.$axios.get("goods/status");
  goodsList.value = res.data;
};
// 向后台服务器发送 get 请求，获取有效客户列表
const getValidateCustomers = async () => {
  const { data: res } = await ctx.$axios.get("customers", {
    params: { status: 0, },
  });
  customers.value = res.data;
};
// 向后台服务器发送 get 请求，通过 redis 获取订单编号
const getCode = async () => {
  const { data: res } = await ctx.$axios.get("getSequence", {
    params: { prefix: "DD", },
  });
  orderNo.value = res.data;
};
// 处理"添加明细"按钮单击事件，添加订单明细录入空白行
const handleAdd = () => {
  const list = { goodsCode: "",goodsName: "",price: "",num: "",quantity: "",
    subtotal: "",gid: "", };
  // 订单明细表单校验
  form3.value.validate((valid) => {
    if (valid) {
      orderTableData.value.push(list);
    } else {
      // 此处为验证失败代码
      ctx.$message.warning("当前行填值有误!");
    }
  });
};
// 处理订单明细行商品编码输入框单击事件，打开商品选择对话框
const selectGoods = (index) => {
  selectGoodsDialogVisible.value = true;
  currentIndex.value = index;
};
// 保存订单
const submitHandler = () => {
  form3.value.validate(async (valid) => {
    if (valid) {
      // 校验订单号
      if (orderNo.value === "")
        return ctx.$message.error("请开启 Redis 服务，生成订单号!");
```

```
        // 校验订单日期
        if (orderTime.value === "")
          return ctx.$message.error("请选择订单日期!");
        // 校验客户名称
        if (customerId.value === "")
          return ctx.$message.error("请选择客户名称!");
        // 校验订单状态
        if (orderStatus.value === "")
          return ctx.$message.error("请选择订单状态!");
        // 校验订单明细记录数
        if (orderTableData.value.length === 0)
          return ctx.$message.error("至少需要添加一条订单明细记录!");
        console.log("校验通过提交了");
        // 提交
        const { data: res } = await ctx.$axios.post("commitOrder", {
          orderNo: orderNo.value,
          ordertime: orderTime.value,
          uid: customerId.value,
          status: orderStatus.value,
          orderprice: totalPrice.value,
          orderDetails: orderTableData.value,
        });
        if (res.code != 0) return ctx.$message.error(res.msg);
        ctx.$message.success(res.msg);
        ctx.$router.push("/home");
      } else {
        ctx.$message({ showClose: true, message: "请完善列表中的数据!",
          type: "warning", });
      }
    });
  };
  // 处理商品选择对话框中表格行的双击事件,
  // 将双击行的商品信息添加到当前订单明细行相应列
  const dataBackFillGoods = async (row, event, column) => {
    orderTableData.value[currentIndex.value].gid = row.id;
    orderTableData.value[currentIndex.value].goodsCode = row.code;
    orderTableData.value[currentIndex.value].goodsName = row.name;
    orderTableData.value[currentIndex.value].price = row.price;
    orderTableData.value[currentIndex.value].num = row.num;
    orderTableData.value[currentIndex.value].quantity = Number(1);
    orderTableData.value[currentIndex.value].subtotal = Number(row.price);
    selectGoodsDialogVisible.value = false;
    // 更新总金额
    var sum = 0;
    orderTableData.value.forEach((item) => {
      sum = Number(sum) + Number(item.subtotal);
    });
    totalPrice.value = Number(sum);
  };
  // 删除订单明细
```

```
const handleDelete = (index) => {
  // 更新总金额
  totalPrice.value =
    totalPrice.value - Number(orderTableData.value[index].subtotal);
  orderTableData.value.splice(index, 1);
};
// 购买数量变化
const quantityChange = (row, index) => {
  if (row.quantity < 1) {
    row.quantity = 1;
  }
  // 控制购买数量
  if (Number(row.quantity) > Number(row.num) || Number(row.quantity) < 1) {
    ctx.$message.warning("购买数量不能超过库存数量,且必须大于0!");
    orderTableData.value[index].quantity = "";
    orderTableData.value[index].subtotal = "";
  }
  // 更新小计
  orderTableData.value[index].subtotal = Number(
    Number(row.price) * Number(row.quantity)
  );
  // 更新总金额
  var sum = 0;
  orderTableData.value.forEach((item) => {
    sum = Number(sum) + Number(item.subtotal);
  });
  totalPrice.value = Number(sum);
};
// 控制购买数量,不允许输入 'E'、'e'、'.'、'-'
const quantityInputLimit = (e) => {
  let key = e.key;
  if (key === "-" ||key === "E" ||key === "e" ||key === "." ||
    Number(key) < 1 ) {
    e.returnValue = false;
    return false;
  }
  return true;
};
```

在上述代码中,getGoodsInfoList、getValidateCustomers 和 getCode 这三个函数都是在组件 CreateOrder.vue 的生命周期钩子函数 onMounted 中被调用。

在 getGoodsInfoList 函数中,使用 axios 发送 get 请求,请求地址为 http://localhost:8888/eshop/goods/status,这个请求将由后端程序 eshop 的控制器类 GoodsInfoController 中的 getValidGoodstInfo()方法处理,获取有效的商品列表。

在 getValidateCustomers 函数中,使用 axios 发送 get 请求,请求地址为 http://localhost:8888/eshop/customers,这个请求将由后端程序 eshop 的控制器类 UserInfoController 中的 getCustomers()方法处理,获取有效客户列表。

在 getCode 函数中，使用 axios 发送 get 请求，请求地址为 http://localhost:8888/eshop/getSequence，这个请求将由后端程序 eshop 的控制器类 SequenceController 中的 getSequence()方法处理，使用 redis 生成订单编号。

handleAdd 函数在单击“添加明细”按钮时调用，用于创建订单明细录入空白行。selectGoods 函数在订单明细行商品编码输入框单击时调用，用于打开选择商品对话框。

submitHandler 函数在单击“保存”按钮时调用，用于保存订单。在 submitHandler 函数中，首先进行表单校验，校验通过后，再使用 axios 发送 post 请求，请求地址为 http://localhost:8888/eshop/commitOrder，并向后台程序传递与订单相关的参数，这个请求将由后端程序 eshop 的控制器类 OrderInfoController 中的 commitOrder()方法处理，将订单及订单明细数据保存到数据库。

dataBackFillGoods 函数在商品选择对话框中表格行被双击时调用，用于将双击行的商品信息填充到当前订单明细行相应列。handleDelete 函数在单击“删除”按钮时调用，用于删除订单明细。quantityChange 函数在订单明细行中购买数量发生变化时调用，用于处理购买数量变化。quantityInputLimit 函数也是在购买数量发生变化时调用，用于控制购买数量，不允许输入 E、e、.、-等字符。在组件 CreateOrder. vue 的生命周期钩子函数 onMounted 中，调用自定义函数 getGoodsInfoList、getValidateCustomers 和 getCode，代码如下。

```
onMounted(() => {
  getGoodsInfoList();
  getValidateCustomers();
  getCode();
});
```

12.12 用户权限管理

用户权限管理包括前台用户管理、后台用户管理和角色管理三个模块，前台用户管理模块包括前台用户列表分页显示与查询、修改和禁用功能。后台用户管理模块包括后台用户列表分页显示与查询、修改、禁用和角色分配等功能。角色管理模块包括角色列表显示、修改、禁用和权限设置等功能。

12.12.1 后台用户管理

在管理首页面中，单击用户权限管理下的后台用户管理菜单，打开后台用户列表页，如图 12-17 所示。

后台用户列表页的所有功能在组件 Backusers. vue 中实现，该组件位于 components/user 目录下。

(1) 页面导航。在组件 Backusers. vue 的 template 模板部分，添加面包屑导航，代码如下。

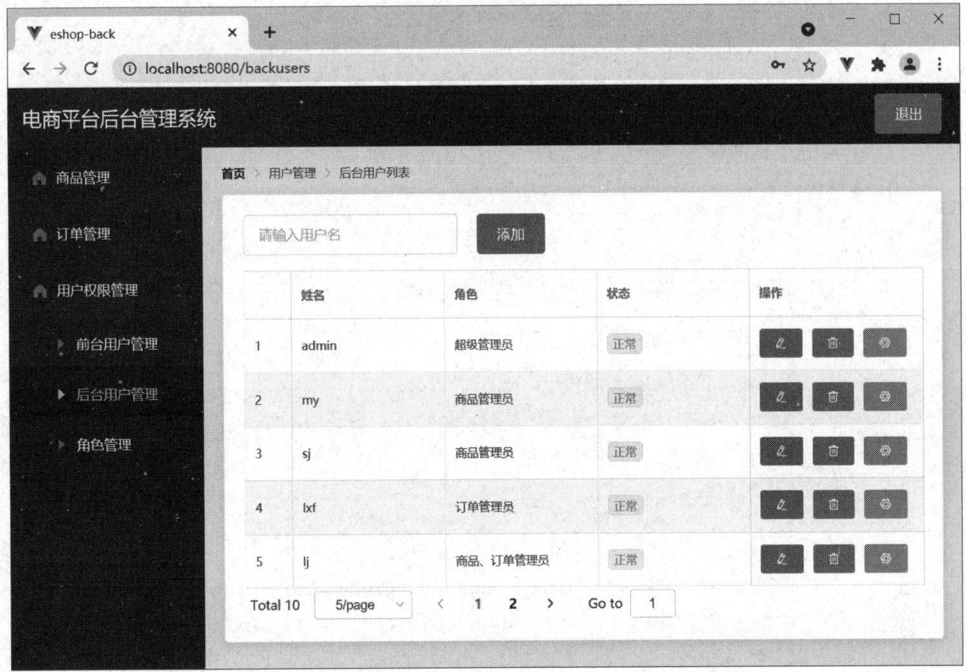

图 12-17　后台用户列表页

```
<el-breadcrumb separator-class="el-icon-arrow-right">
  <el-breadcrumb-item :to="{ path: '/home' }">首页</el-breadcrumb-item>
  <el-breadcrumb-item>用户管理</el-breadcrumb-item>
  <el-breadcrumb-item>后台用户列表</el-breadcrumb-item>
</el-breadcrumb>
```

(2) 后台用户列表分页显示与查询。在组件 Backusers.vue 的 template 模板部分，添加后台用户查询表单、后台用户列表、分页条等内容，代码如下。

```
<el-card>
  <!-- 查询表单 -->
  <el-row :gutter="20">
    <el-col :span="8">
      <el-input
        placeholder="请输入用户名" v-model="queryInfo.name"clearable
        @clear="getAdmininfos" @change="getAdmininfos(-1)">
      </el-input>
    </el-col>
    <el-col :span="4">
      <el-button type="primary" @click="showAddAdminInfoDialog"
        >添加</el-button>
    </el-col>
  </el-row>
  <!-- 管理员列表 -->
```

```
  <el-table :data="adminlist" border stripe>
    <el-table-column type="index"></el-table-column>
    <el-table-column label="姓名" prop="name"></el-table-column>
    <el-table-column label="角色" prop="role.roleName"></el-table-column>
    <el-table-column label="状态">
      <template v-slot="scope">
        <el-tag v-if="scope.row.delState == 0" type="success" size="mini">正常
</el-tag>
        <el-tag v-else type="danger" size="mini">禁用</el-tag>
      </template>
    </el-table-column>
    <el-table-column label="操作" width="180px">
      <template v-slot="scope">
        <div v-if="scope.row.delState === 0">
          <el-tooltip
            class="item" effect="dark" content="修改"
            placement="top":enterable="false">
            <el-button type="primary" icon="el-icon-edit" size="mini"
              @click="openEditAdminInfoDialog(scope.row.id)"></el-button>
          </el-tooltip>
          <el-tooltip class="item" effect="dark" content="禁用"
            placement="top":enterable="false">
            <el-button type="danger" icon="el-icon-delete" size="mini" @click=
"disableAdminInfo(scope.row.id)"></el-button>
          </el-tooltip>
          <el-tooltip class="item" effect="dark" content="分配角色"
            placement="top":enterable="false">
            <el-button type="warning" icon="el-icon-setting"
              size="mini" @click="openSetRoleDialog(scope.row)">
            </el-button>
          </el-tooltip>
        </div>
      </template>
    </el-table-column>
  </el-table>
  <!-- 分页 -->
  <el-pagination
    @size-change="handleSizeChange"@current-change="handleCurrentChange"
    :current-page="queryInfo.curPage" :page-sizes="[5, 10, 20]"
    :page-size="queryInfo.pageSize" layout="total, sizes, prev, pager, next, jumper" :
total="total"></el-pagination>
</el-card>
```

在上述代码中，使用了<el-card>、<el-row>、<el-table>和<el-pagination>等标签。在介绍商品列表功能模块时，对这些标签已做介绍，此处不再赘述。

在组件Backusers.vue的script部分，与后台用户列表分页显示和查询功能相关代码如下。

```
    const { ctx } = getCurrentInstance();
    // 封装查询参数
    const queryInfo = reactive({ name: "", curPage: 1, pageSize: 5, });
    // 保存管理员列表
    const adminlist = ref(null);
    // 保存后台用户总记录数
    const total = ref(0);
    // 向后台程序发送请求,分页获取管理员列表
    const getAdmininfos = async (flag) => {
      if (flag == 0 || flag == -1) {
        queryInfo.curPage = 1;
      }
      // 分页获取管理员列表
      const { data: res } = await ctx.$axios.get("admininfos", {
        params: { name: queryInfo.name, curPage: queryInfo.curPage,
          pageSize: queryInfo.pageSize, },
      });
      if (res.code != 0) return ctx.$message.error("没有管理员记录!");
      adminlist.value = res.adminInfos;
      total.value = res.total;
    };
    // pageSize 改变事件处理函数
    const handleSizeChange = (pageSize) => {
      queryInfo.pageSize = pageSize;
      getAdmininfos();
    };
    // page 改变事件处理函数
    const handleCurrentChange = (curPage) => {
      queryInfo.curPage = curPage;
      getAdmininfos();
    };
    onMounted(() => {
      getAdmininfos();
    });
    return { queryInfo, total, adminlist, getAdmininfos, handleSizeChange,
      handleCurrentChange, ... };
```

在上述代码中,变量 adminlist 用于保存管理员列表,变量 total 用于保存后台用户总记录数,变量 queryInfo 用于封装查询参数。

添加自定义函数 getAdmininfos(),用来向后台服务器发送请求获取后台用户列表。在函数 getAdmininfos 中,使用 axios 发送 get 请求,并传递查询参数。请求地址为 http://localhost:8888/eshop/admininfos,这个请求将由后端程序 eshop 的控制器类 AdminInfoController 中的 getAdmininfos()方法处理,分页获取后台用户列表。返回结果解构后,赋值给 res。如果 res.code 等于 0,表示成功获取后台用户列表,则将后台用户列表赋值给响应式变量 adminlist,将后台用户总数赋值给响应式变量 total。否则,通过消息框提示“没有管理员记录”。添加自定义函数 handleSizeChange,用来处理分页条绑定的 size-change 事件,该事件在分页条上每页显示个数选择器的选项发生变化时触发。添加自定义

函数 handleCurrentChange，用来处理分页条绑定的 current-change 事件，该事件在分页条上当前页页码发生变化时触发。在 Vue 生命周期钩子函数 onMounted 中，调用自定义函数 getAdmininfos。最后，将定义的变量、添加的自定义函数通过 return 返回出去，供 template 模板使用。

（3）添加后台用户。在图 12-17 所示的后台用户列表页中，单击“添加”按钮，可以打开“添加后台管理员”对话框，如图 12-18 所示。

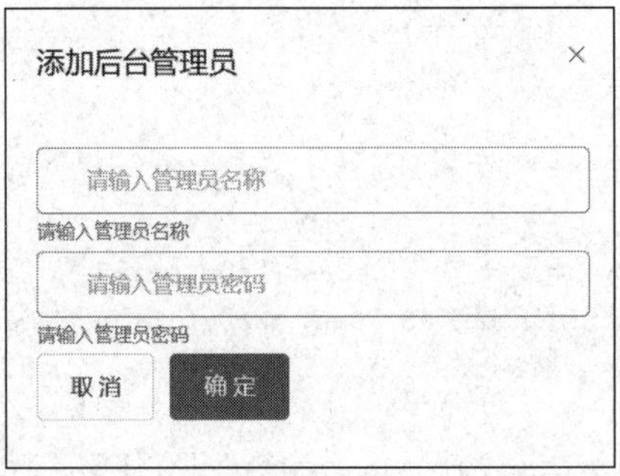

图 12-18 “添加后台管理员”对话框

在组件 Backusers.vue 的 template 模板部分，与“添加后台管理员”对话框相关的代码如下。

```
<el-dialog
  title="添加后台管理员" v-model="addAdminInfoDialogVisible"
  width="30%" @close="resetAddAdminInfoForm()">
  <el-form
    ref="addAdminInfoFormRef" :model="addAdminInfoForm"
    :rules="addAdminInfoFormRules" label-width="0px" class="login-form">
    <!-- 管理员名称 -->
    <el-form-item prop="name">
      <el-input
        v-model="addAdminInfoForm.name"
        prefix-icon="iconfont icon-user"
        placeholder="请输入管理员名称"></el-input>
    </el-form-item>
    <!-- 管理员密码 -->
    <el-form-item prop="pwd">
      <el-input
        v-model="addAdminInfoForm.pwd" prefix-icon="iconfont icon-user"
        placeholder="请输入管理员密码"></el-input>
    </el-form-item>
  </el-form>
  <span slot="footer" class="dialog-footer">
    <el-button @click="addAdminInfoDialogVisible = false">取 消</el-button>
    <el-button type="primary" @click="addAdminInfo()">确 定</el-button>
  </span>
</el-dialog>
```

在上述代码中，使用了< el-dialog >和< el-form >等标签，在介绍商品列表功能模块时，对这些标签已做介绍，此处不再赘述。

在组件 Backusers. vue 的 script 部分，与添加后台管理员功能相关的代码如下。

```
// 控制添加管理员对话框是否可见
const addAdminInfoDialogVisible = ref(false);
// 保存添加管理员表单数据
const addAdminInfoForm = reactive({
  name: "",
  pwd: "",
});
// 指定添加管理员表单被引用时的名称和标识
const addAdminInfoFormRef = ref(null);
// 设置添加管理员表单验证规则
const addAdminInfoFormRules = reactive({
  name: [{ required: true, message: "请输入管理员名称", trigger: "blur" }],
  pwd: [{ required: true, message: "请输入管理员密码", trigger: "blur" }],
});
// 重置添加管理员对话框中的表单
const resetAddAdminInfoForm = () => {
  addAdminInfoFormRef.value.resetFields();
};
// 显示添加管理员对话框
const showAddAdminInfoDialog = () => {
  addAdminInfoDialogVisible.value = true;
};
// 添加管理员
const addAdminInfo = () => {
  addAdminInfoFormRef.value.validate(async (valid) => {
    if (!valid) return;
    var name = addAdminInfoForm.name;
    var pwd = addAdminInfoForm.pwd;
    // 判断管理员名称是否已存在
    const { data: res1 } = await ctx.$axios.get(`admininfos/name/${name}`);
    if (res1.code !== 0) return ctx.$message.error("该管理员名称已存在!");
    const { data: res2 } = await ctx.$axios.post(`admininfos`, {
      name: name,
      pwd: pwd,
    });
    if (res2.code !== 0) return ctx.$message.error("添加失败!");
    ctx.$message.success("添加成功!");
    getAdmininfos();
    addAdminInfoDialogVisible.value = false;
  });
};
return { ...
  addAdminInfoDialogVisible, resetAddAdminInfoForm, addAdminInfoForm,
  addAdminInfoFormRef, addAdminInfoFormRules, showAddAdminInfoDialog,
  addAdminInfo };
```

在上述代码中,变量 addAdminInfoDialogVisible 用于控制添加管理员对话框是否可见。变量 addAdminInfoForm 用于保存添加管理员表单数据,变量 addAdminInfoFormRef 用于指定添加管理员表单被引用时的名称和标识,变量 addAdminInfoFormRules 用于设置添加管理员表单验证规则。

showAddAdminInfoDialog 函数用于显示添加管理员对话框,resetAddAdminInfoForm 函数用于重置添加管理员对话框中的表单,addAdminInfo 函数用于向后台程序发送请求,添加后台管理员。最后,将这些变量和函数 return 出去,供 template 模板使用。

(4) 修改后台用户。在图 12-17 所示的后台用户列表页中,每行记录后面都有一个"修改"按钮。单击"修改"按钮,可以打开"修改后台管理员"对话框,如图 12-19 所示。

图 12-19 "修改后台管理员"对话框

在组件 Backusers. vue 的 template 模板部分,与"修改后台管理员"对话框相关的代码如下。

```
<el-dialog
  title="修改后台管理员" v-model="editAdminInfoDialogVisible" width="30%">
  <el-form
    ref="editAdminInfoFormRef" :model="editAdminInfoForm"
    :rules="editAdminInfoFormRules"label-width="0px" class="login-form">
    <!-- 管理员名称 -->
    <el-form-item prop="name">
      <el-input
        v-model="editAdminInfoForm.name"
        prefix-icon="iconfont icon-user"
        placeholder="请输入管理员名称"></el-input>
    </el-form-item>
    <!-- 管理员密码 -->
    <el-form-item prop="pwd">
      <el-input
        v-model="editAdminInfoForm.pwd"prefix-icon="iconfont icon-user"
        placeholder="请输入管理员密码"></el-input>
    </el-form-item>
  </el-form>
  <span slot="footer" class="dialog-footer">
    <el-button @click="editAdminInfoDialogVisible = false">取 消</el-button>
    <el-button type="primary" @click="editAdminInfo()">确 定</el-button>
  </span>
</el-dialog>
```

在上述代码中,使用了< el-dialog >、< el-form >等标签。在介绍商品列表功能模块时,对这些标签已做介绍,此处不再赘述。

在组件 Backusers. vue 的 script 部分,与修改后台管理员功能相关的代码如下。

```
// 控制修改管理员对话框是否可见
const editAdminInfoDialogVisible = ref(false);
// 保存修改管理员表单数据
const editAdminInfoForm = reactive({ id: "", name: "", pwd: "" });
// 指定修改管理员表单被引用时的名称和标识
const editAdminInfoFormRef = ref(null);
// 设置修改管理员表单验证规则
const editAdminInfoFormRules = reactive({
  name: [{ required: true, message: "请输入管理员名称", trigger: "blur" }],
  pwd: [{ required: true, message: "请输入管理员密码", trigger: "blur" }],
});
// 打开修改对话框
const openEditAdminInfoDialog = async (id) => {
  editAdminInfoDialogVisible.value = true;
  const { data: res } = await ctx.$axios.get(`admininfos/id/${id}`);
  if (res.code != 0) return ctx.$message.error("获取数据失败!");
  editAdminInfoForm.name = res.adminInfo.name;
  editAdminInfoForm.pwd = res.adminInfo.pwd;
  editAdminInfoForm.id = res.adminInfo.id;
};
// 修改
const editAdminInfo = () => {
  editAdminInfoFormRef.value.validate(async (valid) => {
    if (!valid) return;
    var name = editAdminInfoForm.name;
    var pwd = editAdminInfoForm.pwd;
    var id = editAdminInfoForm.id;
    // 判断管理员名称是否已存在
    const { data: res1 } = await ctx.$axios.get(
      `admininfos/name/${name}/id/${id}`
    );
    if (res1.code !== 0) return ctx.$message.error("该管理员名称已存在!");
    const { data: res2 } = await ctx.$axios.put(`admininfos`, {
      name: name, pwd: pwd, id: id,});
    if (res2.code !== 0) return ctx.$message.error("修改失败!");
    ctx.$message.success("修改成功!");
    getAdmininfos();
    editAdminInfoDialogVisible.value = false;
  });
};
return { ..., editAdminInfoForm,editAdminInfoFormRef,editAdminInfoFormRules, editAdminInfoDialogVisible,
openEditAdminInfoDialog,editAdminInfo };
```

在上述代码中，变量 editAdminInfoDialogVisible 用于控制修改管理员对话框是否可见，变量 editAdminInfoForm 用于保存修改管理员表单数据，变量 editAdminInfoFormRef 用于指定修改管理员表单被引用时的名称和标识，变量 editAdminInfoFormRules 用于设置修改管理员表单验证规则，变量 openEditAdminInfoDialog 用于打开修改对话框。

自定义函数 editAdminInfo，先使用 axios 向后台程序发送 get 请求，请求地址为

http://localhost:8888/eshop/admininfos/name/${name}/id/${id}，这个请求将由后端程序 eshop 的控制器类 AdminInfoController 中的 isExistAdminInfoName_Id()方法处理，用于判断管理员名称和 id 是否存在。如果不存在，则再向后台程序发送 put 请求，请求地址为 http://localhost:8888/eshop/admininfos，并传递要修改的管理员对象，该请求将由控制器类 AdminInfoController 中的 editAdminInfo()方法处理，用于修改管理员信息。

最后，将这些变量和函数 return 出去，供 template 模板使用。

(5) 禁用后台用户。在图 12-17 所示的后台用户列表页中，每行记录后面都有一个"禁用"按钮。单击"禁用"按钮，会执行自定义函数 disableAdminInfo，代码如下。

```
const disableAdminInfo = async (id) => {
  const { data: res } = await ctx.$axios.put(`admininfos/id/${id}/delState`);
  if (res.code !== 0) return ctx.$message.error("禁用失败!");
  ctx.$message.success("禁用成功!");
  getAdmininfos();
};
```

在 disableAdminInfo 函数中，使用 axios 向后台程序发送 put 请求，请求地址为 http://localhost:8888/eshop/admininfos/id/${id}/delState，这个请求将由后端程序 eshop 的控制器类 AdminInfoController 中的 disableAdminInfo()方法处理，修改管理员的状态，从而实现禁用管理员功能。

(6) 设置后台用户权限。在图 12-17 所示的后台用户列表页中，每行记录后面都有一个"分配角色"按钮。单击该按钮，会打开"分配角色"对话框，如图 12-20 所示。

图 12-20 "分配角色"对话框

在组件 Backusers.vue 的 template 模板部分，与"分配角色"对话框相关的代码如下。

```
<el-dialog
  title="分配角色" v-model="setRoleDialogVisible" width="40%"
  @close="handleSetRoleDialogClosed">
  <span>
    <div>
      <p v-if="admininfo">用户名：{{ admininfo.name }}</p>
      <p v-if="admininfo && admininfo.role">
        角色名：{{ admininfo.role.roleName }}
```

```
        </p>
        <p v-if="admininfo && !admininfo.role">角色名：无</p>
        <p>
          新角色名：
          <el-select v-model="newRoleId" placeholder="请选择">
            <el-option
              v-for="item in rolelist" :key="item.roleId"
              :label="item.roleName" :value="item.roleId"></el-option>
          </el-select>
        </p>
      </div>
    </span>
    <span slot="footer" class="dialog-footer">
      <el-button @click="powerDialogVisible = false">取 消</el-button>
      <el-button type="primary" @click="saveRole">确 定</el-button>
    </span>
</el-dialog>
```

在上述代码中，使用了<el-dialog>、<el-select>和<el-option>等标签。在介绍商品列表功能模块时，对这些标签已做介绍，此处不再赘述。

在组件 Backusers.vue 的 script 部分，与分配角色功能实现相关的代码如下。

```
// 控制角色设置对话框是否可见
const setRoleDialogVisible = ref(false);
// 待分配角色的管理员对象
const admininfo = ref(null);
// 保存所有角色列表
const rolelist = ref(null);
// 分配的新角色 id
const newRoleId = ref("");
// 打开设置角色对话框
const openSetRoleDialog = async (ai) => {
  admininfo.value = ai;
  // 获取所有角色
  const { data: res } = await ctx.$axios.get("roles/valid");
  rolelist.value = res.data;
  setRoleDialogVisible.value = true;
};
// 处理设置角色对话框关闭事件
const handleSetRoleDialogClosed = () => {
  newRoleId.value = "";
};
// 保存角色更新
const saveRole = async () => {
  if (!newRoleId.value) {
    return ctx.$message.error("请选择一个角色!");
  }
  const { data: res } = await ctx.$axios.put(
```

```
        `admininfos/${admininfo.value.id}/role/${newRoleId.value}`
    );
    if (res.code != 0) return ctx.$message.error("角色更新失败!");
    ctx.$message.success("角色更新成功!");
    getAdmininfos();
    setRoleDialogVisible.value = false;
  };
  return { ..., setRoleDialogVisible, openSetRoleDialog,
  handleSetRoleDialogClosed, admininfo, rolelist, newRoleId, saveRole, ... };
```

在上述代码中，变量 setRoleDialogVisible 用于控制角色设置对话框是否可见，变量 admininfo 用于保存待分配角色的管理员对象，变量 rolelist 用于保存所有角色列表，变量 newRoleId 用于保存分配的新角色 id，变量 openSetRoleDialog 用于打开设置角色对话框，

openSetRoleDialog 函数用于打开设置角色对话框，在 openSetRoleDialog 函数中，通过 axios 向后台程序发送 get 请求，请求地址为 http://localhost:8888/eshop/roles/valid，这个请求将由后端程序 eshop 的控制器类 RoleController 中的 getValidRoles()方法处理，获取未禁用角色列表。handleSetRoleDialogClosed 函数用于处理设置角色对话框关闭事件。saveRole 函数用于保存角色设置，在 saveRole 函数中，通过 axios 向后台程序发送 put 请求，请求地址为 admininfos/${admininfo.value.id}/role/${newRoleId.value}，这个请求将由控制器类 AdminInfoController 中的 saveRole()方法处理，用于设置管理员角色。

最后，将定义的变量和函数 return 出去，供 template 模板使用。

12.12.2 角色管理

在管理首页面中，单击用户权限管理下的角色管理菜单，打开角色列表页，如图 12-21 所示。

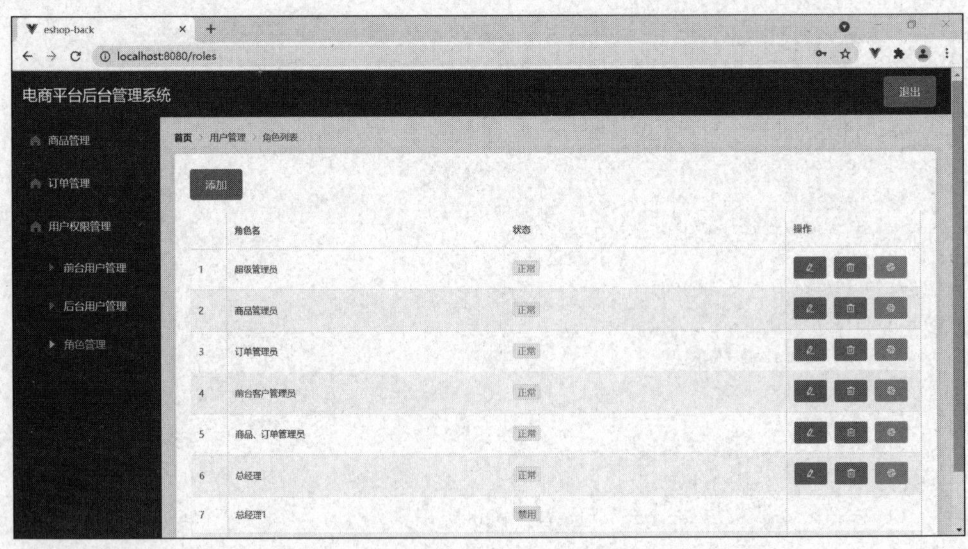

图 12-21 角色列表页

在角色列表页中，可以显示角色、添加角色、修改角色、禁用角色和设置权限。角色列表页的所有功能在组件 Role.vue 中实现，该组件位于 components/user 目录下。

（1）页面导航。在组件 Role.vue 的 template 模板部分，添加面包屑导航，代码如下。

```
<el-breadcrumb separator-class="el-icon-arrow-right">
  <el-breadcrumb-item :to="{ path: '/home' }">首页</el-breadcrumb-item>
  <el-breadcrumb-item>用户管理</el-breadcrumb-item>
  <el-breadcrumb-item>角色列表</el-breadcrumb-item>
</el-breadcrumb>
```

（2）角色显示。在组件 Role.vue 的 template 模板部分，添加用于显示角色列表的标签，代码如下。

```
<el-card>
  <el-row>
    <el-col :span="4">
      <el-button type="primary" @click="openAddRoleDialog()">添加
      </el-button>
    </el-col>
  </el-row>
  <!-- 角色列表 -->
  <el-table :data="rolelist" border stripe>
    <el-table-column type="index"></el-table-column>
    <el-table-column label="角色名" prop="roleName"></el-table-column>
    <el-table-column label="状态">
      <template v-slot="scope">
        <el-tag v-if="scope.row.delState == 0" type="success" size="mini">正常
</el-tag>
        <el-tag v-else type="danger" size="mini">禁用</el-tag>
      </template>
    </el-table-column>
    <el-table-column label="操作" width="180px">
      <template v-slot="scope">
        <div v-if="scope.row.delState === 0">
          <el-tooltip
            class="item" effect="dark" content="修改" placement="top"
            :enterable="false">
            <el-button
              type="primary" icon="el-icon-edit" size="mini"
              @click="openEditRoleDialog(scope.row.roleId)"></el-button>
          </el-tooltip>
          <el-tooltip
            class="item" effect="dark" content="禁用" placement="top"
            :enterable="false">
            <el-button
              type="danger" icon="el-icon-delete" size="mini"
              @click="disableRole(scope.row.roleId)"></el-button>
```

```
            </el-tooltip>
            <el-tooltip
              class="item" effect="dark" content="设置权限" placement="top"
              :enterable="false">
              <el-button
                type="warning" icon="el-icon-setting" size="mini"
                @click="openSettingPowerDialog(scope.row)"></el-button>
            </el-tooltip>
          </div>
        </template>
      </el-table-column>
    </el-table>
  </el-card>
```

在上述代码中,使用了< el-card >、< el-row >和< el-table >等标签,在介绍商品列表功能模块时,对这些标签已做介绍,此处不再赘述。

在组件 Role.vue 的 script 部分,与角色显示功能相关的代码如下。

```
const { ctx } = getCurrentInstance();
// 角色列表
const rolelist = ref(null);
// 向后台程序发送请求,获取角色列表
const getRoles = async () => {
  // 获取角色列表
  const { data: res } = await ctx.$axios.get("roles");
  if (res.code != 0) return ctx.$message.error("没有角色记录!");
  rolelist.value = res.data;
};
onMounted(() => {
  getRoles();
});
return { rolelist, ... };
```

在上述代码中,变量 rolelist 用来保存角色列表。在函数 getRoles 中,使用 axios 发送 get 请求,请求地址为 http://localhost:8888/eshop/roles,这个请求将由后端程序 eshop 的控制器类 RoleController 中的 getAllRoles()方法处理,获取所有角色列表。在 Vue 生命周期钩子函数 onMounted 中,调用自定义函数 getRoles。

最后,将定义的变量 rolelist 通过 return 返回出去,供 template 模板使用。

(3) 添加角色。在图 12-21 所示的角色列表页中,单击"添加"按钮,可以打开"添加角色"对话框,如图 12-22 所示。

图 12-22 "添加角色"对话框

在组件 Role.vue 的 template 模板部分,与"添加角色"对话框相关的代码如下。

```
<el-dialog
  title="添加角色" v-model="addRoleDialogVisible" width="30%"
  @close="resetAddRoleForm()">
  <el-form
    ref="addRoleFormRef" :model="addRoleForm" :rules="addRoleFormRules"
    label-width="0px">
    <!-- 角色名称 -->
    <el-form-item prop="roleName">
      <el-input
        v-model="addRoleForm.roleName" prefix-icon="iconfont icon-user"
        placeholder="请输入角色名称"></el-input>
    </el-form-item>
  </el-form>
  <span slot="footer" class="dialog-footer">
    <el-button @click="addRoleDialogVisible = false">取 消</el-button>
    <el-button type="primary" @click="addRole()">确 定</el-button>
  </span>
</el-dialog>
```

在上述代码中，使用了< el-dialog >和< el-form >等标签，在介绍商品列表功能模块时，对这些标签已做介绍，此处不再赘述。

在组件 GoodsList.vue 的 script 部分，与添加角色功能相关的代码如下。

```
// 封装添加角色表单数据
const addRoleForm = reactive({ roleName: "" });
// 指定添加角色表单被引用时的名称和标识
const addRoleFormRef = ref(null);
// 设置添加角色表单验证规则
const addRoleFormRules = reactive({
  roleName: [{ required: true, message: "请输入角色名称", trigger: "blur" }]
});
// 控制添加角色对话框是否可见
const addRoleDialogVisible = ref(false);
// 打开添加角色对话框
const openAddRoleDialog = () => {
  addRoleDialogVisible.value = true;
};
// 重置添加角色对话框中的表单
const resetAddRoleForm = () => {
  addRoleFormRef.value.resetFields();
};
// 添加角色
const addRole = () => {
  addRoleFormRef.value.validate(async valid => {
    if (!valid) return;
    var roleName = addRoleForm.roleName;
    // 判断角色名称是否已存在
    const { data: res1 } = await ctx.$axios.get(`roles/roleName/${roleName}`);
```

```
        if (res1.code !== 0) return ctx.$message.error("该角色名称已存在!");
        const { data: res2 } = await ctx.$axios.post(`roles`, {
          roleName: roleName
        });
        if (res2.code !== 0) return ctx.$message.error("添加失败!");
        ctx.$message.success("添加成功!");
        getRoles();
        addRoleDialogVisible.value = false;
      });
    };
    return { ..., openAddRoleDialog, addRoleDialogVisible, resetAddRoleForm,
      addRoleFormRef, addRoleForm, addRoleFormRules, addRole };
```

在上述代码中，变量 addRoleForm 用于封装添加角色表单数据，变量 addRoleFormRef 用于指定添加角色表单被引用时的名称和标识，变量 addRoleFormRules 用于设置添加角色表单验证规则，变量 addRoleDialogVisible 用于控制添加角色对话框是否可见。

openAddRoleDialog 函数用于打开添加角色对话框，resetAddRoleForm 函数用于重置添加角色对话框中的表单，addRole 函数用于向后台程序发送请求添加角色。

最后，将变量和函数 return 出去，供 template 模板使用。

(4) 修改角色。在图 12-21 所示的角色列表页中，每行记录后面都有一个"修改"按钮。单击"修改"按钮，可以打开"修改角色"对话框，如图 12-23 所示。

图 12-23 "修改角色"对话框

在组件 Role.vue 的 template 模板部分，与"修改角色"对话框相关的代码如下。

```
<el-dialog
  title="修改角色" v-model="editRoleDialogVisible" width="30%"
  @close="resetEditRoleForm()">
  <el-form
    ref="editRoleFormRef" :model="editRoleForm" :rules="editRoleFormRules"
    label-width="0px">
    <!-- 角色名称 -->
    <el-form-item prop="roleName">
      <el-input
        v-model="editRoleForm.roleName" prefix-icon="iconfont icon-user"
        placeholder="请输入角色名称"></el-input>
    </el-form-item>
  </el-form>
  <span slot="footer" class="dialog-footer">
    <el-button @click="editRoleDialogVisible = false">取 消</el-button>
    <el-button type="primary" @click="editRole()">确 定</el-button>
  </span>
</el-dialog>
```

在组件 Role.vue 的 script 部分，与修改角色功能相关的代码如下。

```
// 封装修改角色表单数据
const editRoleForm = reactive({ roleId: "", roleName: "" });
// 指定修改角色表单被引用时的名称和标识
const editRoleFormRef = ref(null);
// 设置修改角色表单验证规则
const editRoleFormRules = reactive({
  roleName: [{ required: true, message: "请输入角色名称", trigger: "blur" }]
});
// 控制修改角色对话框是否可见
const editRoleDialogVisible = ref(false);
// 重置修改角色对话框中的表单
const resetEditRoleForm = () => {
  editRoleFormRef.value.resetFields();
};
// 打开修改对话框
const openEditRoleDialog = async roleId => {
  editRoleDialogVisible.value = true;
  const { data: res } = await ctx.$axios.get(`roles/roleId/${roleId}`);
  if (res.code != 0) return ctx.$message.error("获取数据失败!");
  editRoleForm.roleName = res.data.roleName;
  editRoleForm.roleId = res.data.roleId;
};
// 修改角色
const editRole = () => {
  editRoleFormRef.value.validate(async valid => {
    if (!valid) return;
    var roleName = editRoleForm.roleName;
    var roleId = editRoleForm.roleId;
    // 判断角色名称是否已存在
    const { data: res1 } = await ctx.$axios.get(
      `roles/roleName/${roleName}/roleId/${roleId}`
    );
    if (res1.code !== 0) return ctx.$message.error("该角色名称已存在!");
    const { data: res2 } = await ctx.$axios.put(`roles`, {
      roleName: roleName, roleId: roleId });
    if (res2.code !== 0) return ctx.$message.error("修改失败!");
    ctx.$message.success("修改成功!");
    getRoles();
    editRoleDialogVisible.value = false;
  });
};
return { ..., editRoleDialogVisible, resetEditRoleForm, editRoleFormRef,
editRoleForm, editRoleFormRules, editRole, openEditRoleDialog };
```

在上述代码中，变量 editRoleForm 用于封装修改角色表单数据，变量 editRoleFormRef 用于指定修改角色表单被引用时的名称和标识，变量 editRoleFormRules 用于设置修改角色表单验证规则，变量 editRoleDialogVisible 用于控制修改角色对话框是否可见，变量 resetEditRoleForm 用于重置修改角色对话框中的表单。

openEditRoleDialog 函数用于打开修改对话框，editRole 函数用于向后台程序发送请求修改角色。最后,将变量和函数 return 出去,供 template 模板使用。

(5) 禁用角色。在图 12-21 所示的角色列表页中,每行记录后面都有一个“禁用”按钮。单击该按钮,会调用自定义函数 disableRole,代码如下。

```
const disableRole = async roleId => {
  const { data: res } = await
  ctx.$axios.put(`roles/roleId/${roleId}/delState`);
  if (res.code !== 0) return ctx.$message.error("禁用失败!");
  ctx.$message.success("禁用成功!");
  getRoles();
};
```

在 disableRole 函数中,使用 axios 向后台程序发送 put 请求,请求地址为 roles/roleId/${roleId}/delState,这个请求将由后端程序 eshop 的控制器类 RoleController 中的 editRole 方法处理,禁用该角色。

(6) 设置权限。在图 12-21 所示的角色列表页中,每行记录后面都有一个“设置权限”按钮。单击该按钮,打开“设置权限”对话框,如图 12-24 所示。

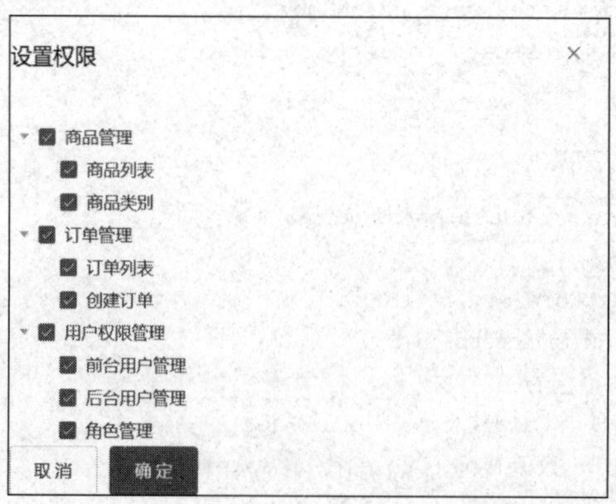

图 12-24 “设置权限”对话框

在组件 Role.vue 的 template 模板部分,与“设置权限”对话框相关的代码如下。

```
<el-dialog
  title="设置权限" v-model="powerDialogVisible" width="40%">
  <span>
    <el-tree
     :data="functions" :props="functionsTreeProps" node-key="id"
     :default-checked-keys="oldLeafPowers" show-checkbox
     default-expand-all ref="functionsTreeRef"></el-tree>
  </span>
  <span slot="footer" class="dialog-footer">
    <el-button @click="powerDialogVisible = false">取 消</el-button>
    <el-button type="primary" @click="savePower">确 定</el-button>
  </span>
</el-dialog>
```

在上述代码中，使用了< el-dialog >和< el-tree >等标签。其中，< el-tree >是 Tree 树形控件，通过层级结构展示数据，可展开或折叠。< el-tree >标签的 data 属性用于指定数据源，这里动态绑定到响应式变量 functions。props 属性用于指定配置选项，这里动态绑定到变量 functionsTreeProps。node-key 属性用于指定每个树节点用来作为唯一标识的属性，整棵树应该是唯一的，这里指定数据源中的 id 属性作为唯一标识。default-checked-keys 属性用于指定默认勾选的节点的 key 的数组，这里动态绑定到变量 oldLeafPowers。show-checkbox 属性用于指定节点是否可被选择，这里设置为 true，表示可被选择。ref 属性用于指定这个树形控件被引用时的名称和标识。

在组件 Role.vue 的 script 部分，与设置角色功能相关的代码如下。

```
// 控制设置权限对话框是否可见
const powerDialogVisible = ref(false);
// 保存系统功能菜单列表
const functions = ref(null);
// 保存用于设置权限的树形控件 prop 属性的配置选项
const functionsTreeProps = reactive({
  label: "text",
  children: "children"
});
// 原有权限叶子节点 id
const oldLeafPowers = ref(null);
// 指定设置权限的树形控件被引用时的名称和标识
const functionsTreeRef = ref(null);
// 角色 id
const roleId = ref(0);
// 打开设置权限对话框
const openSettingPowerDialog = async role => {
  oldLeafPowers.value = [];
  roleId.value = role.roleId;
  // 获取所有功能菜单列表
  const { data: res1 } = await ctx.$axios.get("functions");
  functions.value = res1.data[0].children;
  // 获取该管理员原有的功能菜单列表
  const { data: res2 } = await ctx.$axios.get("getLeafFunctionsByRid", {
    params: { roleId: roleId.value }
  });
  oldLeafPowers.value = res2;
  await nextTick();
  powerDialogVisible.value = true;
};
// 保存权限
const savePower = async () => {
  const newPowers = [
    ...functionsTreeRef.value.getCheckedKeys(),
    ...functionsTreeRef.value.getHalfCheckedKeys()
  ];
  const fids = newPowers.join(",");
  const { data: res } = await ctx.$axios.post(`roles/${roleId.value}`, {
    fids: fids
  });
```

```
    if (res.code !== 0) return ctx.$message.error("设置权限失败!");
    ctx.$message.success("设置权限成功!");
    powerDialogVisible.value = false;
  };
  return { ..., powerDialogVisible, openSettingPowerDialog, functions,
  functionsTreeProps, oldLeafPowers, functionsTreeRef, savePower };
```

在上述代码中，变量 powerDialogVisible 用于控制设置权限对话框是否可见，变量 functions 用于保存系统功能菜单列表，变量 functionsTreeProps 保存用于设置权限的树形控件 prop 属性的配置选项，变量 oldLeafPowers 用于保存原有权限叶子节点 id，变量 functionsTreeRef 用于指定设置权限的树形控件被引用时的名称和标识，变量 roleId 用于保存角色 id。

在 openSettingPowerDialog 函数中，首先使用 axios 发送 get 请求，请求地址为 http://localhost:8888/eshop/functions，这个请求将由后端程序 eshop 的控制器类 FunctionsController 中的 getFunctions()方法处理，获取所有功能菜单列表，作为< el-tree >树形控件的数据源。再使用 axios 发送第二次 get 请求，请求地址为 http://localhost:8888/eshop/ getLeafFunctionsByRid，并传递 roleId 参数，这个请求将由控制器类 RoleController 中的 getLeafFunctionsByRid()方法处理，根据角色 id 获取叶子节点功能菜单 id 列表。

在 savePower 函数中，使用 axios 发送 post 请求，请求地址为 http://localhost:8888/eshop/roles/ ${roleId.value}，这个请求将由控制器类 RoleController 中的 savePowers()方法处理，保存权限。

最后，将定义的变量和函数通过 return 返回出去，供 template 模板使用。

12.12.3 前台用户管理

在管理首页面中，单击用户权限管理下的前台用户管理菜单，打开前台用户列表页，如图 12-25 所示。

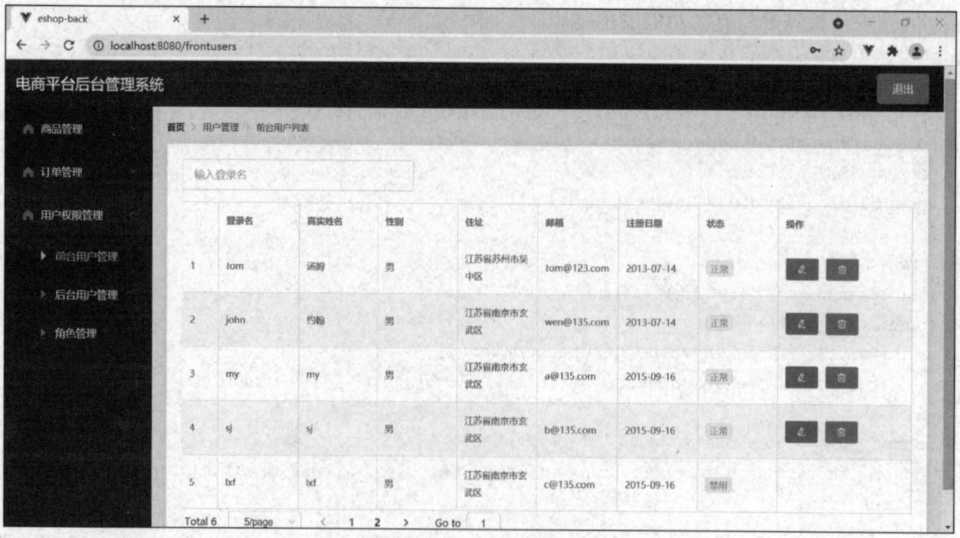

图 12-25 前台用户列表页

在前台用户列表页中，包括前台用户分页显示与查询、修改和禁用前台用户等功能。前台用户列表页的所有功能在组件 Frontusers. vue 中实现，该组件位于 components/user 目录下。

前台用户管理模块功能与后台用户管理类似，读者可以参照理解。此处不再赘述。

12.13　小结

本章基于 Spring Boot，结合 Vue 和 Element Plus 前端框架，详细介绍了一个典型的电商平台后台管理系统的实现过程。系统的主要功能包括商品管理、订单管理和用户权限管理。

通过本章的学习，希望读者能够熟练掌握 Spring Boot、Vue 和 Element Plus 框架整合开发的基本步骤、方法和技巧。

参考文献

[1] 陈恒,楼偶俊,巩庆志,等. Spring Boot 从入门到实战[M]. 北京：清华大学出版社,2020.

[2] 王松. Spring Boot+Vue 全栈开发实战[M]. 北京：清华大学出版社,2018.

[3] 汪云飞. JavaEE 开发的颠覆者：Spring Boot 实战[M]. 北京：电子工业出版社,2016.

[4] 朱建昕. Spring Boot+Vue 开发实战[M]. 北京：电子工业出版社,2021.

[5] 郑天民. Spring 响应式微服务：Spring Boot 2+Spring 5+Spring Cloud 实战[M]. 北京：电子工业出版社,2019.

[6] 黑马程序员. Spring Boot 企业级开发教程[M]. 北京：人民邮电出版社,2019.

[7] 杨洋. Spring Boot 2 实战之旅[M]. 北京：清华大学出版社,2019.

[8] 梁灏. Vue.js 实战[M]. 北京：清华大学出版社,2017.

[9] 申思维. Web 前端技术丛书：Vue.js 快速入门[M]. 北京：清华大学出版社,2018.

[10] 贾志杰. Vue+Spring Boot 前后端分离开发实战[M]. 北京：清华大学出版社,2021.

[11] 李刚. 疯狂 Spring Boot 终极讲义[M]. 北京：电子工业出版社,2021.